1983

APPLICABLE FINITE MATHEMATICS

APPLICABLE FINITE
MATHEMATICS

DAVID S. MOORE, ASSOCIATE PROFESSOR OF STATISTICS, PURDUE UNIVERSITY

JAMES W. YACKEL, ASSOCIATE PROFESSOR OF MATHEMATICS, PURDUE UNIVERSITY

HOUGHTON MIFFLIN COMPANY   BOSTON

Atlanta   Dallas   Geneva, Illinois   Hopewell, New Jersey   Palo Alto   London

Library of Congress Catalog Card Number: 73-9397

ISBN: 0-395-17771-5

CONTENTS

PREFACE

In recent years the use of mathematical models in economics and in the social, managerial, and behavioral sciences has become increasingly common. The modern computer is in part responsible for this; its data-handling capacity allows the use and testing of ever more complex models. Many of the mathematical techniques used in these models belong to the area of "finite mathematics." In writing this book we have deviated from the common outline of other finite mathematics texts in order to present a unified body of material chosen for applicability to model building in the social and managerial sciences. This material is, of course, interesting mathematics in its own right and is widely applied in the natural sciences and engineering as well. We have emphasized the applicability of our material throughout, but this remains a book of mathematics rather than a book of methods. Our stress is on formulation of problems and conceptual understanding of their solutions rather than on detailed solution methods which in practice are implemented by computer routines.

The book opens with a discussion of probability—the mathematics of uncertainty. The concepts and tools of probability are used in Chapter 2 to study Markov chains and their applications. Chapter 3 concerns the apparently unrelated subject of linear programing—the mathematics of best allocation of limited resources. Both probability and linear programing contribute to the presentation of game theory in Chapter 4—the mathematics of competitive situations. Chapter 5 uses probability and game theory to discuss decision theory—the mathematics of decision making in the face of uncertainty. We have stressed the game-theoretic and decision-making aspects of this subject rather than its close connection with statistics.

The mathematical prerequisite for reading this text is a good knowledge of high school algebra. To be sure, a successful reader will acquire other tools as he progresses. In particular, we have introduced basic material on sets, summation notation, vectors, matrices, and the n-dimensional geometry of lines and planes as needed. On the other hand, we have not included a treatment of combinatorial problems, and we have kept the use of combinatorics to a minimum in our presentation of probability theory. We believe that the difficulties students encounter in combinatorial problems impede their understanding of the concepts of probability.

Several features of the text deserve mention. The sections entitled "Applications" contain extended illustrations of the use of the ideas in each chapter. The applications given in these sections vary in mathematical complexity, as well as in the field to which the mathematics is applied, so that each instructor can choose those examples most appropriate to the students. The sections entitled "Some Theory" are optional and can be omitted without loss of continuity. They give proofs (or near proofs) of some harder results, such as the fundamental theorem of game theory. A few other sections are starred, indicating that they contain material not essential to the remainder of the text and of somewhat greater difficulty. The concluding section of each chapter, entitled "Food for Thought," is also optional. These sections suggest, in a nontechnical way, further developments which are amusing and instructive to think about.

Each chapter also contains several projects suggested for students with access to a digital computer and knowledge of a language such as Basic or Fortran. In some cases the results of these projects will lend plausibility to theoretical results stated in the text, while other projects provide more realistic experience in the use of text material. We encourage instructors to assign these projects where possible; our usual practice is to assign each project to a group of students who work together.

We have used this material at Purdue University for a two-semester course taught primarily to underclass students in industrial management. Chapters 1 and 2 are covered in the first semester; Chapters 3, 4, and 5 in the second. In teaching this course we omit the starred sections as well as Secs. 2.11 to 2.13 on absorbing Markov chains. A higher level course can be taught by including the starred sections and discussing in detail several of the more difficult examples from the "Applications" sections.

We have also planned the text to allow a choice of any of 3 one-semester courses in applicable finite mathematics. These are as follows.

1. *Probability and decision making* (Chapters 1, 4, and 5, omitting Sec. 4.8) Chapter 5 is the easiest chapter in the text. In budgeting time for this course, note that Sec. 1.14 on expected value is essential for

Chapters 4 and 5, while Application 1.3 is strongly recommended. Sections 1.9, 1.12, 1.15, 1.16, 1.18, and 1.19 should be omitted, and the starred sections of Chapters 4 and 5 (which use material from Chapters 2 and 3) must be omitted.

2. *Probability and Markov chains* (Chapters 1 and 2) Most instructors will find it difficult to cover both the material on limiting probabilities (Secs. 2.8 to 2.10) and the material on absorbing chains (Secs. 2.11 to 2.13). Either may be included in the course, but the latter is more difficult.

3. *Probability and linear Programing* (Chapter 1, Sec. 2.3 and 2.4, Chapter 3) Since these topics are independent, any choice of material in Chapter 1 is possible, and in fact linear programing can be taught first if desired. We recommend beginning in Chapter 1, then teaching Sec. 3.1, followed by Secs. 2.3 and 2.4 and the remainder of Chapter 3. Most instructors will require slightly less than half a semester to teach Chapter 3, so careful allocation of time in Chapter 1 is necessary.

We recommend the first of these options for most students in the managerial sciences and the second for students interested in probabilistic models in the social sciences. Either is an excellent prelude to a course in statistical methods.

We are grateful to several colleagues who read the manuscript and made many helpful suggestions: Professors Herman Chernoff of Stanford University; William Fulton of Brown University; Donald E. Myers of the University of Arizona; Arlo Schurle of the University of North Carolina; and Frank Wolf of Carleton College. Dorothy Penner and Edna Hicks provided rapid and accurate translation of our work into typed copy.

1 PROBABILITY

1.1 INTRODUCTION

Most parts of mathematics have roots in the tangible world around us, though the roots may seem hidden by the subsequent growth. In particular, all the areas of mathematics discussed in this book center on attempts to describe observable phenomena and to solve practical problems. A mathematical description is often called a *mathematical model*. Probability theory is the branch of mathematics which tries to provide descriptions or models for "chance behavior." Gambling provides many simple examples of chance behavior, such as rolling a die, spinning a roulette wheel, or dealing a poker hand. In fact probability theory was born when certain 17th century French aristocrats worried about the practical problem of why they were losing their fortunes gambling. They solved the problem not by ceasing to gamble, but by asking their more intelligent friends to calculate odds for them. The mathematics which resulted was applied in the following centuries to the study of errors in careful measurements and (by life insurance companies) to the study of human mortality. But only in the 20th century has probability become a major branch of mathematics with widespread applications in all the sciences.

Probability theory is used today to describe such things as spread of infectious diseases, growth of animal populations, waiting times of cars at a tollbooth, lifetime in service of electronic components, movement of stock market indices, and effectiveness of new teaching methods. The common factor here is randomness or uncertainty. We cannot predict in advance whether a person exposed to an infectious disease will be infected or how

1

long a car will wait in line at a tollbooth any more than we can predict in advance what number will come up when a die is rolled. This unpredictability is the essence of chance behavior. Yet we do have some intuition about chance behavior, at least in simple situations such as rolling a die. Our first goal is to develop that intuition and show that everyday notions of probability do lead to a precise mathematical description. The mathematics developed in this chapter is not the only possible way of analyzing our intuitive notion of probability as a measure of uncertainty. We claim only that it is reasonable (you will have to decide this for yourself) and that it has proved very useful in applications to real problems.

When we speak of probability in everyday conversation, we may mean several things. Most likely we are simply protecting ourselves against the chance that we may be wrong ("I will probably be in my office tomorrow at 9"). Or we may be giving our personal assessment of the likelihood of an event ("he has not got 1 chance in 10 of meeting his deadline"). To introduce the mathematical formulation of probability, we concentrate on another intuitive notion of probability. When we say that the probability of getting a head on a coin toss is $1/2$, what do we mean? "Why, that the coin will come up heads half the time." The idea here is that in a very long sequence of tosses of the coin about half will be heads. More generally, suppose we have any experiment with several possible outcomes. Suppose further that we cannot predict which outcome will occur, so that we are faced with uncertainty. We now repeat the experiment many times. If the proportion of occurrences of a given outcome approaches a number p as we continue to repeat the experiment, it is reasonable to call p the probability of that outcome. This is the "relative frequency" interpretation of probability.

Our intuition is based on our experience. If you are a bridge or poker addict, the relative frequency interpretation of probability will seem natural, for you will have observed that important configurations of cards recur with fixed frequencies in the long run. If you were a life insurance executive, you would know that a predictable proportion of 18-year-olds will die this year, even though the death of an individual 18-year-old is unpredictable. But you may well have little experience of repeated trials under uncertainty, and hence little intuition about probability. To remedy this, there follow some exercises which invite you to experiment with repeated trials, and also some questions to probe your intuition.

Exercise 1.1   1.   Toss a penny 20 times and record the number of heads. In class, pool your results as follows: Go down the class roll in some order. After each student reports the results of his 20 tosses, compute the proportion of heads observed in all tosses reported up to that point. Make a graph of proportion of heads versus number of tosses. (This experiment is equivalent to a very long sequence of tosses of a single penny, if we assume that all the pennies tossed are identical.)

Study the graph. Does the proportion of heads appear to settle down near some fixed value? Do any properties of the graph surprise you? Based on this experiment, what is the probability of a head in tossing pennies?

2. Take 2 pennies and toss them both 20 times. The possible results of each toss are 2 heads, 2 tails, or one of each. What would you guess to be the probability of 2 heads? Of 2 tails? Of one of each? Pool your results in class as in Exercise 1.1.1. Graph the proportion of outcomes equal to each of the three possible results against the number of tosses. Do the proportions settle down to fixed numbers? Are the proportions of the various outcomes as you guessed they would be?

Interpreting probability as relative frequency, what is the sum of the probabilities of the three possible outcomes? Why is this result expected?

3. A stock speculator theorizes that trends tend to persist in the stock market. Thus he feels that an "up" day tends to be followed by another "up" day, and a "down" day by another "down" day. Before using this theory as a guide for his trading, he wants to know the probability of such continuations and the probability of reversal. Look up the closing Dow Jones Industrial Average (DJIA) for every trading day of a month. (The last interior page of the daily *Wall Street Journal* is one source for this information. Each student should be assigned a different month.) With these data at hand, count the continuations (up following up or down following down) and reversals from day to day. Graph the proportion of continuations against the number of days observed cumulatively as each student presents data.

Does the proportion of continuations approach a fixed number as the data accumulate? Is the speculator right in guessing that the probability of a continuation exceeds the probability of a reversal? Do you think the probability of continuation is large enough to support the "buy today because the market was up yesterday" strategy?

4. The six letters occurring most frequently in English prose are (in order) e, t, a, o, i, n. Estimate the relative frequency of each of these letters by counting the letters on a page of English text and keeping track of the number of e's, t's, and so on. If you opened a book blindly, stabbed at the page with a pencil, and recorded the letter nearest the pencil point, what would you estimate to be the probability of an e? What is the probability of recording one of the six letters e, t, a, o, i, n? What is the probability of getting a letter other than these six?

1.2 FIRST STEPS

If you have done the exercises above, you should have some feeling for how probability (in the sense of relative frequency) behaves. Our goal is to give a precise mathematical description which reflects that feeling. Suppose that we have a repeatable experiment with uncertain outcome. We described four such experiments in the exercises: Tossing a coin, tossing two coins simultaneously, computing the Dow Jones Industrial Average at the end of tomorrow's trading, and choosing a printed letter blindly from a book. We will call these *random experiments*. Other random experiments are counting the number of cars arriving at a tollbooth in a 5-min. period, measuring the number of hours a light bulb lasts before burning out, asking an individual if she favors legalizing marijuana, and asking a consumer which brand of beer he plans to buy next.

In a random experiment we do not know in advance what the outcome will be. Yet we usually can say what the *possible* outcomes are. If we toss a coin, we get either heads or tails. If we ask an individual if she favors legalizing marijuana, she says either "yes," "no," or "undecided." If we count the cars arriving at a tollbooth in a 5-min period, the result must be some whole number greater than or equal to zero. So the mathematical description of uncertainty begins by specifying the collection or set of all possible outcomes. This is called the *sample space* of the random experiment. In setting up a mathematical model of a random experiment, we often have some freedom in choosing the sample space. In tossing a coin we usually say that "heads" and "tails" are the only possible outcomes. Yet it is conceivable that the coin might stand on edge or roll down a grate. In counting cars at a tollbooth we may allow any nonnegative whole number of cars as a possible outcome. Yet someone may object that surely no more than 1000 cars can pass the booth in 5 min, so that we could allow only whole numbers from 0 to 1000. When we toss two coins we might take the possible outcomes to be "one head and one tail," "two heads," and "two tails." Alternatively, we might distinguish between the first and second coins and say that the possible outcomes are "heads on the first, tails on the second," "tails on the first, heads on the second," "two heads," and "two tails." Which choice of sample space is most convenient depends on the particular questions we wish to answer. The goal is to choose a sample space which adequately describes the experiment and is easy to work with. The art of doing this is learned by experience.

What about probability? We want to be able to talk about the probability of getting a head on a coin toss, or about the probability that more than 25 cars will arrive at the tollbooth in 5 min. In general terms, we want to be able to attach probabilities to sets of possible outcomes, such as "25 or more cars," as well as to individual possible outcomes, such as "heads." If we call any subset of the sample space an *event*, then we want to assign probabilities to events. Intuitively, the probability of an event is the proportion of the time that event occurs in a very long sequence of repetitions of the experiment. Mathematically, an event is a set of possible outcomes or a subset of the sample space. In discussing events and their probabilities, we therefore use the mathematical language of set theory. The small amount of set theory needed is presented in Sec. 1.3, which may be reviewed even if the subject is known to you.

The basic notion of the probability of an event is that it is a measure of the likelihood that the event will occur. The probability of an event is therefore a number attached to the event. Let us guide our thinking by supposing that the probability of an event is the relative frequency of the event in a long sequence of repetitions of the random experiment. The following three statements are true.

1. The probability of any event must be a number between 0 and 1.
2. The probability of the entire sample space must be 1.
3. If two events have no outcomes in common (i.e., both cannot occur simultaneously), then the probability that one or the other occurs must be the sum of their individual probabilities.

In Sec. 1.4 we will formally define a legitimate assignment of probabilities to events as one satisfying statements 1, 2, and 3. This approach is based on our intuition concerning relative frequencies, but is nonetheless intellectually subtle. We begin by ignoring the particular random experiment we want to describe. Instead, we spell out the properties that probability ought to have in any situation. This rather abstract approach arose only in the 1930s, after particular assignments of probability had been studied for several centuries. One advantage of this method is that we will be able to build up a body of facts and techniques which applies to any assignment of probabilities—we need not begin anew in describing each new random experiment. Another advantage is that the resulting body of mathematics stands by itself—you need not always interpret probability as relative frequency. In studying decision making, for example, it is often convenient to think of probability as representing the decision maker's personal assessment of the likelihood of events. ("Let's see, I think we have probability $3/4$ of meeting our production schedule. If we don't make it, I think there is only a probability of $1/10$ that they will cancel the order.") In this case, properties 1, 2, and 3 still ought to hold if the decision maker's assignment of probabilities makes sense. We will understand probability better if we try to see as we go along that mathematical statements make sense in terms of relative frequencies or in terms of personal assessments of likelihood. But the mathematics does not depend on which interpretation we give it.

Exercise 1.2

1. Explain carefully why long-run relative frequencies have properties 1, 2, and 3.
2. What is a reasonable sample space for each of the following random experiments?

a. Roll a single standard die.
b. Roll a pair of dice.
c. Deal a 13-card bridge hand.
d. Record whether or not a certain drug arrests the growth of cancer in the liver of a white rat.

e. Count the number of electric toenail cleaners we will sell in a year if we begin to market them.

f. Ask a TV viewer if he remembers a particular commercial.

g. Measure the lifetime in service of a TV picture tube.

3.   Sometimes the assignment of probabilities to outcomes can be obtained empirically. Here is an example in which the random experiment is choosing a word at random and recording its length. Divide the class into two groups of about equal size. Each member of the first group is to visit the library and select a paragraph from a novel of Henry James, while each member of the second group is to select a paragraph from a back issue of *Popular Science*. Record the total number of words in each paragraph and the number of words of each length (one letter, two letters, and so on).

Pool your results in class to get two assignments of probabilities to word length, one a model for Henry James' novels, the other for *Popular Science*. Compare the two—for example, which source tends to have longer words?

4.   Sometimes a reasonable assignment of probabilities to outcomes can be made a priori, on the basis of some balance or symmetry. If we bought a new die, what probabilities would be likely a priori for the possible outcomes 1, 2, 3, 4, 5, 6? What is the probability of an odd number under this assignment? Do *all* dice behave in the manner described by the assignment?

5.   For some purposes an assignment of probabilities can be made by the personal judgment of someone who knows the situation. What is your personal assignment of probabilities to the possible grades you might receive in this course? According to that assignment, what is the probability that you will pass the course?

1.3   SETS

We have seen that events are collections of possible outcomes of a random experiment. A *set* is a collection of objects; the objects are called *members* of the set. The members of a set need not be mathematical objects—we can speak of the set of left-handed first basemen or of the set of unicorns. When we are discussing events, all the members of all the sets we encounter are members of the sample space for the random experiment in question. We will assume whenever we discuss relations among sets that all the members of all the sets we talk about are members of a given set S. It is simplest to think of sets as events and S as the sample space. We use the language of sets because we will want to apply this language to subjects other than probability theory.

Here is some basic notation. We usually use capital letters to denote sets. A set can be specified by listing the objects it contains. The statement "A is the set of whole numbers between 0 and 5 inclusive" can be written as

$$A = \{0,1,2,3,4,5\}$$

Here we described A by listing its members. The curly brackets { } stand for "the set whose members are." It is often inconvenient to list the members of a set, and impossible to do so if the set has infinitely many members. Fortunately, we can also specify a set by simply describing its members, i.e., giving a rule for deciding whether or not any particular object is in the set. Examples of such descriptions are "the set of whole numbers" and "the set of American land-grant universities." The same set can often be described in more than one way. For example, the set of solutions of the equation $x^2 - 1 = 0$ is the same as the set $\{-1,1\}$ because both sets have exactly the same members. To be formal, we write $A = B$ when the sets A and B have exactly the same members.

It is convenient to be able to talk about sets having *no* members. The set of numbers which are both less than 0 and greater than 0 is such a set, and the set of unicorns is probably another. Since these sets have the same members (namely, none), they are just different descriptions of the same set. The set having no members is called the *empty set*, and is denoted by $\emptyset$.

Our primary purpose in discussing sets is to provide a language for expressing relations among events, especially for describing complicated events in terms of simpler ones. We often find ourselves saying that both of two events occurred, or that one occurred and the other did not, and so on.

The formal terms used to express such relations are the *set operations* defined in Definition 1.1. Each set operation produces a new set from sets already given.

DEFINITION 1.1

a.   The *union $A \cup B$* of sets A and B is the set of all objects which are members of A or members of B or both.

b.   The *intersection $A \cap B$* of sets A and B is the set of all objects which are members of both A and B.

c.   The *complement A^c* of a set A is the set of all objects in S which are not members of A.

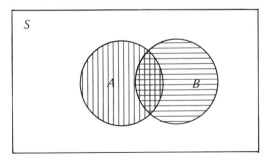

Figure 1.1   Set A is vertically scored, set B is horizontally scored. The entire scored region is $A \cup B$, while the doubly scored region is $A \cap B$.

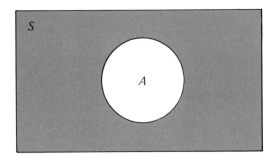

Figure 1.2   The region shaded is A^c.

There are two important ways of clarifying these operations on sets. The first is to pictorially represent the operations for the case in which S is the plane or a subset of it. Then all sets are collections of points in the plane. Figure 1.1 represents union and intersection, and Fig. 1.2 complementation.

These pictures are called *Venn diagrams*. They are invaluable as visual proof of the truth of set-theoretic statements. For example, we claim that for any two sets A and B

$$(A \cup B)^c = A^c \cap B^c \tag{1.1}$$

To see that this is true, we draw Venn diagrams of each side of the equation, being careful to apply the set operations in the order specified. It can be seen from Figs. 1.3 and 1.4 that the blue regions at the end of each sequence of operations are identical.

Since we claimed that Eq. (1.1) is true no matter what the sets A and B are, we were careful to start Figs. 1.3 and 1.4 with sets which overlapped in part but not altogether. We avoided two special situations: that A and B do not overlap at all, or that one entirely contain the other. These two special situations are important enough to merit another definition.

DEFINITION 1.2

a.   A set B *is contained in* a set A ($B \subset A$ or, equivalently, $A \supset B$) if every

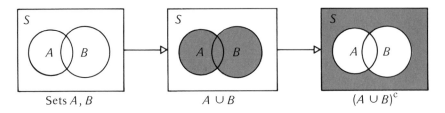

Figure 1.3   Venn diagram of $(A \cup B)^c$

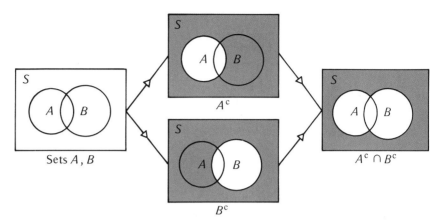

Figure 1.4 Venn diagram of $A^c \cap B^c$

member of B is also a member of A. In this case we say that B is a *subset* of A.
b. Sets A and B are *disjoint* if they have no members in common, i.e., if $A \cap B = \emptyset$.

These special relations between two sets are illustrated in Fig. 1.5. In such cases results may hold which are not true for all sets, as Exercise 1.3.4 demonstrates.

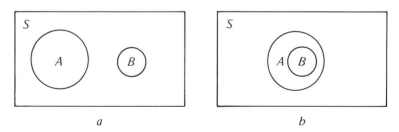

Figure 1.5 (a) $A \cap B = \emptyset$; (b) $A \supset B$

There is another interpretation of set operations which is more important than Venn diagrams. If S is the sample space for a random experiment, we say that an event A *occurs* if the outcome is a member of A. Set operations can then be read as statements about the occurrence of events. From Definition 1.1 you can see that if A and B are events, $A \cap B$ is the event that A *and* B both occur. Similarly A^c is the event that A does *not* occur, and $A \cup B$ is the event that *A or B* or both occur. Thus intersection, complement, and union express "and," "not," and "or" respectively. (Note that the English word *or* is ambiguous. It can be used *exclusively*, in which case "*A or B*" means "*A or B* but not both." We use the exclusive or when we say, "I will rent a car from Hertz or Avis." If or is used *inclusively*, "*A or B*" means "*A or B* or both" as in "Hertz or Avis should start renting Ferraris." Union is not ambiguous. It expresses the inclusive or.)

This interpretation of set operations in terms of logical connectives is important because it helps us to express complicated events in terms of simpler events. We will see later that this often allows us to find the probabilities of complicated events.

Example 1.1   Schwarz, Thompson, and Williams are workers eligible for promotion. Management may promote none of the three, any one of them, any two of them, or all of them. If we let *A*, *B*, and *C* denote respectively the events that Schwarz is promoted, Thompson is promoted, and Williams is promoted, then

$B \cap C$ is the event that Thompson and Williams are both promoted.

$B \cup C$ is the event that at least one of Thompson and Williams is promoted.

The event that Thompson or Williams, but not both (exclusive or), is promoted is more complicated. If we think of this as the event "Thompson is promoted and Williams is not, or Williams is promoted and Thompson is not," a little thought should convince us that in terms of the sets *B* and *C* this is

$$\{B \cap C^c\} \cup \{B^c \cap C\}$$

If we think of this event as "Thompson or Williams is promoted, and not both of them," we get

$$\{B \cup C\} \cap \{B \cap C\}^c$$

These expressions both represent the same set, of course. You should check that both have the Venn diagram given in Fig. 1.6.

If we can think logically, we can prove set identities by simply applying the definitions. For example, we claim that for any set *A*

$$(A^c)^c = A$$

To see this, note that by definition A^c is the set of those objects in *S* that are not in *A*. The complement of A^c is the set of those objects in *S* that are not in A^c, which is just the set of objects in *A*. We won't dwell on such proofs, relying on Venn diagrams to establish the few facts we need. Some are very simple. For example, it should be obvious to you that

$$S^c = \emptyset$$

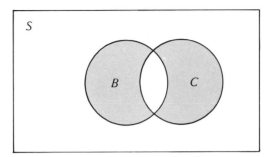

Figure 1.6   Venn diagram for exclusive *or*

The set relations of Definition 1.2 also have interpretations as statements about events. Events A and B are disjoint ($A \cap B = \emptyset$) if it is impossible for both to occur simultaneously. Event B is a subset of A if A must occur whenever B occurs. We then say that event B implies event A.

Example 1.2   For the experiment of rolling a single die, the sample space is $S = \{1,2,3,4,5,6\}$. Let A be the event that we roll an odd number, B the event that we roll an even number, and C the event that we roll a 3. Then A and B are disjoint because we cannot simultaneously roll an even number and an odd number. In symbols, $A \cap B = \emptyset$. C is a subset of A because 3 is odd, so that whenever C occurs, we know that A also occurs. Therefore C implies A, or in symbols, $C \subset A$.

Exercise 1.3   1.   Suppose that S is the set of all whole numbers, A is the set of all whole numbers less than or equal to 100, and B is the set of all whole numbers greater than or equal to zero and less than or equal to 150. What numbers are members of the following sets?

a. A^c d. $(A \cap B)^c$
b. $A \cap B$ e. $A^c \cap B^c$
c. $A \cup B$

Is A a subset of B? Is B a subset of A? Are A and B disjoint?
2.   Draw Venn diagrams representing $(A \cap B)^c$, $A^c \cap B^c$, and $A^c \cup B^c$. Are any of these sets equal?
3.   Convince yourself that for any sets A, B, and C

$$A \cup (B \cap C) = (A \cup B) \cap (A \cup C)$$

by drawing Venn diagrams of both sets. (It is best to draw a sequence of several Venn diagrams for each side of the equality, showing the effect of applying each set operation in turn.)

4.   Demonstrate by Venn diagrams that if $A \supset C$, then $A \cap C = C$ and $A \cup C = A$. If A and C are as in Example 1.2, state the verbal equivalents of these symbolic statements.

5.   Which of the following statements are always true and why? (Try to argue in terms of the logical equivalents of set operations.)

a.   If $A \subset B$, then $A^c \supset B^c$.
b. $A \cap A^c = \emptyset$.
c.   If A and B are disjoint, so are A^c and B^c.

6.   Unions and intersections can be applied to more than two sets. What should the definitions of $A \cup B \cup C$ and $A \cap B \cap C$ be? Illustrate the definitions by properly shading the Venn diagram below.

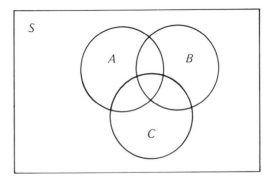

7.   In Example 1.1, the event that at least one of the three is promoted is $A \cup B \cup C$. The event that all three are promoted is $A \cap B \cap C$. What are set expressions for the following events?

a.   Thompson and Williams are promoted, and Schwarz is not.
b.   Schwarz is promoted, and Thompson and Williams are not.
c.   Exactly one of the three is promoted.
d.   Exactly two of the three are promoted.

8.   A stock market speculator bases his decisions on the movement of the market (up, down, or unchanged) on each of the last 3 trading days. Number these days 1 through 3 (day 3 being most recent) and let

    A_i be the event that the market was up on day i
    B_i be the event that the market was down on day i
    C_i be the event that the market was unchanged on day i

Answer the following, using set notation.

a.   Express C_i in terms of A_i and B_i.
b.   What is the event "up on the most recent trading day"?

c. What is the event "up on each of the last 3 trading days"?
d. What is the event "up on a majority of the last 3 trading days"?
e. What is the event "down exactly once in the last 3 trading days"?

9. Suppose A and B are disjoint and $A \supset C$. Show that in this case C and B are also disjoint by interpreting these set expressions as statements about events.

10. Show by Venn diagram that for any events A and B

$$A \cup B = A \cup (B \cap A^c)$$

and that A and $B \cap A^c$ are disjoint. Show also that for any events A and B

$$B = (A \cap B) \cup (B \cap A^c)$$

and that $A \cap B$ and $B \cap A^c$ are disjoint.

1.4 PROBABILITY MEASURES

We are now ready to give the formal mathematical framework of probability theory. As outlined in Sec. 1.2, the elements of a mathematical description (or mathematical model) of a random experiment are a sample space, events, and an assignment of probabilities to events. Although we are already familiar with sample spaces and events, we will give a definition for completeness.

DEFINITION 1.3 The *sample space* of a random experiment is a set such that each possible outcome of the experiment is represented by exactly one member of the set. Any subset of the sample space is an *event*.

The exact wording of Definition 1.3 is intended to allow us to be flexible in the choice of a sample space. If the experiment involves measuring the aptitude of a prospective employee by a test scaled from 0 to 100 in 1-point steps, there is no question what the sample space ought to be. But if we are counting the calls arriving at a telephone switchboard in an hour, it is difficult to say exactly what the largest possible number of calls is. Definition 1.3 allows us to use the set of all nonnegative whole numbers as the sample space, since each possible number of calls is certainly such a number. Again, if we are training a laboratory animal to perform a certain task, each trial results in success or failure. If the experiment consists of 10 trials, it may be convenient to label successes by 1 and failures by 0 so that each outcome of the experiment is a sequence of 10 numbers, each either a 1 or a 0. The sample space can be taken to be the set of all such sequences, since each possible outcome is represented by a member of this set.

The next step in building a probability model is to assign a probability to each event. To do this we give a rule which assigns to each event (subset of the sample space) its probability (a number). Since such a rule assigns to every event a measure of the likelihood that the event will occur (its probability), we call the rule a *probability measure*.

Example 1.3   A "fair" coin is tossed once. A reasonable probability model for this random experiment is to take $S = \{H,T\}$ and to assign a probability of $\frac{1}{2}$ to each of the two possible outcomes. The only subsets of S are $\{H\}$, $\{T\}$, S itself, and the empty set $\varnothing$. Here is a complete description of the assignment of probabilities to events.

S is assigned probability $P(S) = 1$.
$\{H\}$ is assigned probability $P(\{H\}) = \frac{1}{2}$.
$\{T\}$ is assigned probability $P(\{T\}) = \frac{1}{2}$.
$\varnothing$ is assigned probability $P(\varnothing) = 0$.

Assignments of probability should satisfy the requirements 1, 2, and 3 discussed in Sec. 1.2. We therefore give the following definition.

DEFINITION 1.4   A *probability measure* P on a sample space S assigns to every event A a number $P(A)$ (the probability of A) in such a way that

a. (positivity) For every event A, $P(A) \geq 0$.
b. (totality) $P(S) = 1$.
c. (additivity) If events A and B are disjoint, then

$$P(A \cup B) = P(A) + P(B)$$

Definition 1.4 spells out the requirements for a legitimate assignment of probabilities to events. It is simply a formalization of the properties 1, 2, and 3 for relative frequencies, except that Definition 1.4a only requires that the probability of any event be greater than or equal to 0. We will shortly see that Definition 1.4a, b, and c imply that the probability of any event is also less than or equal to 1, so that we get property 1 of Sec. 1.2. We use Definition 1.4a because it is easier to check. A *probability model* for a random experiment consists of a sample space S and a probability measure P on S. Definitions 1.3 and 1.4 do not tell us which probability model is correct for a particular random experiment; they only describe the properties which any legitimate probability model must have. The examples which follow will help us grasp this important point.

Example 1.4   Our random experiment is to choose a woman in the American civilian labor force and record her marital status. This experiment has four possible outcomes, and we assign a probability to each outcome equal to the proportion of women in the labor force having that marital status in 1970.

Outcome	Probability
Single	0.22
Married, husband present	0.59
Married, husband not present	0.05
Widowed or divorced	0.14

Having assigned probabilities to individual outcomes, we can define a probability measure P on S by taking $P(A)$ to be the sum of the probabilities of the outcomes which are members of A. For example, if A is the event that the woman chosen is married,

$$P(A) = 0.59 + 0.05 = 0.64$$

It is easy to check that this P *is* a probability measure. $P(A) \geq 0$ for any A, because each individual outcome has positive probability, and $P(A)$ is a sum of these probabilities. $P(S) = 1$ because the probabilities of all the outcomes add to 1. To see that P is additive, let us look at a particular case. If A is the event that the woman chosen is married, and B is the event that she is single, then A and B are disjoint. $A \cup B$ contains all individual outcomes except "widowed or divorced," so

$$P(A \cup B) = 0.22 + 0.59 + 0.05 = 0.86$$

The same result is obtained by finding $P(A)$ and $P(B)$ individually (by adding the probabilities of the individual outcomes in these events) and then computing $P(A) + P(B)$.

Example 1.5   Consider a sociologist's data on social mobility. He can place each individual into one of five social classes on the basis of such factors as education and occupation. Long-term frequency data in Denmark and England give the following proportions of sons of lower-class (class 1) fathers ending up in each social class. For example, 46 percent of sons of class 1 fathers in Denmark remain in class 1.

	Denmark				
Class	1	2	3	4	5
Relative frequency	0.46	0.38	0.13	0.02	0.01

	England				
Class	1	2	3	4	5
Relative frequency	0.48	0.38	0.08	0.05	0.01

These data can be used to set up probability models for the experiment of observing the son of a lower-class father and recording his social class as an adult. The sample space is $S = \{1,2,3,4,5\}$. We assign to each subset of S the sum of the frequencies of its members. For example, the probability that the son ends up in one of the upper two classes is $P(\{4,5\}) = 0.03$ using the Denmark data or $P(\{4,5\}) = 0.06$ using the England data. Since each complete set of frequencies sums to 1 we have $P(S) = 1$ in both cases. Just as in Example 1.4, this method of assigning probabilities also satisfies Definition 1.4a and c. Here we have two probability models for social mobility of sons of lower-class fathers. Both are legitimate. The first is "correct" in Denmark, in the sense that the probabilities agree with observed relative frequencies. The second is correct in England. Either or neither might be correct in the United States—this must be decided by collecting data in the United States.

Example 1.6   Return to the experiment of tossing two coins performed in Exercise 1.1.2. Some people think that an adequate probability model is to take $S = \{0,1,2\}$ corresponding to "zero heads," "one head," and "two heads" and to assign a probability of $\frac{1}{3}$ to each of these possible outcomes. Others say we ought to keep the coins distinct and use $S = \{(H, H),(H, T),(T, H),(T, T)\}$. Here (H, T) stands for "heads on the first coin and tails on the second" and so on. They would assign a probability of $\frac{1}{4}$ to each of these four possible outcomes. In each case the probability of any event can be found by adding up the probabilities of the individual outcomes which are members of the event. For example, if A is the event "one head and one tail," then under the first model $P(A) = \frac{1}{3}$; but under the second model, $A = \{(H, T),(T, H)\}$, and $P(A) = \frac{1}{2}$. Both models are legitimate in the sense that they satisfy the conditions of Definition 1.4. There are good reasons to think that the second model comes much closer to describing normal coin tossing. But these reasons must be checked against actual experience. What was the relative frequency of event A in the experiment with real coins in Exercise 1.1.2?

Example 1.7   Schwarz, Williams, and Thompson (Example 1.1) are awaiting word on which of them has been promoted. Williams remarks that, from what he knows of their work records, he guesses that he and Thompson each have a probability of $\frac{1}{2}$ of being promoted while Schwarz has a probability of $\frac{2}{3}$. (These are "personal probabilities" reflecting Williams' assessment of likelihood.) Williams has assigned probabilities to the three specific events A, B, and C on page 10, but he has not given a probability measure. (*Note: A, B*, and *C* are *not* individual outcomes since each specifies the fate of only one of the workers.) We cannot assign a probability to the event $A \cup B$ (at least one of Schwarz and Thompson is promoted) on the basis of what he has said. A probability measure must assign probabilities to *all* events. Williams might do this if we asked, but he has not yet done so. Notice that it is *not* true that

$$P(A \cup B) = P(A) + P(B)$$
$$= \frac{2}{3} + \frac{1}{2}$$
$$= \frac{7}{6}$$

In fact, the result is ridiculous, since it says that the probability that either Schwarz or Thompson will be promoted exceeds 1. We cannot find $P(A \cup B)$ in this way because A and B are not disjoint—it is possible for both men to be promoted.

Exercise 1.4

1.   A student announces to his roommate that he has probability 90 percent of getting an A on an upcoming quiz, probability 50 percent of a B, and only probability 30 percent of a C or lower. Criticize his assignment of probabilities.

2.   Using the data on Danish social mobility, what is the probability that the son of a lower-class (class 1) father

a.   remains in class 1?
b.   ends up in one of the bottom three classes?
c.   ends up in one of the top two *or* one of the bottom two classes?
d.   does not remain in his father's class?

3.   If a "fair" die is rolled once, it is reasonable to assign probability $\frac{1}{6}$ to each outcome in the sample space $S = \{1,2,3,4,5,6\}$. As in Examples 1.4 and 1.5, a probability measure P on S can be defined by assigning to any event A the sum of the probabilities of the outcomes which are members of A. Find the probabilities of the following events using this probability model.

    A: Roll a 3.
    B: Roll an odd number.
    C: Roll an even number.
    D: Roll a number less than 3.
    E: $A \cup D$.
    F: $B \cup C$.
    G: $A \cap B$.
    H: A^c.

4.   Our car is parked in a towaway zone which will be checked by the police during the next hour. In an attempt to give a mathematical model which reflects our feeling that the check will be made at a random time during the hour, we assign probabilities as follows. Measuring time in

minutes, we take the sample space to be

$$S = \{1,2,3, \ldots, 60\}$$

where outcome i means that the police arrive during minute i. We assign any event (subset of S) A the probability

$$P(A) = \frac{1}{60} \times \{\text{number of minutes in } A\}$$

Answer the following questions:

a. Have we given a probability measure?
b. What is the probability that the police will arrive during the first 15 min of the hour?
c. What is the probability that the police will arrive during the first 15 min or the last 15 min of the hour?
d. What is the probability that the police will arrive during an odd-numbered minute (1, 3, 5, etc.)?

5.   The proportions of American workers in 1970 falling in each of four categories by race and job description were as follows.

	White	Nonwhite
White collar	0.45	0.03
Blue collar	0.44	0.08

a. What is a sample space for the experiment of choosing a worker and classifying him by race and job category as in the table?
b. Using the national proportions given as probabilities, what is the probability that the worker is nonwhite? That he holds a white-collar job?

6.   Salesmen of the Fairhair Wig Company call on five customers per day. The sales manager claims that the performance of an average salesman can be summarized by the following assignment of probabilities. Let A_i be the event that the salesman makes exactly i sales in a day. Then

$$P(A_0) = 0.3 \qquad P(A_3) = 0.1$$
$$P(A_1) = 0.3 \qquad P(A_4) = 0.1$$
$$P(A_2) = 0.2 \qquad P(A_5) = 0.1$$

Has the manager given a probability measure?

7.   To aid in deciding how many of a certain replacement part to keep in stock, a plant manager records requests per week for the part over a long period of time. If A_i is the event that i requests are received in a week, he observes the following relative frequencies.

$$P(A_1) = 0.3 \qquad P(A_3) = 0.1$$
$$P(A_2) = 0.1 \qquad P(A_4) = 0.05$$

a. If no more than four requests per week are ever received, what must be the value of $P(A_0)$, the probability of receiving no requests in a week, if we are to have a probability measure?

b. Express the event "no more than two requests" in terms of the A_i, and find its probability.
c. Express the event "at least three requests" in terms of the A_i and find its probability.
d. Explain why the sum of your answers to Exercise 1.4.7b and c should be 1.

8.   Recall that in Example 1.7 any of Schwarz, Thompson, and Williams can be promoted, and that A, B, C stand for the events that (in order) Schwarz is promoted, Thompson is promoted, Williams is promoted.

a. What is a sample space (list of all possible outcomes) for this experiment?
b. Give a probability measure on this sample space which assigns $P(A) = \frac{2}{3}$, $P(B) = P(C) = \frac{1}{2}$. (There are many correct ways of doing this. The first step is to express A, B, and C as collections of individual possible outcomes listed in Exercise 1.4.8a.)

9.   Applicants for an outside sales position are given an aptitude test before being hired. Experience shows that only 10 percent of all applicants both pass the test and do badly on the job, while 30 percent pass the test and do well on the job.

a. What percentage of applicants pass the test?
b. What percentage of applicants either fail the test or do badly on the job?
c. What percentage of applicants do well on the job?

10.   A dart player throws a dart at the dartboard below and receives a score equal to the number marked in the region he hits.

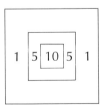

The area of the region marked 5 is 8 times that of the center region (marked 10), and the area of the region marked 1 is 16 times that of the center region.
    Suppose that the player always hits the board and that the probability that he hits any region is proportional to the area of that region.

a. What is the probability he scores 1?
b. What is the probability he scores 5?
c. What is the probability he scores 10?
d. What is the probability he scores 5 or more?

1.5   PROPERTIES OF PROBABILITY MEASURES

Probability measures have many properties other than positivity, totality, and additivity. We are now going to deduce some of these properties from Definition 1.4. The point of this work is that we will then know that any assignment of probabilities to events having the three properties in Definition 1.4 automatically has all these other properties. Each theorem we prove states an intuitively reasonable property of probability. Only if we thus grasp the meaning of these statements will we be able to apply them easily.

The first property we wish to establish is that the probability that an event does not occur is 1 minus the probability that it does occur. This is surely true if we interpret probability as the proportion of repetitions of a random experiment on which the event occurs. We now show that Definition 1.4 implies this reasonable result.

THEOREM 1.1   For any event A, $P(A^c) = 1 - P(A)$.

PROOF   Any event A satisfies $A \cup A^c = S$ (any outcome is either in A or not in A) and $A \cap A^c = \varnothing$ (no outcome is both in A and not in A). So by additivity

$$P(A \cup A^c) = P(A) + P(A^c)$$

and by totality

$$P(A \cup A^c) = P(S) = 1$$

So we see that $P(A) + P(A^c) = 1$, which yields the desired result.

Theorem 1.1 allows us to find the probability of an event by first finding the probability of its complement. This sometimes saves a good deal of work.

Example 1.8   In Example 1.5 on Danish social mobility, let A be the event "the son of a class 1 father changes class." Then A^c is the event "the son of a class 1 father remains in class 1" and

$$P(A) = 1 - P(A^c) = 1 - 0.46 = 0.54$$

We could also have computed $P(A)$ directly by adding the probabilities that the son ends up in classes 2, 3, 4, and 5. In this simple model, the use of Theorem 1.1 saved us only a minor amount of arithmetic.

From Theorem 1.1 we can deduce a corollary which formalizes our feeling that an "impossible" event (one containing no outcomes) should have probability zero.

COROLLARY 1.1 $P(\varnothing) = 0$.

PROOF Since $\varnothing = S^c$, Theorem 1.1 and totality state that

$$P(\varnothing) = 1 - P(S) = 1 - 1 = 0$$

A second important property of probability measures is that if event B implies event A, then $P(A)$ is greater than or equal to $P(B)$. This is reasonable, for if A must occur whenever B does, we feel that A is "larger" than B (look at Fig. 1.5) and must have greater probability. Thus in Example 1.4, the probability that a working woman is married is surely at least as large as the probability that she is married and her husband is present.

THEOREM 1.2 If $A \supset B$, then $P(A) \geq P(B)$.

PROOF If $A \supset B$, we can write

$$A = B \cup (A \cap B^c)$$

and the sets B and $A \cap B^c$ are disjoint. (Draw a Venn diagram to check the truth of that statement.) So by additivity

$$P(A) = P(B) + P(A \cap B^c)$$

But by positivity $P(A \cap B^c) \geq 0$, so that $P(A)$, being equal to $P(B)$ plus a nonnegative number, is at least as great as $P(B)$.

Theorem 1.2 is sometimes useful when an event B is so complicated that we cannot easily calculate $P(B)$. It may be that we can calculate $P(A)$ for an event A containing B. We then have an upper bound for $P(B)$, i.e., a number $P(A)$ which is at least as great as $P(B)$. We will use this idea in Sec. 1.18. For now we will use Theorem 1.2 only to show that any event must have probability no greater than 1 when the assignment of probabilities satisfies Definition 1.4. This fact often helps us detect errors in computing probabilities, for if we obtain a probability greater than 1, we know that we have made a mistake.

COROLLARY 1.2 For any event A, $P(A) \leq 1$.

PROOF Any event A satisfies $A \subset S$, and since $P(S) = 1$, Theorem 1.2 gives

$$P(A) \leq P(S) = 1$$

To this point we have concentrated on properties common to all legitimate assignments of probability. It is time to say something about the construction of actual probability models, though the presentation of any complicated models must be postponed to the following sections. A fundamental distinction among probability models is based on the nature of the sample space. The simplest sample spaces are those which contain only finitely many individual outcomes. Such finite sample spaces are adequate for describing many interesting random experiments. We confine ourselves almost entirely to finite sample spaces. But in some cases, such as measuring the lifetime in service of a transistor or the reaction time of a subject who has been given a certain drug, it is convenient to use infinite sample spaces. Two types of infinite sample space are discussed in Secs. 1.9 and 1.19, along with examples of their use. The most important point to remember if we encounter infinite sample spaces in the future is that the rules of probability which we have developed apply to probability measures on any sample space.

When the sample space is finite, we can begin by assigning probabilities to individual possible outcomes and then obtain the probability of any event by adding up the probabilities of the individual outcomes which are members of the event. This simplifies our work (at least conceptually) since we need not assign probabilities directly to events, but only to individual outcomes. We have used this idea in a number of examples and exercises, beginning with Examples 1.4 and 1.5. We now state formally which assignments of probabilities to individual outcomes produce a probability measure. The requirements are that the probability of each individual outcome be nonnegative and that the sum of the probabilities of all the individual outcomes be 1. The proof is just a formalization of the reasoning used in Example 1.4, so we will omit it. As with Definition 1.4, this theorem does not help you to assign correct probabilities in a particular case. It only says which assignments are mathematically legitimate. Exercises 1.2.3 to 1.2.5 indicate three ways to assign probabilities to individual outcomes in practice: use of actual frequency data, a priori reasoning, and personal assessment of likelihood.

THEOREM 1.3 Suppose $S = \{s_1, s_2, \ldots, s_n\}$ is a finite sample space, that p_i is the probability of outcome s_i for each $i = 1, \ldots, n$, and that

a.   all $p_i \geq 0$
b. $p_1 + p_2 + \cdots + p_n = 1$

For any event $A = \{s_k, s_l, \ldots, s_t\}$ define

$$P(A) = p_k + p_l + \cdots + p_t$$

Then P is a probability measure on S.

Example 1.9 Here are data on the proportions of Americans who entered fifth grade in 1962 for whom each successive grade was the last completed. (Some, of course, went on to college.)

Grade	4	5	6	7	8	9	10	11	12
Proportion	0.010	0.007	0.007	0.013	0.032	0.068	0.070	0.041	0.752

Do these numbers give a legitimate probability assignment for the experiment of choosing an American who entered fifth grade in 1962 and asking what his last completed grade was? By Theorem 1.3 they do, for the numbers given are all nonnegative and add up to 1. The probability of any event can be easily found by addition. For example, if A is the event "completed at least 1 year of high school," then

$$P(A) = 0.068 + 0.070 + 0.041 + 0.752 = 0.931$$

Exercise 1.5

1.   In Example 1.9 what is the probability that a person who entered fifth grade in 1962 did not graduate from high school (i.e., did not complete twelfth grade)?

2.   The distribution of blood types among the American population is: 40 percent type A, 7 percent type B, 43 percent type O, 10 percent type AB. Show by using Theorem 1.3 that these frequency data give a probability measure on the sample space {A,B,O,AB} for the experiment of choosing an individual at random and typing his blood. An individual with type-B blood can safely receive transfusions only from persons with type B or AB. What is the probability that a randomly chosen individual is an acceptable blood donor for a type-B patient?

3.   Here are data on the proportions of white and nonwhite employed persons whose occupation fell in various job categories in 1970.

Job Category	White	Nonwhite
Professional	0.147	0.091
Managerial	0.113	0.035
Clerical	0.180	0.132
Salesworkers	0.067	0.021
Craftsmen	0.135	0.082
Operatives	0.170	0.237
Non-farm laborers	0.041	0.103
Service industries	0.107	0.260
Farmhands	0.040	0.039

Show that these figures give two legitimate assignments of probability—one for the experiment of choosing a white worker and categorizing his occupation, the second for choosing a nonwhite worker. The first four categories make up the "white-collar" occupations. What is the probability that a white worker will have a white-collar job? That a nonwhite worker will have a white-collar job?

4.   Zeke the Greek, asked by his friends to predict the Big Ten basketball champion, follows the modern practice of giving probabilistic predictions. He says, "Purdue's probability of winning is twice Ohio State's. Michigan and Illinois each have probability 0.1 of winning, but Ohio State's probability is 3 times that. Nobody else has a chance." Has Zeke

given a probability model for this random experiment with 10 possible outcomes?

5.   Andrew, Bob, and Carol are applying for a job. Exactly one of the three will be hired.

a. Bob says, "I think that Andrew and I have equal probability of being hired, but that, because of pressure to hire women, Carol is twice as likely to be hired as either of us." What probabilities has Bob assigned to the three possible outcomes?
b. Carol says, "Since Bob is unqualified, he has only probability 0.1 of being hired. Because of prejudice against women, my probability of being hired is 0.1 less than Andrew's." What probabilities has Carol assigned to the outcomes?

6.   It is reasonable (as we will see in Sec. 1.7) to expect that when we throw two balanced dice, each pair of faces has the same probability of coming up as any other pair.

a. List the members of the sample space for this experiment. What is the probability of each individual outcome?
b. Let A_i be the event that the sum of the two faces is i. Find $P(A_i)$ for $i = 2, 3, \ldots , 12$.
c. Some people are very interested in the event B that the sum of the two faces is 7 or 11. Find $P(B)$ by expressing B in terms of the events A_i and using the results of part b. Find $P(B)$ also by listing the individual outcomes in B and using Theorem 1.3.

7.   License plates of cars assigned to a certain government agency are numbered CIA 1, CIA 2, . . . , CIA 1000. You are assigned one of these 1000 cars "at random" so that you are equally likely to receive any of them.

a. What is the probability that the first number on the license plate is each of $1, 2, \ldots , 9$?
b. What is the probability that the last number on the license plate is each of $0, 1, 2, \ldots , 9$?

8.   Example 1.7 shows that it is *not* always true that

$$P(A \cup B) = P(A) + P(B)$$

We claim that it *is* always true that

$$P(A \cup B) = P(A) + P(B) - P(A \cap B)$$

Prove this result by applying additivity and the facts about sets derived in Exercise 1.3.10.

9.   Consolidated Builders has bid on two construction projects. The company president feels that the probability of being awarded the first is 0.4, the probability of being awarded the second is 0.6, and the probability

of being awarded both is 0.2. Use the result of Exercise 1.5.8 to find the probability that Consolidated is awarded at least one of the contracts. Find also the probability of being awarded neither contract.

10. Suppose that events A, B, and C are disjoint in pairs. Apply additivity twice to show that

$$P(A \cup B \cup C) = P(A) + P(B) + P(C)$$

Give an example (a Venn diagram will do) to show that events for which $A \cap B \cap C = \emptyset$ need not be disjoint in pairs.

11. Give an example to show that it is *not* always true that

$$P(A \cap B) = P(A)P(B)$$

(*Hint:* Very easy examples can be given using the probability model of Example 1.3.2)

12. A psychologist is arranging an experiment in which a group of people collaborate in making a decision which is rewarded with 10 cents, 20 cents, . . . , 100 cents depending on its "correctness." Actually the psychologist will decide the degree of "correctness" randomly—he wants to observe the effect of the resulting frustration on the group. So the reward for any decision is the outcome of a random experiment with $S = \{10, 20, . . . , 100\}$. To avoid paying his subjects too much, the psychologist decides to assign probabilities inversely proportional to the size of the reward. Thus $p_{10} = k/10$, $p_{20} = k/20$, and so on, for some constant k. What probabilities does the psychologist assign to each possible reward?

1.6 CONDITIONAL PROBABILITY

Although we have discussed properties of probability measures and have seen how to give probability measures on finite sample spaces by assigning probabilities to individual outcomes, we have not yet given any complicated probability models. Most complicated random experiments are such that we find it hard to describe them without referring to the influence or lack of influence of one event on another. For example, suppose that a firm rates the efficiency of each of its departments at the end of each month as "inefficient," "acceptable," or "efficient." It is very likely that the probability of being efficient next month given the information that the department in question was efficient this month is not the same as the probability of being efficient next month given the information that the department was inefficient this month. To take another example, in conducting an opinion poll to determine what proportion of the population believes marijuana should be legalized, great pains are taken to see that respondents' answers do not influence each other. This might not be the case if, for example, the pollster stupidly told a respondent what the last respondent's opinion was. Again a reasonable probability model must reflect the lack of dependence of each response on previous responses. In this section we will concentrate on a mathematical formulation of dependence of events. Lack of dependence will be studied in the following section.

Consider two events *A* and *B*. We want to make formal the idea that the probability of *A given the knowledge that B has occurred* need not be the same as *P(A)*. To clarify this, let us look at a simple probability model.

Example 1.10   We roll a balanced die once so that $S = \{1,2,3,4,5,6\}$, and each individual outcome has probability $\frac{1}{6}$. Let *A* be the event that 1 is rolled, and *B* the event that an odd number is rolled. Suppose we are told that *B* occurred, but are not told the actual outcome of the roll. The only possible outcomes are now 1, 3, and 5—these are just the members of *B*. What probabilities should we assign to these outcomes now that we know that one of them has occurred? They are equally likely under the original assignment of probabilities, and we have no reason to think that the relative chance of the odd outcomes has changed. So it seems reasonable to say that if we know that *B* has occurred, the probabilities of 1, 3, and 5 are each $\frac{1}{3}$. We say that the *conditional probability* of *A*, given *B*, is $\frac{1}{3}$. Notice that this is not the same as *P(A)* (which is $\frac{1}{6}$) either numerically or conceptually. If *C* is the event that a 2 is rolled, what probability should be assigned to *C* if we know that *B* occurred? If the outcome was odd (event *B*), it is impossible that a 2 was rolled. So the conditional probability of *C* given *B* is 0. Of course, $P(C) = \frac{1}{6}$ is not 0, which reminds us again that conditional probabilities can be very different from the original assignment of probabilities.

This example illustrates what we expect in general. If we know that *B* has occurred, the sample space is in effect reduced from *S* to *B*. Therefore the conditional probability of any event *A* given the occurrence of *B* is the same as the conditional probability of $A \cap B$. The part of *A* not contained in *B* (namely $A \cap B^c$) has conditional probability 0. Figure 1.7 illustrates the sets involved.

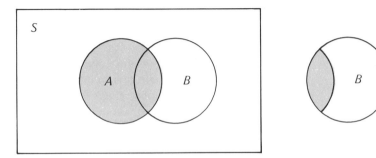

Figure 1.7   Original sample space *S* with event *A* shaded, and sample space reduced to *B* with $A \cap B$ shaded

What is more, knowing only that *B* has occurred does not give any information about the relative chances of the various events contained in *B*. They are still as given by the original probability measure. This means that conditional on *B* occurring, any event *A* has probability equal to the proportion of the original probability of *B* assigned to *A* ∩ *B*. This proportion is $P(A \cap B)/P(B)$. Notice that this ratio equals 1 for $A = B$ and 0 for $A = B^c$, as should be the case. In terms of relative frequencies $P(A|B)$ is the proportion of occurrences of *B* on which *A* also occurs. You should check that this is $P(A \cap B)/P(B)$ since $P(B)$ is the proportion of occurrences of *B* and $P(A \cap B)$ the proportion of occurrences of *A* and *B* simultaneously. We have arrived at the following definition.

DEFINITION 1.5   Suppose *P* is a probability measure on a sample space *S*, and *B* is an event with $P(B) > 0$. The *conditional probability of A, given B*, is defined for any event *A* by

$$P(A|B) = \frac{P(A \cap B)}{P(B)}$$

It is very important not to be confused by the notation $P(A|B)$. The two events *A* and *B* play very different roles. *A* is the event whose probability we are calculating. *B* is the event which we know occurred—it represents partial knowledge of the outcome of the random experiment. It makes no sense to ask for probabilities conditional on the occurrence of an event which has probability 0. That is why Definition 1.5 requires that $P(B)$ be positive.

Example 1.11   In Example 1.9, let *B* be the event "completed at least 1 year of high school" and *A* the event "completed twelfth grade." Then $A \subset B$ (to complete twelfth grade one must complete at least 1 year) so that $A \cap B = A$. The conditional probability that a person known to have finished at least 1 year of high school finished twelfth grade is therefore

$$P(A|B) = \frac{P(A \cap B)}{P(B)} = \frac{P(A)}{P(B)} = \frac{0.752}{0.931} = 0.81$$

[$P(B)$ was computed in Example 1.9.] The calculation of conditional probability in this case reduces to concluding that 81 percent of all students who completed at least 1 year of high school graduated. Notice that $P(A) = 0.752$ means that 75 percent of all students (we are only speaking of those who entered fifth grade in 1962) graduated from high school. Knowledge of event *B* raised the percentage of graduates, as we would expect.

In some probability models (especially those used for various games of chance), conditional probabilities can be computed using Definition 1.5. When probability models are constructed for social, economic, or biological phenomena, the artificial symmetry present in games of chance is often missing. On the other hand, we often know conditional probabilities in advance and wish to take advantage of that knowledge in building the model. Definition 1.5 is then most often used in this form: If $P(B)$ and $P(A|B)$ are known, then we can find $P(A \cap B)$ from the *multiplication rule*

$$P(A \cap B) = P(A|B)P(B)$$

Example 1.12   An international corporation purchases electrical equipment from a Swiss supplier. The controller of the corporation feels that the probability is 0.7 that the dollar will be devalued next week (event B). He also feels that *if* the dollar is devalued, the conditional probability is 0.6 that the supplier will demand renegotiation of the current contract (event A). What personal probability has the controller assigned to the event that the dollar is de-valued *and* renegotiation is demanded? This is exactly the form

$$P(A \cap B) = P(A|B)P(B)$$
$$= (0.6)(0.7) = 0.42$$

Example 1.13   A psychologist trains rats to move from one compartment to another by ringing a bell and administering a shock to the rat if it does not respond. Long-term data show that the probability that a rat responds on the first trial is 0.2, and that the probability that a rat responds on the second trial given that it responded on the first is 0.3. The multiplication rule therefore tells us that the probability is $(0.2)(0.3) = 0.06$ that a rat responds on both of the first two trials.

Exercise 1.6   1.   In Example 1.10, find the conditional probability of the event $C = \{1,2,3\}$ given that event B occurred.

2.   Mortality tables show that of 100,000 people at age 1, 95,594 live to be 18 and 94,917 live to be 21. What is the conditional probability that a person who reaches 18 will live to be 21?

3.   Use the probability distributions of Exercise 1.5.3 to answer the following questions.

a.   Given the knowledge that a white worker holds a white-collar job, what is the conditional probability that the worker is a professional?

b.   Answer the same question for nonwhite workers.

c.   In 1970, 11 percent of the labor force was composed of nonwhites. The second column in Exercise 1.5.3 can be interpreted as giving condi-tional probabilities that a worker's job falls in certain categories given that he is nonwhite. What is the probability that a worker is a nonwhite professional?

4.   Use the model for social mobility in Denmark (Example 1.5) to find the conditional probability that a son of a class 1 father ends up in each of classes 1, 2, 3, 4, 5 given that he leaves his father's class (class 1). Is this new assignment of probabilities on $S = \{1,2,3,4,5\}$ a probability measure?

5.   Refer to Exercise 1.5.2. What is the conditional probability that a person has type-B blood given that he is an acceptable donor for a type-B patient?

6.   Let us suppose (as is not quite true) that among families having two children the four possible outcomes

   $\{BB,BG,GB,GG\}$

are equally likely. (Here BG stands for "elder child is a boy, younger child is a girl," and so on.)

a.   If a family with two children has a boy, what is the conditional probability that their elder child is a boy?
b.   If the elder child is a boy, what is the probability that they have one child of each sex?

7.   A congressman's poll revealed that of his male constituents 62 percent would be willing to go to war over Berlin, while only 48 percent of his female constituents would be willing to do so. He knows that 53 percent of his constituents are male.

a.   What percentage of the Congressman's constituents are male *and* ready to fight for Berlin?
b.   What percentage are female and ready to fight?
c.   Show from the additivity property of probability measures that for any sets A and B,

   $$P(A) = P(A \cap B) + P(A \cap B^c)$$

d.   Combine the results of Exercise 1.6.7a, b, and c to find the percentage of the congressman's constituents who are willing to go to war over Berlin.

8.   In Example 1.12, suppose the controller knows that the Swiss supplier will demand renegotiation of the contract only if the dollar is devalued. Show that he can now say that 0.42 is the probability that the supplier will demand renegotiation. (*Hint:* Use the result of Exercise 1.6.7c.)

9.   Use the model and results of Exercise 1.5.6 to find the conditional probability that the face of the first die is odd given that the sum of the two faces is odd.

10.   Probabilities are assigned to certain events as indicated in the following Venn diagram. For example, $P(A \cap B \cap C) = \frac{1}{16}$ and $P(A \cap B^c \cap C^c) = \frac{1}{8}$.

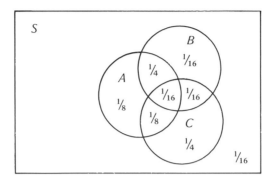

Use this information to compute the following:

a. $P(B)$ d. $P(B|A \cup C)$
b. $P(A \cup B)$ e. $P(A \cup B|C)$
c. $P(B|A)$

11.   Use Definition 1.5 and additivity of P to show that if $A \cap B = \emptyset$ and $P(C) > 0$, then

$$P(A \cup B|C) = P(A|C) + P(B|C)$$

This says that conditional probability given a fixed set C is additive. Give an example to show that it is *not* always true that

$$P(C|A \cup B) = P(C|A) + P(C|B)$$

12.   Use Definition 1.5 twice to show that

$$P(A \cap B \cap C) = P(A|B \cap C)P(B|C)P(C)$$

when all sets have positive probabilities.

13.   Suppose that the probability is $1/2$ that the Dow Jones Industrial Average advances on a fixed trading day. Suppose further that the probability of an advance given an advance on the previous day is $2/3$, and that knowledge of market behavior on yet earlier days gives no additional information. Use the result of Exercise 1.6.12 to find the probability that three consecutive advances occur.

1.7   INDEPENDENCE

The definition of the conditional probability $P(A|B)$ is intended to make precise the idea that the fact that event B occurs may cause a reassignment of probability to event A differing from the original assignment $P(A)$. Very often we are trying to construct a probability model for a random experiment in which we feel that the fact that B occurs gives no information about

A. We can make this precise in the model by saying that the new assignment of probability $P(A|B)$ is the same as the original assignment $P(A)$.

DEFINITION 1.6 Suppose that events A and B both have positive probability. A and B are *independent* if

$$P(A|B) = P(A)$$

It is natural to ask about $P(B|A)$ also, so we immediately give the following important fact.

THEOREM 1.4 Suppose that $P(A) > 0$ and $P(B) > 0$. Then the following three statements are equivalent:

a. $P(A|B) = P(A)$
b. $P(A \cap B) = P(A)P(B)$
c. $P(B|A) = P(B)$

PROOF To show that Theorem 1.4a, b, and c all mean the same thing, we will show that a is equivalent to b. The proof that c is also equivalent to b is exactly the same.

We know that by the multiplication rule

$$P(A \cap B) = P(A|B)P(B)$$

So if a holds, we get

$$P(A \cap B) = P(A)P(B)$$

which is b. If, on the other hand, b is true, then by Definition 1.5 and b we see that

$$P(A|B) = \frac{P(A \cap B)}{P(B)}$$
$$= \frac{P(A)P(B)}{P(B)} = P(A)$$

So a holds. Thus a and b are equivalent.

Theorem 1.4 gives us three ways of stating that A and B are independent. Statements a and c connect independence with the idea of conditional probability. Statement b is especially important in model building. You learned earlier that probabilities of unions can be found by using

$$P(A \cup B) = P(A) + P(B)$$

if A and B are disjoint, but not otherwise. Theorem 1.4 now says that probabilities of intersections can be found by using

$$P(A \cap B) = P(A)P(B)$$

if A and B are independent, but not otherwise. You must resist the tempta-
tion to use these simple formulas when conditions needed are not present.
(See Exercises 1.5.8 and 1.5.11.) You must also be clear that disjointness and
independence are quite different. If A and B are disjoint, then knowing that
B occurs tells you that A cannot occur. So A and B ought not to be independent if they are disjoint. More formally, suppose $P(A) > 0$ and $P(B) > 0$. If A
and B are disjoint, $A \cap B = \emptyset$ and $P(A \cap B) = P(\emptyset) = 0$. But $P(A)P(B) > 0$,
so Theorem 1.4b does not hold, and these events are not independent.

   The definition of independence gives us a way to build into a probability model our feeling that certain events are physically independent—do
not give information about each other. Sometimes events which do not
seem to be physically independent turn out to be independent in the sense
of Definition 1.6. So independence in the probability model and physical
or "real-world" independence are not always the same thing. But when
physical independence seems to be present in a random experiment, we
usually assume independence in building the probability model. An example will illustrate this process.

Example 1.14   Suppose an experiment consists of throwing two dice and recording (in
order) the two numbers on the upward faces. The sample space, as we know
from Exercise 1.5.6, is the set of all pairs (x, y), where x is the number showing on the first die and y the number showing on the second. Both x and y
can be any of the numbers 1, 2, 3, 4, 5, 6, and so S has 36 members. What
probability measure should we use on S? We will make two reasonable
assumptions about the experiment. The first is that the dice are both balanced. That means that each of the six possible outcomes of either die has
probability $1/6$ (each face is equally likely to come up). The second assumption is that the dice fall independently. This physical independence
suggests that knowledge of which face came up on one gives no information
about which face came up on the other. These assumptions completely
specify the probability model, for we can now compute the probability of
any individual outcome. For example, using Theorem 1.4b,

P(1 on first die and 6 on second die)
$$= P(1 \text{ on first die})P(6 \text{ on second die})$$
$$= 1/6 \cdot 1/6 = 1/36$$

Similarly, each of the 36 possible outcomes has probability $1/36$. By Theorem
1.3, this determines a probability measure.

   This example is important, because it illustrates a basic method used in
constructing probability models. We began with reasonable assumptions
about the experiment—balanced dice thrown independently. These we
made precise by translating them into mathematical language—equally
likely outcomes for any single die, independence of outcomes of the first

and second dice. If the assumptions about the physical experiment are wrong, the resulting model will not describe the experiment. (By the way, actual frequency data show that the assumption of independence in experiments such as this is justified if professional manipulators are not throwing the dice.)

To construct more complicated models, it is necessary to discuss independence of many events. This is not as simple as it might seem. It is possible (see Exercise 1.7.2) to find events A, B, and C such that any two are independent, but it does not seem reasonable to say that all three together are. We therefore define independence of three or more events to be stronger than independence of each pair. Theorem 1.4*b* is our guide.

DEFINITION 1.7 Events A_1, A_2, ..., A_n (all having positive probability) are *independent* if the probability of the intersection of any collection of the A_i is the product of the probabilities of the events in the collection.

Independence of two events is the same using either Definition 1.6 or Definition 1.7, by Theorem 1.4. Independence of three events A, B, and C means that all the following are true:

$$P(A \cap B) = P(A)P(B)$$

$$P(A \cap C) = P(A)P(C)$$

$$P(B \cap C) = P(B)P(C)$$

$$P(A \cap B \cap C) = P(A)P(B)P(C)$$

Independence is often *assumed* in setting up models, as in the following examples.

Example 1.15   A manufactured product contains five subassemblies, and quality control procedures guarantee that only 2 percent of each type of subassembly will be defective. The finished product will be defective if any of the subassemblies are. What is the probability that the finished product will not be defective? Let this event be B, and A_i be the event that the ith subassembly is not defective. Then

$$B = A_1 \cap A_2 \cap A_3 \cap A_4 \cap A_5$$

It is reasonable to assume that whether or not one subassembly is defective tells us nothing about whether or not another subassembly going into the same finished product is defective. We therefore assume that the A_i are independent. So

$$P(B) = P(A_1)P(A_2)P(A_3)P(A_4)P(A_5)$$
$$= (0.98)^5$$
$$= 0.904$$

[Observe that we used Theorem 1.1 to get $P(A_i) = 0.98$.] It is worth noticing that although 98 percent of the subassemblies are perfect, only about 90 percent of the finished products are. What is more, a product with more subassemblies, each with probability 0.98 of being perfect, is even less likely to work. There is a moral here somewhere.

Example 1.16   As another example of model building using independence, consider this simple model for group problem solving. A group of n individuals works on a problem, which must be solved by a given deadline. Suppose the ith individual has probability p_i of solving the problem before the deadline. If the n individuals work on the problem independently, what is the probability that the group solves the problem? Let B be the event that the group solves the problem, A_i the event that the ith individual solves it. Then

$$B = A_1 \cup A_2 \cup \cdots \cup A_n$$

The events A_i are *not* disjoint (several individuals may succeed), so this is little help. However,

$$B^c = A_1{}^c \cap A_2{}^c \cap \cdots \cap A_n{}^c$$

[This is a consequence of the set-theoretic fact that

$$(A_1 \cup A_2 \cup \cdots \cup A_n)^c = A_1{}^c \cap A_2{}^c \cap \cdots \cap A_n{}^c$$

but we should be able to see that it simply says, "the group fails exactly when all individuals fail."]

Now if the A_i are independent, so are the events $A_i{}^c$ (see Exercise 1.7.9). So

$$\begin{aligned} P(B^c) &= P(A_1{}^c)P(A_2{}^c) \cdots P(A_n{}^c) \\ &= (1 - p_1)(1 - p_2) \cdots (1 - p_n) \end{aligned}$$

and

$$\begin{aligned} P(B) &= 1 - P(B^c) \\ &= 1 - (1 - p_1)(1 - p_2) \cdots (1 - p_n) \end{aligned}$$

If all n individuals are of equal ability, so that all p_i are equal (say p is the common value), then the probability of a solution is

$$P(B) = 1 - (1 - p)^n$$

If $p = 0.10$, then with two people working $P(B) = 1 - (0.9)^2 = 0.19$, with five people $P(B) = 0.41$, and with ten individuals $P(B) = 0.65$.

If you are properly alert, you will object that the assumption that individuals work independently is not realistic in Example 1.16. Group problem solving usually involves collaboration, so this model seems more appropriate as a model for (say) whether any of *n* students working individually will solve a certain examination problem. That is quite right. But a model for collaboration would be complicated, perhaps hopelessly so. Literature in all the sciences is full of oversimplified probability models. Understanding concepts such as independence will help us judge how appropriate a particular model is.

Exercise 1.7   1.   The table below shows the distribution of "top wealthholders" in the United States by age and sex in 1962. For example, the top left entry says the probability is 0.24 that a top wealthholder is a male under 50.

	Male	Female
Under 50 years	0.24	0.12
50 to 69 years	0.29	0.19
70 years and over	0.08	0.08

Use this probability distribution for the experiment of choosing a top wealthholder at random. Are the events

A: wealthholder chosen is a male

B: wealthholder chosen is under 50

independent?

2.   In the model for throwing two balanced dice independently discussed in Example 1.14, define the events

A: odd face on first die

B: odd face on second die

C: sum of the faces is odd

Show that each pair of events from these three is independent (it may help to notice that C is the same as "one face even, one face odd"). Since $A \cap B \cap C = \varnothing$, we feel that there is nonetheless some dependence among A, B, and C. Show that they are not independent according to Definition 1.7.

3.   A general can plan his campaign to fight one decisive battle, which he estimates he has probability 0.80 of winning, or to fight three smaller battles, each of which he has probability 0.90 of winning. To win the campaign by the second plan, he must win all three smaller battles. He believes the outcomes of the small battles to be independent. Which plan should he choose?

4.   A local farmer claims to be able to detect the presence or absence of water by using an old coathanger and his extraordinary abilities. A skeptic confronts him with four sealed tins and asks him whether each contains water. The farmer is right all four times. Assuming his four answers to be independent, what is the probability of this string of successes if he is right

only 50 percent of the time? (If he is "just guessing," he'll be right about half the time by chance.) Do you think the skeptic should concede the farmer's ability?

5.   In a certain school district, 28 percent of the population is black. All five school administrators are white. Some local leaders charge discrimination, and support their charge by a computation of the probability of five whites if choice were made at random from the district's population. Make such a computation, being careful to state your assumptions.

The school board replies that administrators are chosen from individuals having master's degrees in school administration, and that only 6 percent of these are black. They then compute the probability of five whites if the choice were made without discrimination from those holding this degree. What is the board's result? (The local leaders now discover that two of the five earned their degrees after being appointed to their jobs. Both sides now cease computing probabilities.)

6.   An industrial safety device contains a servomechanism and three sensors. The backup device contains a servomechanism and one sensor. Each servomechanism has probability 0.95 of remaining in operation for a month, and each sensor has probability 0.90 of remaining in operation during that time. All components operate (or fail to operate) independently.

a.   What is the probability that all components of the safety device remain in operation for a month?
b.   What is the probability that all components of the backup device remain in operation for a month?
c.   What is the probability that all components of at least one of the two devices remain in operation for a month?

7.   A maze contains the paths sketched below (movement is possible only in the direction of the arrows).

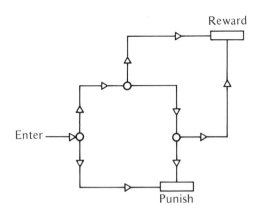

When a rat running this maze reaches a node (marked O), it makes a decision to follow one of the two paths other than the one he is on. If both paths are at right angles to the path he is on, he chooses each with probability $1/2$. If one path is straight ahead and the other is at a right angle, he chooses the straight path with probability $3/4$. All decisions are independent. Find the probabilities that the rat

a. reaches the "punish" square
b. reaches the "reward" square

8.   Let our experiment be to choose an American worker at random and note his race and job category. W will be the event that the worker is white. Then the data of Exercise 1.5.3 give conditional probabilities of various job categories given W or W^c. If A is the event that the worker is a professional, we see that

$$P(A|W) = 0.147 \qquad P(A|W^c) = 0.091$$

We feel that if A and W were independent, these conditional probabilities should be equal. Therefore we conclude that events A and W are *not* independent here.

Show that this reasoning is correct by proving: if events A and B are independent, and if $P(B)$ is not equal to 0 or 1, then

$$P(A|B) = P(A|B^c)$$

9.   Suppose that events A and B are independent

a.   Use additivity to show that

$$P(A \cap B^c) = P(A) - P(A \cap B)$$

b.   Use (a) and independence of A and B to show that A and B^c are also independent.
c.   Use independence of A and B^c and

$$P(A^c|B^c) = 1 - P(A|B^c)$$

to show that A^c and B^c are independent.

10.   A graduate student of management wanted to use the following model to construct a rating of executives. An executive, in the course of the job, makes one decision after another. Assume that each decision can be classified as correct or incorrect, that any executive has a fixed probability p of making a correct decision, and that the decisions are independent. Discuss the adequacy of this model.

1.8   BINOMIAL PROBABILITY DISTRIBUTIONS

Very often we must construct probability models for random experiments in which we observe the outcome of a fixed number of trials. This is the case if we record the scores of 35 students on a standardized test, observe the closing price of wheat futures for 10 days, record whether or not a rat responds to a stimulus on each of 20 trials, or check whether each of the 8000 transistors coming from an assembly line in a day passes inspection. Many different probability models can be constructed for such experiments, depending on the characteristics of the particular experiment. One model of great importance is appropriate when the following hold.

1. Each trial has only two possible outcomes. Call them "success" and "failure."
2. Outcomes of different trials are independent.
3. The probability of "success" is the same on each trial.

In other words, our random experiment consists of a fixed number (say n) of independent repetitions of the same experiment, each repetition resulting in a success with probability p (or in a failure with probability $1 - p$). This is the case if a coin is tossed n times (say "success" is a head), if a basketball player shoots n foul shots, if n patients suffering headache pains are given a sugar pill and asked in an hour if their headaches are gone, if n television viewers are asked if they recall a Zipf Beer commercial, and so on. Most experiments in which a phenomenon is followed over time (stock or commodity prices, responses of a rat as he becomes conditioned to respond to a stimulus, etc.) do *not* satisfy item 2, for successive trials are dependent. They may also fail to satisfy item 3, as in the case of the rat's responses or other learning situations. In such cases more elaborate probability models are required. Some of these are presented in Chapter 2.

The probability model for an experiment satisfying items 1, 2, and 3 is not hard to derive. The sample space S consists of all possible outcomes of the n trials. Each trial results in a success or a failure, so an outcome of the experiment is a sequence of n trial outcomes, each a success or failure. For example, if the experiment consists of tossing a coin five times, the outcome might be

HTHHT

standing for head on first toss, tail on second, head on third, and so on. S is a finite sample space. (*Challenge*: Can you see how many possible outcomes there are for n trials?) We can therefore give a probability measure on S by assigning a probability to each individual outcome. This is easy by the assumption of independence if we remember Definition 1.7. In the example of tossing a coin five times, suppose that the probability of a head is $p = 1/3$. (This is a very unbalanced coin.) Then by independence the probability of the outcome HTHHT is

$$\frac{1}{3} \cdot \frac{2}{3} \cdot \frac{1}{3} \cdot \frac{1}{3} \cdot \frac{2}{3} = \left(\frac{1}{3}\right)^3 \left(\frac{2}{3}\right)^2$$

Notice that *any individual outcome having 3 heads and 2 tails in some order has the same probability*. This is *not* to say that HTHHT and HHHTT are the same outcome—only that these different outcomes have the same probability. You can see that the general situation is the same: in n independent trials with probability p of success on each trial, any sequence of k successes and $n - k$ failures in some order has probability $p^k(1 - p)^{n-k}$. This gives us the probability of any individual outcome, and hence a probability measure on S. (We are confident that these assignments are correct, and therefore that they satisfy Theorem 1.3b. If you are worried, you can try to show mathematically that the sum of all these individual probabilities is 1.)

Example 1.17   A new spray is claimed to be 80 percent successful in preventing black pod disease of cocoa plants. A grower treats four plants in an infected area and finds that the first, second, and fourth plants do not contract black pod, while the third plant does. What is the probability of this result if the claim is true?

We assume that the four plants are well separated and that other precautions are taken to ensure that we have four independent trials. On any single trial the claim is that the probability is 0.8 of not being infected and 0.2 of being infected. So the outcome given has probability

$$(0.8)(0.8)(0.2)(0.8) = (0.8)^3(0.2) = 0.1024$$

[In this case $n = 4$, $p = 0.8$, and $k = 3$. Note that the resulting probability is $p^k(1 - p)^{n-k}$ as claimed.]

The grower is not particularly concerned with which particular plants were infected, but rather with the total number which were infected. So we might ask what the probability of exactly one infected plant is.

There are four possible outcomes in this event: Only the first plant is infected, only the second, only the third, and only the fourth. Each of these four outcomes has probability $(0.8)^3(0.2) = 0.1024$. The event "exactly one infected plant" has probability equal to the sum of the probabilities assigned to these individual outcomes, or

$$4(0.8)^3(0.2) = 4(0.1024) = 0.4096$$

Example 1.17 gives a pattern for our next step. In most experiments consisting of repeated independent trials, we are not interested in which particular trials resulted in successes, only in the number of successes obtained. We do not care which tosses of the coin gave heads, only how many heads we got in five tosses. To assess the performance of the transistor assemblyline, we can report that only 38 of 8000 were defective. We do not care that the 127th, 216th, etc., were the defective ones. If this is the case, we can let our experiment consist of running the n independent trials and recording only the number of successes. The sample space for this experiment consists of all possible total numbers of successes and is just

$$S^* = \{0,1,2, \ldots, n\}$$

because we can get any number of successes between 0 and n. Now this experiment is clearly very closely related to the original experiment. Only the particular data recorded were changed—we now record only the number of successes, not the exact sequence of successes and failures. This change cannot be ignored, for it changes the sample space. Section 1.13 is devoted to exploring situations such as this. For now we are content to find the probability measure on our new S^*. Example 1.17 illustrates how this is done.

The occurrence of k successes is an individual outcome in S^*, but is an event which contains many individual outcomes in S. We know that the probability of any event in a finite sample space such as S is the sum of the probabilities of the individual outcomes making up the event. In this case those individual outcomes are all sequences of k successes and $n - k$ failures in some order. What is more, if p is the probability of success on an individual trial, we know that each such outcome has probability $p^k(1 - p)^{n-k}$. So adding up these probabilities amounts to multiplying $p^k(1 - p)^{n-k}$ by the number of different arrangements of k successes and $n - k$ failures among n trials. This leaves us the problem of finding that number. This we now do.

Notice that any arrangement of k successes among the n trials amounts to choosing a subset of size k from the set of n trials to be successful. Thus the number of ways of having k successes and $n - k$ failures in n trials is the same as the number of ways that we can choose a subset of k out of the n trials. So we want to know how many different subsets of size k a set of size n has. Here is the answer (the proof is not required reading).

LEMMA 1.1 Given a set of n objects, the number of different subsets containing k of these objects is

$$\frac{n(n - 1)(n - 2) \cdots (n - k + 1)}{k(k - 1)(k - 2) \cdots 3 \cdot 2 \cdot 1}$$

for any whole number k, $1 \leq k \leq n$.

*PROOF Let us ask first how many ways there are to choose k of the n objects in order. Think of this as filling k bins

$$\underset{1}{\sqcup} \; \underset{2}{\sqcup} \; \underset{3}{\sqcup} \; \cdots \; \underset{k}{\sqcup}$$

labeled 1 to k. The first can be filled with any of the n objects in the set. When that is done, the second bin can be filled with any of the remaining $n - 1$ objects, the third with any of the remaining $n - 2$, and so on. The last bin (number k) can be filled with any of the $n - k + 1$ objects which still remain. So there are

$$n(n - 1)(n - 2) \cdots (n - k + 1)$$

ways of choosing k out of n objects in a particular order.

Now a subset of size k is a collection of k objects without regard to their order. Any fixed subset of k objects gives rise to

$$k(k-1)(k-2) \cdots 3 \cdot 2$$

different ordered arrangements of these objects. (The argument is as above: Any of the k can go in bin 1, any of the remaining $k-1$ in bin 2, and so on.)

So the number of different subsets of size k is the number of different ordered choices of k out of n objects divided by the number of these choices which come from a single subset. The resulting quotient is the result stated by Lemma 1.1.

The quantities which appear in Lemma 1.1 are useful enough to deserve a name. Let us also notice that the formula given does not allow us to count the subsets of size 0. There is only one such subset, the empty set. This leads to the following definition.

DEFINITION 1.8 If n is any whole number, and k is a whole number with $1 \le k \le n$, the *binomial coefficient* of n over k is

$$\binom{n}{k} = \frac{n(n-1)(n-2) \cdots (n-k+1)}{k(k-1)(k-2) \cdots 3 \cdot 2 \cdot 1}$$

In addition,

$$\binom{n}{0} = 1$$

Example 1.18 A committee of three can be chosen from a group of eight people in

$$\binom{8}{3} = \frac{8 \cdot 7 \cdot 6}{3 \cdot 2 \cdot 1} = 56$$

ways. This is true because the committee is just a subset of size 3 of the set of eight people.

The number of ways three of eight horses in a race can be awarded win, place, and show is *not* $\binom{8}{3}$. This is because order matters—the same three horses can win, place, and show in several orders (six orders, in fact; try to see that by writing down the possible orders of finish).

Assembling our results, we get a very important theorem. The proof has been given over the past several pages, but we review it to remind you of the principles involved.

THEOREM 1.5 The probability of obtaining exactly k successes in n independent trials, where on each trial there is probability p of a success, is

$$\binom{n}{k} p^k (1-p)^{n-k} \qquad k = 0, 1, 2, \ldots, n$$

PROOF Any sequence of k successes and $n - k$ failures in n trials has probability $p^k(1 - p)^{n-k}$ by independence. Each such sequence corresponds to a way of choosing k of the n trials and placing the k successes on the trials chosen. By Lemma 1.1, this can be done in $\binom{n}{k}$ ways. So the sum of the probabilities assigned to outcomes having exactly k successes is $\binom{n}{k}p^k(1 - p)^{n-k}$.

The family of probability models (depending on n and p) appearing in Theorem 1.5 is called binomial probability models or *binomial distributions*. The word *distribution* is descriptive, for the formula of Theorem 1.5 describes the allocation or distribution of the total probability 1 over the sample space $S^* = \{0,1,2, \ldots , n\}$. *Learn this fact: The probability distribution of the number of successes in n independent trials with the same probability p of success on each trial is a binomial distribution.* Computing the actual values of the binomial probabilities $\binom{n}{k}p^k(1 - p)^{n-k}$ for various n, p, and k is tedious. In practical work tables are used to eliminate this arithmetic. A short table of binomial probabilities appears as Table A at the end of the book.

Example 1.19 The assumptions of Example 1.17 tell us that a binomial distribution with $n = 4$ and $p = 0.8$ applies to the number of uninfected plants. So the probability of exactly one infected plant—that is, exactly three uninfected plants—is

$$\binom{4}{3}(0.8)^3(0.2)^1 = \frac{4 \cdot 3 \cdot 2}{3 \cdot 2 \cdot 1}(0.8)^3(0.2)^1$$
$$= 4(0.8)^3(0.2)^1 = 0.4096$$

Here it was easy to evaluate the binomial coefficient $\binom{4}{3} = 4$. Notice that in Example 1.17 we found without using binomial coefficients that 4 was the number of ways in which exactly one uninfected plant could occur. Also check Table A to see that it gives the answer 0.4096 without calculation.

Example 1.20 A student takes a true-false exam of 20 questions. We will adopt a simple model which assumes that there is some probability p that the student answers any individual question correctly, and that the exam amounts to 20 independent trials. Then the number of correct answers has the binomial distribution with $n = 20$ and p. Suppose $p = 0.9$; i.e., in the long run the student would answer correctly 90 percent of questions appropriate for such an exam. Then the probability that he gets at least 18 correct is the sum of the probabilities of exactly 18, 19, and 20 correct answers. This is

$$\binom{20}{18}(0.9)^{18}(0.1)^2 + \binom{20}{19}(0.9)^{19}(0.1)^1 + \binom{20}{20}(0.19)^{20}(0.1)^0$$

and Table A reduces this to

$$0.285 + 0.270 + 0.122 = 0.677$$

So the probability that the student will do at least as well as his ability indicates is about $2/3$.

On the other hand, if the student has only $p = 0.5$ (which corresponds to blind guessing), then the probability he gets at least 18 correct is only

$$\binom{20}{18}(0.5)^{18}(0.5)^2 + \binom{20}{19}(0.5)^{19}(0.5)^1 + \binom{20}{20}(0.5)^{20}(0.5)^0$$
$$= 0.003 + 0.000 + 0.000 = 0.003$$

(The last two binomial probabilities are not exactly 0, of course. But they are so small that they are 0 to 3 decimal places.)

Example 1.21   A polygraph operator must decide whether the subject is telling the truth or lying. The operator is right with probability 0.99. A firm examines 20 employees concerning a theft. If all are in fact innocent, what is the probability that at least one fails the polygraph test?

This is the probability that the polygraph operator is wrong at least once in 20 trials. A binomial model with $n = 20$ and $p = 0.99$ seems appropriate for the number of correct decisions he makes. So, using Table A,

$$P(\text{at least 1 failure}) = 1 - P(\text{no failures})$$
$$= 1 - P(20 \text{ successes})$$
$$= 1 - \binom{20}{20}(0.99)^{20}(0.01)^0$$
$$= 1 - 0.818 = 0.182$$

The practical conclusion is that the probability of at least one error in a long series of trials is sizable even if the probability of error on an individual trial is very small.

You should notice that Example 1.21 could have been done with no knowledge of the binomial distributions. By independence, the probability of 20 consecutive successes is $(0.99)^{20}$, so the probability of at least 1 failure is $1 - (0.99)^{20}$. Since $\binom{20}{20} = 1$ (see Exercise 1.8.4) and $(0.01)^0 = 1$, this *is* the same as the result in Example 1.21. Table A in effect saved us the work of computing $(0.99)^{20}$.

Exercise 1.8   1.   In each of the following cases, decide whether or not a binomial distribution is an appropriate model, and give your reasons.

a.   An employer records how many of the 50 disadvantaged youths in a federally funded training program successfully complete the program.

b. A teaching machine instructs a child how to do a certain type of mathematical problem, then presents six such problems with additional instruction between problems if the student gets a wrong answer. The number of problems correctly solved is recorded.

c. A student counts how many years between 1960 and 1970 the DJIA ended the year higher than it began.

d. In a test for the presence of ESP, a subject is asked to identify which of four possible patterns is on a card visible to the experimenter but not to the subject. This is repeated for 100 cards, and the number of correct identifications is recorded.

e. A firm gives applicants a personality test designed to detect managerial talent. The firm keeps track of how many of 16 applicants singled out by the test in 1970 had achieved managerial status by 1975.

f. A chemist wishes to determine solubility of a substance, so he repeats a solubility test 10 times. Each successive test is conducted at an increase of 10°C (celsius) of the solute.

g. In checking for defective items on a production line, a worker continues to test items until he finds a defective one, then he stops.

h. Fifty students are taught by a new method. At the end of the trial an exam is given, and the number of students scoring more than 60 percent are considered successes; the others are considered failures.

2. State whether each of the following is true or false and explain why.

a. The number of 5-letter code words (such as brxbu) which can be formed from the 26-letter alphabet is $\binom{26}{5}$.

b. The number of ways a group of 15 workmen can choose a two-man bargaining team is $\binom{15}{2}$.

c. Of 30 graduates of a certain law school, 21 pass the bar exam on the first try. There are $\binom{30}{9}$ possibilities for the list of the 9 who failed.

3. To gain some experience with the general form of binomial distributions, graph the binomial probabilities as follows: Mark off $k = 0, 1, 2, \ldots, n$ on the horizontal axis, then above each k draw a vertical line of length $\binom{n}{k}p^k(1 - p)^{n-k}$ on a convenient vertical scale. Using Table A, do this for

a. $n = 8, p = 0.5$ c. $n = 20, p = 0.5$
b. $n = 8, p = 0.9$ d. $n = 20, p = 0.9$

What is the effect of increasing p from 0.5 to 0.9?

4. Show that each of the following is true, and (more important) explain *why* each is true in the light of Lemma 1.1.

a. $\binom{n}{n} = 1$ for any whole number $n \geq 1$
b. $\binom{n}{k} = \binom{n}{n-k}$ for all n and k

5. A fuse manufacturer tests his product by subjecting 20 fuses each

day (produced independently) to a specified overload and counting the number that blow. If 90 percent of all the fuses being produced blow under this overload, what is the probability that 18 or more of the 20 tested blow?

6.   A factory employs 3500 workmen, 30 percent of whom are black. If the fifteen members of the union executive council are chosen without discrimination from the work force, what is the probability that three or fewer are black?

Here is a rewording of the same problem: 70 percent of the workers are white. What is the probability that twelve or more of the fifteen members of the council are white? Show that you get the same result whichever way you state the problem.

7.   A student asked 10 people if they favored legalizing marijuana, and 7 said they did. What is the probability of at least this many favorable responses if the persons interviewed were selected from a population which is in fact evenly divided on the subject?

It turned out that the student only asked other students. If 60 percent of the students are in favor of legalizing marijuana, what is the probability of at least 7 out of 10 responding favorably?

8.   Of the women in the American labor force, 20 percent are single. If you choose five working women at random, what is the probability that at least one is single? That all five are single?

9.   A graduate student plans to ask faculty members at his school whether they agree with a language requirement for graduate degrees. He is going to report his results to the campus paper if at least half are opposed to the requirement. Suppose that in fact 60 percent of the faculty support the language requirement. What are the probabilities that of those the student asks

a. at least 3 out of 5 oppose the requirement?
b. at least 5 out of 10 oppose the requirement?
c. at least 8 out of 15 oppose the requirement?
d. at least 10 out of 20 oppose the requirement?

Will asking a larger number of faculty tend to protect the student against reporting faculty opinion wrongly?

10.   In a test for the presence of ESP, a subject is told that each of 15 cards visible to the experimenter but not to the subject may contain a star, a square, or a circle. As the experimenter looks at each card in turn, the subject is asked to name the shape on the card.

a. If the subject is just guessing, what should the probability of success on a single trial be?
b. What is the probability that the subject will get at least 8 out of 15 cards right if he is just guessing?
c. What is the conditional probability that he will get the sixth card right given that he got the first five wrong if he is just guessing?

d. You are willing to believe that the subject has ESP if he gets at least k out of 15 cards right, where k is the smallest integer such that the probability of getting k or more right by guessing is no more than 0.1. What is the number k?

e. Suppose the subject is told in advance that the 15 cards contain five stars, five squares, and five circles. Is a binomial model still appropriate for this experiment?

*1.9   GEOMETRIC PROBABILITY DISTRIBUTIONS

The binomial distributions form one of many families of probability models which receive special attention because they are models for situations which arise frequently in practice. In the binomial case, that situation is counting the number of successes in a fixed number of independent repetitions of the same experiment. Another common situation arises when we count the number of independent trials needed to achieve the first success. Once again the assumption of independence makes it easy to construct a probability model.

To be specific, let us say that we are again faced with independent trials, that each trial results in one of two outcomes which we call "success" and "failure," and that the probability of success is the same number p on each trial. Then

P(first success on trial k)
$\quad = P$(failure on trial 1, 2, . . . , $k-1$, and success on trial k)
$\quad = (1-p)(1-p) \cdots (1-p)p$
$\quad = (1-p)^{k-1}p$

If instead of asking the probability that the first success occur exactly on trial k we ask the probability that the first success occur on or before trial k, simple reasoning with independence succeeds again. The event "first success on or before trial k" is the same as "at least one success in k trials." The complementary event is therefore "no successes in k trials." So

P(first success on or before trial k) $= P$(at least 1 success in k trials)
$\quad = 1 - P(k \text{ consecutive failures})$
$\quad = 1 - (1-p)^k$

Example 1.22   A process for individually manufacturing synthetic gemstones produces industrial quality stones 80 percent of the time and jewelry quality stones 20 percent of the time. Experience shows that producing successive gemstones results in independent trials. Then

P(jewel on trial 1) $= 0.2$

P(first jewel on trial 2) = (0.8)(0.2) = 0.16
P(first jewel on trial 3) = (0.8)(0.8)(0.2) = 0.128
P(first jewel on trial 4) = (0.8)(0.8)(0.8)(0.2) = 0.1024

The probability of producing at least one jewel in four trials is

$$1 - (0.8)^4 = 1 - (0.4096) = 0.5904$$

We could have found this last answer by observing that the event "at least one jewel in four trials" is the union of the four events "first jewel on trial k" for $k = 1, 2, 3, 4$. These events are disjoint, and we found their probabilities just above, so

$$P\text{(at least one jewel in four trials)} = 0.2 + 0.16 + 0.128 + 0.1024$$
$$= 0.5904$$

just as before.

Example 1.23   A somewhat shady company wants to advertise that, "Laboratory studies show that in 9 out of 10 cases, Exache cured throbbing pain of headache in less than 15 min." In fact the company knows that cures result in only 50 percent of patients, not 90 percent. To get a laboratory study in their files to back their false claim, they decide to administer Exache to groups of 10 patients until they get at least 9 cures in a group. Then they will throw out the records of all the other groups and collect statements from the 10 patients in the successful group. The company president applauds this brilliant idea, but is worried about the costs of the study. He wants to know for any k the probability that k groups must be used to get a success.

This is moderately complicated, but we can now do it. First we need to know the probability of a success (9 or 10 cures out of 10 patients) in a single group. Since the true probability of an individual cure is 0.5, the number of cures in a group follows a binomial distribution with $n = 10$ and $p = 0.5$. From Table A,

$$P\text{(9 or 10 cures)} = P\text{(9 cures)} + P\text{(10 cures)}$$
$$= 0.010 + 0.001$$
$$= 0.011$$

Therefore the probability that the first success occurs on the kth group of patients is

$$(1 - 0.011)^{k-1}(0.011) = (0.989)^{k-1}(0.011)$$

These numbers are distressingly small. The probability of needing more than 20 groups is

$$P\text{(no success in 20 trials)} = (0.989)^{20}$$
$$= 0.801$$

On hearing this, the company president declares that it would be a dis-service to the public to justify advertising claims in this way, and the project is abandoned.

Although we have successfully used our result on the probability that k trials are required to achieve a success, we have avoided carefully specify-ing a probability model. Our random experiment consists of repeating independent trials each having probability p of success until the first success occurs, then recording the number of trials needed. Any positive whole number of trials may be needed, so the sample space is

$$S = \{1,2,3,\ldots\}$$

We are faced with an *infinite* discrete sample space. How shall we give a probability measure on such a sample space? We know that the proba-bility of individual outcome k (that is, first success on trial k) is

$$p_k = (1-p)^{k-1}p$$

Can we assign probabilities to any event (subset of S) by adding the proba-bilities of the individual outcomes in S, as we did when S was finite? The answer is yes, provided that the individual probabilities p_k are nonnegative and sum to 1. The difficulty is that there are infinitely many p_k.

This difficulty vanishes if you are acquainted with the formula for the sum of a *geometric series*. For any number r such that $0 < r < 1$

$$1 + r + r^2 + r^3 + \cdots = \frac{1}{1-r}$$

Here the $\cdots$ means "keep on adding r^4, r^5, and so on." If you accept this result, we can reason as follows:

$$
\begin{aligned}
p_1 + p_2 &+ p_3 + p_4 + \cdots \\
&= p + (1-p)p + (1-p)^2p + (1-p)^3p + \cdots \\
&= p[1 + (1-p) + (1-p)^2 + (1-p)^3 + \cdots] \\
&= p\frac{1}{1-(1-p)} \quad \text{(using the geometric series formula)} \\
&= p \cdot \frac{1}{p} = 1
\end{aligned}
$$

So the p_k *do* sum to 1, and we can give a probability measure on S by adding probabilities of individual outcomes. Alas, this addition may require sum-mation of infinitely many p_k.

The probability models with $S = \{1,2,3,\ldots\}$ and

$$p_k = (1-p)^{k-1}p \qquad k = 1, 2, 3, \ldots$$

for some $0 \le p \le 1$ are called *geometric distributions* from their connec-tion with geometric series.

In discussing geometric distributions we have strayed from finite mathematics by introducing the sum of an infinite series. It is nonetheless valuable to take a glance at how probabilities are assigned on infinite sample spaces. Section 1.19 describes a quite different class of probability models with infinite sample spaces. (That section can be read now if so desired.)

Exercise 1.9

1.   An oil exploration team estimates that they drill four dry holes for every producing well. That is, they hit oil 20 percent of the time. What is the probability that they will drill three consecutive dry holes? What is the probability that their first producing well will result from the third hole they drill?

2.   A possible scheme for assessing quality of a manufactured product is to inspect randomly chosen items and record the number of items inspected before the first defective item is found. Suppose 80 percent of the items are acceptable. What is the probability that the fourth item inspected is the first defective?

3.   A graphologist claims to be able to distinguish physicians from others by their handwriting. In fact, he is a fraud and simply guesses blindly. In an exhibition he is presented with five handwriting samples.

a.   What is the probability that he classifies all five correctly?
b.   What is the probability that he classifies four of the five correctly?
c.   What is the probability that his first failure occurs on the fifth sample?
d.   What is the conditional probability that he is wrong on the fifth sample given that he is right on the first four?

4.   A space experiment requires launching two rockets. The launch crew attempts to launch the second rocket on the day following the launching of the first. A launch attempt requires 1 day and succeeds 90 percent of the time. If a launch attempt fails, an independent attempt is made the following day. The experiment will be successful only if the second rocket is launched within 3 or fewer days after the first. What is the probability that the experiment succeeds?

5.   A balanced coin is tossed until the first head appears. The event "first head occurs on an odd-numbered toss" contains infinitely many individual outcomes. Find its probability by summing the probabilities of these outcomes and using your knowledge of the sum of a geometric series.

1.10   PARTITIONS

We have seen how the assumption of independence allows us to build models for random experiments in which physical independence is present. The more general ideas of dependence of events and of conditional probability are also very useful in model building and calculation. Recall that we thought of the conditional probability given that an event B has occurred as the original assignment of probability altered only by being restricted to the set B. Thus all events outside B have conditional probability 0; while those contained in B have conditional probability proportional to their original probability. This is the sense of Definition 1.5. This suggests that when event B is fixed, the assignment of probability $P(A|B)$ to events A is a probability measure on S. To prove this we need only verify positivity, totality, and additivity. You were asked to check additivity in Exercise 1.6.11 and will find that positivity and totality are not hard to establish. Here is the theorem which results.

THEOREM 1.6   Let B be an event in S with $P(B) > 0$. Then the assignment of probabilities $P(A|B)$ to events A in S is a probability measure.

Example 1.24   Using the distribution given in Exercise 1.5.3, find the conditional probability that a white worker is not a professional given that he holds a white-collar job.

If B is the event that the worker holds a white-collar job, we see that

$$P(B) = 0.147 + 0.113 + 0.180 + 0.067 = 0.507$$

Taking A to be the event that the worker is a professional, we see that A is a subset of B, so that

$$P(A|B) = \frac{P(A \text{ and } B)}{P(B)}$$
$$= \frac{P(A)}{P(B)} = \frac{0.147}{0.507} = 0.29$$

We were asked to find $P(A^c|B)$. Theorem 1.6 assures us that conditional probabilities given B have all the properties of probability measures. In particular,

$$P(A^c|B) = 1 - P(A|B)$$
$$= 1 - 0.29 = 0.71$$

In plain language, 71 percent of white workers holding white-collar jobs are not professionals.

Example 1.25   Here is a question which is typical of many we must be able to use conditional probabilities to answer. A maker of racing bicycles has a long-term agreement to buy Adams gears, and has found that only 2 percent of the gears supplied are defective. Due to rapidly rising bicycle sales, Adams can now supply only 70 percent of the gears needed. So 20 percent are purchased from Baker and 10 percent from Charlie. These are less reliable suppliers: 4 percent of Baker's gears and 5 percent of Charlie's are defective. What is the overall proportion of defective gears supplied? These proportions become probabilities if the random experiment is drawing a gear from the entire stockpile in such a way that each gear is equally likely to be chosen. This is called *drawing at random*. If events A, B, C are (in order) the events that the gear drawn is supplied by Adams, Baker, or Charlie, and D is the event that it is defective, then we are given the information

$$P(A) = 0.70 \qquad P(D|A) = 0.02$$
$$P(B) = 0.20 \qquad P(D|B) = 0.04$$
$$P(C) = 0.10 \qquad P(D|C) = 0.05$$

The problem is to compute $P(D)$.

We must develop the tools to handle problems such as Example 1.25. The central idea is that each gear in the stockpile comes from exactly one of the suppliers. We say that events A, B, C partition the sample space S (the set of all gears in the stockpile). A formal definition of a partition and a basic property of probabilities of partitions follow.

DEFINITION 1.9 The events $B_1, B_2, \ldots, B_n$ *partition* the sample space S if

a. $B_1 \cup B_2 \cup \cdots \cup B_n = S$
b.   The B_i are pairwise disjoint, that is

$$B_i \cap B_j = \varnothing$$

whenever $i \neq j$.

THEOREM 1.7 If $B_1, B_2, \ldots, B_n$ partition S, then

$$P(B_1) + P(B_2) + \cdots + P(B_n) = 1$$

PROOF Since $P(S) = 1$, we know by Definition 1.9a that

$$P(B_1 \cup B_2 \cup \cdots \cup B_n) = 1$$

The additivity property of P holds for any number of pairwise disjoint sets. (This can be seen by repeatedly applying the additivity property for two sets, as in Exercise 1.5.10.) Therefore

$$P(B_1 \cup B_2 \cup \cdots \cup B_n) = P(B_1) + P(B_2) + \cdots + P(B_n)$$

which completes the proof.

Definition 1.9 says that events $B_1, \ldots, B_n$ partition S when each possible outcome belongs to exactly one of the events B_i. Theorem 1.7 says that in this case the probabilities of these events must add to 1. Notice that this is true of the events A, B, C in Example 1.25. The next theorem will be used often in this chapter and in Chapter 2. It says that the probability of any event A is a weighted average of the conditional probabilities $P(A|B_i)$, the weights being the probabilities $P(B_i)$.

THEOREM 1.8   The Partition Equation Suppose that events B_1, B_2, ..., B_n partition S and that each $P(B_i) > 0$. Then for any event A

$$P(A) = P(A|B_1)P(B_1) + P(A|B_2)P(B_2) + \cdots + P(A|B_n)P(B_n)$$

PROOF   (See Fig. 1.8) We can write any set A as

$$A = (A \cap B_1) \cup (A \cap B_2) \cup \cdots \cup (A \cap B_n)$$

and the sets $A \cap B_i$ are pairwise disjoint. Therefore by additivity

$$P(A) = P(A \cap B_1) + P(A \cap B_2) + \cdots + P(A \cap B_n)$$

The partition equation follows from this by using the multiplication rule $P(A \cap B) = P(A|B)P(B)$ on each summand.

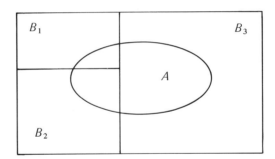

Figure 1.8 B_1, B_2, and B_3 partition S. Event A is partitioned into $A \cap B_1$, $A \cap B_2$, and $A \cap B_3$.

EXAMPLE 1.25 Continued

The proportion of defective gears in the stockpile is given by the partition equation.

$$\begin{aligned}
P(D) &= P(D|A)P(A) + P(D|B)P(B) + P(D|C)P(C) \\
&= (0.02)(0.70) + (0.04)(0.20) + (0.05)(0.10) \\
&= 0.027
\end{aligned}$$

If you have fully understood the meaning of partitioning and of conditional probability, you should be able to read the probability equation above ("The probability of a defective gear is the conditional probability that a gear made by Adams is defective times the proportion of gears supplied by Adams, plus . . .") and be confident that it is true.

Example 1.26   A New York bank recently advertised, "We give you two chances to get a car loan." They explained that if the first loan officer approves your application, you get the loan. If he rejects the application, it goes to a second loan officer and you get the loan if this gentleman approves. The extent to which this policy increases your chances of getting a loan depends on several assumptions. We illustrate one possibility. Let the experiment consist of a customer applying for a loan, and let A be the event that he gets it. Let B be the event that the first officer approves. The bank has kept records which show that in the past 75 percent of auto loan applications have been approved when reviewed by a single officer. So we take $P(B) = 0.75$. The bank's training program attempts to teach loan officers to make similar judgments, and records show that a trained loan officer will approve a loan which (unknown to him) was refused by another officer only 5 percent of the time. So if a rejected application is given without comment to a second officer, $P(A|B^c) = 0.05$. (B^c expresses our knowledge that the application was rejected; the second loan officer does not know this.) Now B and B^c partition S, since every application is either approved or rejected by the first officer. So by the partition equation,

$$P(A) = P(A|B)P(B) + P(A|B^c)P(B^c)$$
$$= (1)(0.75) + (0.05)(0.25)$$
$$= 0.7625$$

Thus the bank's scheme increases the proportion of applicants given loans by only 0.0125, for a single officer grants loans to 75 percent of the applicants. (The bank may well reply that it wishes to give a second chance to a good risk who is mistakenly turned down the first time. Our calculations do not refute this advantage.)

Exercise 1.10   1.   A pollster knows that 55 percent of the voters in Congressman Klimko's district are Republicans, 30 percent are Democrats, and 15 percent are Independents. He polls voters from these three groups separately and discovers that 80 percent of the Republicans, 5 percent of the Democrats, and 10 percent of the Independents plan to vote for Klimko in the forthcoming election. What does the pollster report about the overall proportion of voters favoring Klimko?
2.   Interpret Exercise 1.5.3 as stating the conditional probabilities of various job categories given that a worker is either white or nonwhite. In 1970, 11 percent of the labor force were nonwhites. What proportion of the total labor force held professional jobs in 1970? What proportion held white-collar jobs?
3.   In Exercise 1.6.7 find the proportion of the Congressman's constituents who are willing to go to war over Berlin by using the partition equation.

4.   In Example 1.26 suppose that the first loan officer, knowing that the applicant has a second chance, unconsciously raises his standards so that he approves only 70 percent of loan applications. Suppose also that the second officer knows that the application has been rejected once. This makes him cautious, so he approves only 4 percent of such applications. What proportion of applicants receive loans?

5.   An investor gives his personal assessment of next year's business conditions as follows:

$$P(\text{weak}) = 0.2 \qquad P(\text{moderate}) = 0.4 \qquad P(\text{strong}) = 0.4$$

He further feels that the stock of Nanosecond Computers, Inc., has probability 0 of paying a dividend in weak business conditions, probability 0.2 in moderate conditions, and probability 0.9 of paying a dividend in strong business conditions. What probability has the investor assigned to the event that Nanosecond pays a dividend next year?

6.   We have used a binomial model (with success = right, and failure = wrong) in several situations in which two kinds of error are possible. Examples are Exercise 1.7.4, Exercise 1.9.3, and Example 1.21. A more realistic model would allow the two kinds of possible errors to have different probabilities.

Suppose, for example, that a lie detector operator decides "lie" 1 percent of the time when the subject is telling the truth and 95 percent of the time when the subject is really lying. Experience shows that 98 percent of the time subjects tell the truth when taking a lie detector test. What is the operator's overall probability of error? Would using this model change the work we did in Example 1.21?

7.   Complete the proof of Theorem 1.6 by proving that conditional probability given a fixed event B satisfies $P(A|B) \geq 0$ for any event A and $P(S|B) = 1$.

1.11   BAYES' FORMULA

The partition equation is useful in computing the probability of an event A when the conditional probabilities of A given each event B_i of a partition, and the (unconditional) probabilities of events B_i, are known. It often happens that from this same information we want to find the conditional probability $P(B_i|A)$. In some examples we can think of A as an "effect" and of B_i as the possible "causes" of A; the partition equation then gives the probability of the effect A. With little additional effort we can compute $P(B_i|A)$, the probability of a specific cause given that we observed the effect. An example will show you the procedure involved.

Example 1.27   It is difficult to detect cancer of the pancreas and primary cancer of the liver (hepatoma) in their early stages, and more difficult to distinguish between them without surgery. A medical researcher has developed a diagnostic test based on blood chemistry. He finds that the test registers positive in 95 percent of the cases of hepatoma, 50 percent of pancreatic cancer cases, and in only 5 percent of cases where neither malignancy is present. The researcher proposes making this technique a standard part of blood testing, so that it would be given to the general population. Now if only 0.001 of the population has cancer of the liver and 0.0005 has pancreatic cancer, what is the probability that an individual has cancer of the liver if his test response is positive?

Here the sample space consists of the population and is partitioned into

B_1:   persons with hepatoma
B_2:   persons with pancreatic cancer
B_3:   persons with neither

We know that $P(B_1) = 0.001$, $P(B_2) = 0.0005$, and, by Theorem 1.7, this means that $P(B_3) = 0.9985$. The events B_i are the possible "causes" of the "effect" that the diagnostic test registers positive. Call this effect event A. We were told the conditional probability of the effect given each cause

$P(A|B_1) = 0.95$
$P(A|B_2) = 0.50$
$P(A|B_3) = 0.05$

The problem is to compute the probability of the cause B_1 (hepatoma) given that we observe the effect A (positive test reaction). This is

$$P(B_1|A) = \frac{P(A \cap B_1)}{P(A)}$$

We now know how to find the probabilities on the right side. By the partition equation the denominator is

$$\begin{aligned}
P(A) &= P(A|B_1)P(B_1) + P(A|B_2)P(B_2) + P(A|B_3)P(B_3) \\
&= (0.95)(0.001) + (0.50)(0.0005) + (0.05)(0.9985) \\
&= 0.00095 + 0.00025 + 0.04992 \\
&= 0.0511
\end{aligned}$$

By the multiplication rule the numerator is the first of these summands

$$P(A \cap B_1) = P(A|B_1)P(B_1) = 0.00095$$

Therefore

$$P(B_1|A) = \frac{0.00095}{0.0511} = 0.019$$

Notice in Example 1.27 that even though the diagnostic test picks up 95 percent of those with hepatoma and a much smaller proportion of those without, only 1.9 percent of persons with positive test reactions have hepatoma. This illustrates the difficulty of effectively detecting rare conditions. From another point of view, however, the test is quite successful. The probability that an individual in the general population has hepatoma is only 0.001, and a positive test reaction increases that probability by a factor of 19. This leads to a second way of grasping the meaning of the computations we have done: The $P(B_i)$ are our initial assessments of the probabilities of these events, the positive reaction A is a piece of evidence, and the conditional probabilities $P(B_i|A)$ show how this evidence has changed our initial assessment of probability.

The method used in Example 1.27 to compute probabilities of causes given the observed effect can be reduced to a formula. When this is done, the result is called Bayes' formula after its discoverer, an 18th century English clergyman. To prove that the formula is true, simply repeat the argument given in Example 1.27 without using the particular numbers given there. (This is Exercise 1.11.7.) It is far better to learn the method than to memorize the formula.

THEOREM 1.9 Bayes' Formula Suppose that events $B_1, B_2, \ldots, B_n$ partition S and that each $P(B_i) > 0$. Then if $P(A) > 0$, we have for each $i = 1, 2, \ldots, n$ that

$$P(B_i|A) = \frac{P(A|B_i)P(B_i)}{P(A|B_1)P(B_1) + \cdots + P(A|B_n)P(B_n)}$$

So if we know each probability $P(B_i)$, and each conditional probability $P(A|B_i)$, Bayes' formula tells us how to find $P(B_i|A)$.

Example 1.28   Here is a probabilistic model which might be used to correct for guessing in scoring multiple-choice examinations. Let A be the event that a student gets the right answer on a k-choice multiple-choice question. Either the student knows the answer (event B), in which case he gets it right [$P(A|B) = 1$], or he does not know (event B^c) and guesses blindly [$P(A|B^c) = 1/k$]. If the proportion of students who know the answer is p, then the proportion who give the correct answer is

$$P(A) = P(A|B)P(B) + P(A|B^c)P(B^c)$$
$$= (1)(p) + \frac{1}{k}(1 - p)$$
$$= p + (1 - p)\frac{1}{k}$$

We see that the proportion who get the question right is greater than the proportion who know the answer because of the effect of guessing. What is the probability that a student who gave the right answer actually knew the answer?

This is

$$P(B|A) = \frac{P(A \cap B)}{P(A)}$$
$$= \frac{P(A|B)P(B)}{P(A)}$$
$$= \frac{(1)(p)}{p + (1 - p)(1/k)} = \frac{kp}{1 + (k - 1)p}$$

For example, if 80 percent of the students know the answer to a question, the probability of a correct answer if there are four choices is

$$0.8 + (0.2)\frac{1}{4} = 0.85$$

and if a student gives the correct answer, the conditional probability that he knew it is

$$\frac{(4)(0.8)}{1 + (3)(0.8)} = 0.94$$

Exercise 1.11

1.   If the pollster in Exercise 1.10.1 tells you that a certain voter plans to vote for Klimko, what is the probability that the voter is a Republican? A Democrat? An Independent? Why should these conditional probabilities sum to 1?

2.   Aluminum ingots coming to a plant for extrusion into aircraft wing supports are high-grade or low-grade. Only high-grade ingots can be successfully extruded, but unfortunately the grade cannot be ascertained until extrusion is attempted. Experience shows that 60 percent of the ingots are high-grade. A preliminary test is used which passes 90 percent of high-grade ingots and 30 percent of low-grade ingots.

What proportion of ingots passes the test? If an ingot passes the test, what is the probability that it is high-grade?

3.   A study shows that 3 percent of an axle plant's work force are alcoholics and 2 percent are drug addicts and that no one is both. In the course of a year 58 percent of the alcoholics are involved in plant accidents, as are 40 percent of the addicts and 5 percent of the remaining workmen. Management asks that you compute the probability that a workman is either an alcoholic or an addict, given that he was involved in an accident this past year.

4.   Refer to Exercise 1.10.2. What proportion of workers in white-collar jobs in 1970 were nonwhite?

5.   Refer to Example 1.25. What is the probability that a gear was supplied by each of Adam, Baker and Charlie, given that the gear is defective?

6.   A lie detector operator can declare a subject to be lying either because of error or because the subject really is lying. Find the probability of each of these causes given the effect (operator says he is lying) using the model of Exercise 1.10.6.

7.   Write out a formal proof of Theorem 1.9.

1.12   APPLICATIONS

In this section we present two applications of probability theory which are more extensive than those we have encountered to this point. The first application is a probabilistic analysis of a scheme for deciding whether a company should accept a shipment of parts from a supplier. Most companies of any size are partners to such a scheme, either as purchasers of parts or as suppliers. The second, and mathematically more difficult, application is a probability model for a learning experiment with rats. It is an example of the elaborate probability models now being used in the social sciences.

Application 1.1   Acceptance Sampling

An industrial concern, which we will call the consumer, contracts to buy parts from another concern, the producer. The contract states the specifications to be met by the parts, so that inspection of a given part can determine whether it meets the contract specifications. If not, we say that the part is defective. The contract also specifies that parts are to be shipped in lots of a fixed size, and almost certainly states the maximum allowable number of defective parts in a lot. Why not ask for perfect lots—lots with no defectives? Because the producer cannot mass-produce parts without having some defectives, and if the consumer wants only perfect lots shipped, the producer will have to inspect each individual part. He will naturally charge the consumer for this. What is more, the inspectors are human, so that some defective parts will slip through. Unless the consumer is NASA, he probably cannot afford to ask for perfect lots.

How is the consumer to determine if a given lot is acceptable? He can of course inspect every part and count the defectives. If there are more defectives than the contract allows, the producer pays a penalty. But complete inspection is usually not practical for several reasons. First, it is expensive. Second, if the inspection is of fuses or of artillery shells, it may destroy the item inspected. Third, inspectors charged with looking at each of a lot of 3000 bearings make a good many mistakes, so that complete inspection is not always reliable.

This is where *acceptance sampling* comes in. *A sample is a subset of the lot.* The consumer can carefully inspect a sample of the parts in the lot, usually a small number of parts compared with the size of the lot. He then decides to accept or reject the lot on the basis of the number of defective parts he finds in the sample. This idea avoids all three objections to complete inspection.

How is the sample to be chosen? We want a subset of the lot which is representative of the complete lot. Allowing the inspector to select whatever parts are handy is no way to get a representative sample, if only because the producer might pack the defective parts in the bottom of the crate. In fact, there is no way to guarantee a representative sample without knowing the composition of the entire lot. The solution to this dilemma is to draw samples in a way which can be analyzed by probability theory. We will see that while we cannot guarantee that a decision made on the basis of such a sample is right, we can compute the probability of making an error. The producer and the consumer often write a particular sampling plan into the contract. We will describe a simple and widely used plan for acceptance sampling.

Suppose that the size of each lot is N and it is agreed that a sample of size n (where $n < N$) will be drawn from the lot and inspected. A *simple random sample is a sample drawn so that every possible sample of size n is equally likely to be drawn.* Giving every possible sample an equal chance to be the one we actually draw seems "fair." We hope that such a sample will be representative of the entire lot. The word *random* here has the same meaning as in our earlier usage *random experiment.* It means that the particular sample we draw is not predictable in advance. But a random sample is *not* a haphazard sample—it is drawn so that a particular probability model applies. There are $\binom{N}{n}$ possible samples of size n which could be drawn from a lot of size N. (This is a direct application of the basic meaning of the binomial coefficients as stated in Lemma 1.1.) So the sample space S in our model for simple random sampling consists of all $\binom{N}{n}$ possible sets (samples) of n parts chosen from the lot of N parts. If each sample is equally likely to be chosen, each must have probability $1/\binom{N}{n}$. This assignment of probabilities to individual outcomes gives a probability measure on the sample space by Theorem 1.3.

Obtaining a simple random sample may not be easy. We are, after all, starting with a probability model and attempting to organize a random experiment which the model describes. A naive but effective method is to label the N parts in the lot 1 through N, put identical tabs numbered 1 through N in a box, mix thoroughly, and draw n tabs. The sample consists of the n parts labeled with the numbers on these tabs. Naturally, this is rarely done in practice. We will say no more, except that simple random samples are not available unless someone works at obtaining them.

Returning to the contract signed by the producer and the consumer, it describes the acceptance sampling plan using simple random samples by three numbers:

N: the lot size

n: the sample size

c: the acceptance number

This means that the consumer draws a simple random sample of size n, inspects each part in the sample, and accepts the lot if he finds c or fewer defective parts. If he finds $c + 1$ or more defectives, he rejects the lot. What happens then is spelled out in the contract, and is likely to cost the producer money.

To investigate the effectiveness of such a sampling plan, we must answer this question: *Suppose that the lot in fact contains D defective parts. What is the probability that the lot will be accepted?*

The lot is accepted if the sample contains no more than c defective parts. So to compute the probability of accepting the lot, we first ask another question: What is the probability that the sample contains a specified number (say k) of defective parts? Each possible sample has probability $1/\binom{N}{n}$. The probability of any event is therefore the number of samples falling in that event divided by $\binom{N}{n}$. So we must count the number of samples containing exactly k defective parts and $n - k$ nondefective parts. To do this, imagine that the sample is drawn in two steps. First draw any k of the D defective parts. This can be done in $\binom{D}{k}$ different ways. Then draw any $n - k$ of the $N - D$ nondefective parts, which can be done in $\binom{N-D}{n-k}$ different ways. Any one choice of k defective parts can be combined with the $\binom{N-D}{n-k}$ possible choices of nondefective parts to form $\binom{N-D}{n-k}$ possible samples. Since there are $\binom{D}{k}$ ways to choose the defective parts, and each way produces $\binom{N-D}{n-k}$ samples having exactly k defective parts, the total number of possible samples having exactly k defective parts is $\binom{D}{k}\binom{N-D}{n-k}$. Therefore the probability of such a sample is

$$P(\text{exactly } k \text{ defectives}) = \frac{\binom{D}{k}\binom{N - D}{n - k}}{\binom{N}{n}}$$

The possible values of k are the whole numbers between 0 and the *smaller* of the sample size n and the total number D of defectives in the lot.

We have just derived a probability distribution which, like the binomial and geometric distributions, is sufficiently important to merit a name of its own. It is called a *hypergeometric distribution*. The hypergeometric distributions form a family of probability models in which a particular distribution is determined by the values of N, D, and n. Such distributions arise whenever a simple random sample of size n is drawn from a population of N objects, D of which are of some special kind—defective in our case. The number of "special" objects in the *sample* has a hypergeometric distribution. Exercise 1.12.2 provides some examples of this.

We now know that if the lot is accepted when the sample contains no more than c defective parts, and the lot in fact contains D defectives, the probability of accepting the lot is

$$P_a = P(0 \text{ defectives}) + P(1 \text{ defective}) + \cdots + P(c \text{ defectives})$$

$$= \frac{\binom{D}{0}\binom{N-D}{n}}{\binom{N}{n}} + \frac{\binom{D}{1}\binom{N-D}{n-1}}{\binom{N}{n}} + \cdots + \frac{\binom{D}{c}\binom{N-D}{n-c}}{\binom{N}{n}}$$

To give a concrete example, suppose that roller bearings are shipped in lots of $N = 3000$. The contract specifies that a simple random sample of size $n = 150$ will be drawn, and that the lot will be accepted if no more than $c = 4$ defective bearings are found in the sample. If the lot in fact contains $D = 120$ defective bearings, the probability of accepting the lot is

$$P_a = \frac{\binom{120}{0}\binom{2880}{150}}{\binom{3000}{150}} + \frac{\binom{120}{1}\binom{2880}{149}}{\binom{3000}{150}} + \frac{\binom{120}{2}\binom{2880}{148}}{\binom{3000}{150}}$$

$$+ \frac{\binom{120}{3}\binom{2880}{147}}{\binom{3000}{150}} + \frac{\binom{120}{4}\binom{2880}{146}}{\binom{3000}{150}}$$

It should be clear that finding the actual numerical value of P_a in an example such as this will involve an unpleasant amount of arithmetic. There are various tricks and approximations which are helpful, but rather than discuss them, we will just give the results for the example of a lot of size $N = 3000$, a sample of size $n = 150$, and an acceptance number $c = 4$. It is convenient to use the *proportion of defectives* $p = D/N$ rather than the actual number of defectives D. So Fig. 1.9 is a graph of the probability of accepting the lot P_a against the proportion p of defective parts in the lot. Understand that these probabilities are correct only for the specific sampling plan specified by $N = 3000$, $n = 150$, and $c = 4$. A different choice of N, n, and c would give a different graph, though one of the same general shape. If we are alert, we will notice that in fact the only possible values of p in this example are 0, $1/3000$, $2/3000$, and so on. We drew a continuous curve in Fig. 1.9 to make it easier to read. In the example above, the proportion of defectives is $p = 120/3000 = 0.04$, and from Fig. 1.9 it can be seen that the probability of accepting such a lot is approximately $P_a = 0.28$.

In practice, the proportion p of defectives in the lot is unknown—if it were known, there would be no need to inspect a sample. Figure 1.9 is valuable because it gives the probability P_a of accepting a lot with any proportion p of defectives. We can therefore tell what kind of lots this

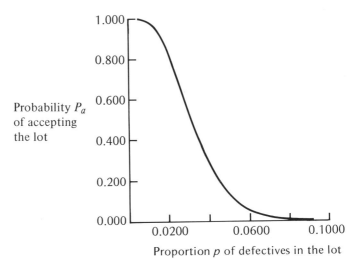

Figure 1.9 The operating characteristic (OC) curve for the acceptance sampling plan with $N = 3000$, $n = 150$, and $c = 4$

sampling plan will probably accept and what kind of lots it will probably reject. For example, if the proportion of defectives is around 10 percent, we will almost never accept the lot. But if only 2 percent of the parts in a lot are defective, the probability of accepting that lot is 0.81. If only 1 percent of the parts in a lot are defective, the probability of acceptance increases to 0.98. So "bad" lots are usually rejected, and "good" lots are usually accepted. The graph of P_a against p shown in Fig. 1.9 is called the *operating characteristic (OC) curve* for this sampling plan. We must examine in more detail how OC curves are used.

Suppose it is agreed that a lot is "good" if it contains 60 or fewer defective parts and "bad" if it contains more. This means that, if D is no greater than 60 ($p \leq 0.02$), the correct decision is to accept the lot. If D is 61 or more ($p > 0.02$), the correct decision is to reject the lot. Our sampling plan can make two kinds of error:

1. reject the lot when in fact the lot is good
2. accept the lot when in fact the lot is bad

For any particular value of D only one of these errors is possible—type 1 if $D \leq 60$ or type 2 if $D > 60$. The probability of an error of type 1 is called the *producer's risk*, since the producer suffers when a good lot is rejected. The consumer suffers if he accepts a bad lot, and so the probability of an error of type 2 is the *consumer's risk*. The OC curve gives us both: If the lot is bad ($p > 0.02$), then P_a is the consumer's risk. If the lot is good ($p \leq 0.02$), then $1 - P_a$ is the producer's risk. These probabilities change with the true

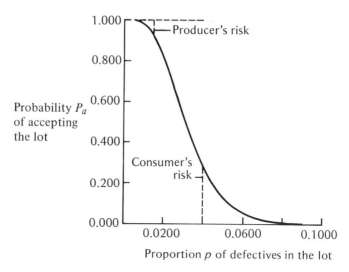

Figure 1.10   Producer's risk when $p = 0.015$ and consumer's risk when $p = 0.04$

value of p. Figure 1.10 is a copy of the OC curve of Fig. 1.9 which indicates the producer's risk when in fact $p = 0.015$ ($D = 45$ defective parts in the lot) and the consumer's risk when in fact $p = 0.04$ ($D = 120$ defective parts in the lot).

The particular sampling plan we have been discussing favors the producer, in the sense that for any good lot—any $p \leq 0.02$—the probability of rejection is never more than 0.18. For lots which are bad, but not very bad—p just a little greater than 0.02—the probability of acceptance is almost 0.82. So a small producer's risk goes along with a high consumer's risk for lots with D slightly above 60, the maximum allowable number. It is certainly reasonable that a sampling plan has a hard time telling the difference between lots with $D = 60$ and $D = 61$. This particular plan usually accepts when the true D is anywhere near 60.

How could we change the plan to favor the consumer? If we decrease c, the maximum allowable number of defectives in the sample, we make it easier to reject the lot. That decreases P_a for all values of p. So the consumer's risk decreases for any $p > 0.02$, and the producer's risk for any $p \leq 0.02$ increases. The producer's and consumer's risks are on a seesaw—decreasing one increases the other. That is true as long as the lot size N and the sample size n are fixed. In practice, the producer and consumer agree on values of N, n, and c which give an OC curve acceptable to both. In general, they must increase the sample size n if they wish to decrease both risks. That means that they must pay to have more parts inspected. The final sampling plan must therefore take into account the cost of accepting a bad lot, the cost of rejecting a good lot, and the cost of drawing and inspecting the sample.

Exercise 1.12

1.   Here is an experiment to illustrate simple random samples. Number 50 identical tags 1 to 50, place in a box, and mix well. Suppose that tags 1 through 5 represent the defective parts in this lot of 50. Draw a sample of size 10, and record the number of defectives. Return the tags drawn to the box, mix well, and repeat this random experiment 20 times. Graph the relative frequency of the possible numbers 0, 1, 2, 3, 4, 5 of defectives.

2.   Which of these events have probability given by a hypergeometric distribution? Explain your answers.

a.   A bargaining committee of 4 men and 1 woman represents a union with 250 male and 200 female members. A women's caucus computes the probability that a simple random sample of size 5 from the membership would give this composition.

b.   A producer corporation bases its product quality control on inspecting 100 parts produced independently on its assemblyline and counting the number which are defective.

c.   Another producer corporation examines the parts being produced independently on its assemblyline beginning each hour and counts the number of parts until the first defective appears.

d.   A lottery sells 1000 tickets and 10 are drawn at random to receive prizes. Relatives of the organizer bought 30 tickets and received 4 prizes. A suspicious ticketholder computes the probability of this event.

3.   The experiment of counting the defectives in a simple random sample of size n from a lot of size N having D defectives is described by a hypergeometric distribution. Explain carefully why the sample space consists of the whole numbers 0, 1, 2, . . . up to the smaller of n and D.

4.   Find from Fig. 1.9 the producer's or consumer's risk (whichever is appropriate) for a lot containing each of the following proportions of defective parts:

a. $p = 0.06$ b. $p = 0.03$ c. $p = 0.01$

5.   Complete inspection by inspectors who never make an error results in both risks being 0. If the contract defines a good lot as one with no more than proportion $p = 0.02$ of defective parts, draw the OC curve of this "perfect" inspection plan.

6.   Based on the OC curve of Fig. 1.9, which situation do you think best justifies the use of that particular acceptance sampling plan, and why?

a.   A producer who in the past has often shipped lots with more than 60 defective parts

b. A new producer about whom the consumer knows little

c. A trusted producer who in the past has very rarely shipped lots with more than 60 defective parts

7. The reasoning used in acceptance sampling applies to many other situations in which any of a family of probability distributions may describe a random experiment, and we must come to a decision concerning which is true. Here is an example.

An individual claims that his ESP allows him to identify the shapes on cards invisible to him more often than someone who simply guesses. Suppose there are five possible shapes, so that a guesser would be right about $1/5$ of the time. To test this, he is asked to identify the shapes on 20 cards, and a record of his success is kept. The number of successes is described by a binomial distribution with $n = 20$ and p unknown. The possible decisions are: (1) he has ESP (correct if in fact $p > 1/5$), and (2) he does not have ESP (correct if in fact $p \leq 1/5$).

a. Describe the two types of possible errors we can make when we reach a decision.

b. Suppose that we agree to decide that the subject has ESP if he gets 10 or more cards correct out of 20. What is the probability of error if in fact the subject is just guessing ($p = 1/5$)?

c. What is the probability of error using this procedure if in fact $p = 0.5$, so that the subject has strong ESP?

8. Explain why acceptance sampling cannot guarantee even a high probability of obtaining a good quality lot. (*Hint*: What will the results be if only lots with many defectives are shipped to the consumer?)

Application 1.2 A Model from Learning Theory

This model for "avoidance conditioning" in rats is an example of the many probability models which have been developed by psychologists studying learning in animals. Some additional aspects of the model are discussed in Section 6.1 of Atkinson, Bower, and Crothers, *An Introduction to Mathematical Learning Theory* (Wiley, 1965). Yet other aspects will be studied in Sec. 2.13 after we have learned more probability theory.

An experiment is performed as follows: The apparatus consists of a cage with two compartments separated by a door. One compartment is painted black and is unlighted, while the other is painted white and is lighted. A rat is placed in the black compartment, the door is opened, and a bell begins to ring. If the rat is still in the black compartment after 5 sec, it is shocked through the metal floor until it moves to the white compartment. After a minute's rest the rat is returned to the black compartment, and the experiment is repeated. Since the rat prefers the dark, it will not go to the white compartment when the bell rings until it has "learned" to avoid the shock. The rat usually learns within 20 trials to leave the black compartment when the bell rings. It then avoids all future shocks.

Some psychological theorists believe that the rat undergoes a two-stage learning process, first being conditioned to fear the bell and then learning to avoid the shock by deliberately entering the white chamber. Here is a mathematical model based on this theory. When the rat is first placed in the cage, it is in "state 0" and has probability 1 of receiving a shock. When it has been conditioned to fear the bell, it has passed to "state 1" and now has probability p of receiving a shock. Here $p > 0$, since the rat has not yet learned what specific action will avoid the shock. When it has learned this, it passes to "state 2" and has probability 0 of receiving a shock. Finally, the model assumes that whenever a rat in states 0 or 1 is shocked, it has probability α of "learning," i.e., of passing to the next state. If the rat avoids shock, it cannot learn on that trial. Notice that the probability of learning is the same (namely, α) whether the rat is in state 0 or state 1.

You should realize that *the experimenter cannot tell which state the rat is in* at any point during the experiment. If it is shocked, it could be in either state 0 or state 1. If it avoids shock, it could be in either state 1 or state 2. (Of course, if the rat is shocked after a trial on which it avoided shock, we then know that it was in state 1 on all the trials between these two.) How then are we to test the usefulness of the model by comparing it with experimental data? We will deduce from the model the probability distribution of the *total number of shocks* the rat receives. This can be observed by the experimenter, so that the model can be tested against his observations. One theme of this application is that some thought may be needed to produce from a mathematical model predictions which can be experimentally tested.

Even though the state of the rat is not observable, we must use the fact that the rat can receive a shock in either state 0 or state 1. Define B_j as the event that exactly j shocks are received while in state 0, and C_j as the event that exactly j shocks are received while in state 1. We will find the probabilities of these events. The rat is shocked on each trial as long as it stays in state 0, so the event B_j is the same as the event that the rat stays in state 0 for $j - 1$ trials and advances to state 1 after the jth trial. The probability of advancing after any trial is α. So we have a geometric probability (see Sec. 1.9)

$$P(B_j) = (1 - \alpha)^{j-1}\alpha \qquad \text{for} \quad j = 1, 2, \ldots$$

Finding $P(C_j)$, the probability of j shocks while in state 1, is more subtle. Once in state 1, the rat can advance to state 2 only after receiving a shock. This does not happen on every trial—the probability of a shock on any trial while in state 1 is only p—so the number of shocks received in state 1 is not the same as the total number of trials spent in that state. What we do is simply ignore trials on which the rat avoids shock. What is left is a sequence of shocks, and each shock gives the rat probability α of advancing to state 2. Just as before, the probability of receiving exactly j shocks is the probability of failing to advance after each of $j - 1$ shocks, then advancing after the jth shock. So

$$P(C_j) = (1 - \alpha)^{j-1}\alpha \qquad \text{for} \quad j = 1, 2, \ldots$$

Now we are ready to find the probability of the event D_k that the rat receives exactly k shocks in all. To receive a total of k shocks, the rat must receive 1 in state 0 and $k - 1$ in state 1, or 2 in state 0 and $k - 2$ in state 1, or 3 in state 0 and $k - 3$ in state 1, and so on. In symbols

$$D_k = \bigcup_{j=1}^{k-1} (B_j \cap C_{k-j}) \qquad k = 2, 3, \ldots$$

Notice that j cannot be 0 or k because the rat must receive at least one shock in each state. Now the events $B_j \cap C_{k-j}$ are disjoint for different j, and the events B_j and C_{k-j} are independent. Therefore

$$
\begin{aligned}
P(D_k) &= P(B_1 \cap C_{k-1}) + P(B_2 \cap C_{k-2}) + \cdots + P(B_{k-1} \cap C_1) \\
&= P(B_1)P(C_{k-1}) + P(B_2)P(C_{k-2}) + \cdots + P(B_{k-1})P(C_1) \\
&= \alpha \cdot (1 - \alpha)^{k-2}\alpha + (1 - \alpha)\alpha \cdot (1 - \alpha)^{k-3}\alpha + \cdots \\
&\qquad\qquad\qquad\qquad\qquad\qquad + (1 - \alpha)^{k-2}\alpha \cdot \alpha \\
&= \alpha^2(1 - \alpha)^{k-2} + \cdots + \alpha^2(1 - \alpha)^{k-2} \qquad (k - 1 \text{ terms}) \\
&= \alpha^2(k - 1)(1 - \alpha)^{k-2}
\end{aligned}
$$

For any α, the probabilities $P(D_k)$ which we have computed can be calculated and compared with the relative frequencies observed in experimentation with many rats. Atkinson, Bower, and Crothers cite some experimental work indicating that the model fits real rats quite well. They also discuss the related question of deciding from the data what the probability α of learning is.

The remainder of this section is optional. We want to compute the probabilities of advancing from one state to another on a given trial. These are conditional probabilities given the present state. We will see in Chapter 2 that they have many uses. For convenience we introduce a shorthand notation. Let X_n stand for the rat's state at the *end* of the nth trial. Thus $\{X_5 = 1\}$ is the event "the rat is in state 1 after five trials."

The rat begins in state 0 and is shocked on every trial as long as it remains there. So if the rat is in state 0 at the beginning of any trial, it has probability α of advancing to state 1. In symbols,

$$P(X_n = 1 \mid X_{n-1} = 0) = \alpha$$

The rat can only go to states 0 or 1 from state 0, so

$$P(X_n = 0 \mid X_{n-1} = 0) = 1 - \alpha$$

Suppose next that the rat begins trial n in state 1. What is the probability

$$P(X_n = 2 | X_{n-1} = 1)$$

that it advances to state 2 on that trial? It can advance only if shocked, so let A_n be the event that a shock is administered on trial n. Then the model states that

$$P(A_n | X_{n-1} = 1) = p$$

What is more, we know that if the rat begins trial n in state 1 *and* is shocked on that trial, it advances to state 2 with probability α. In symbols

$$P(X_n = 2 | X_{n-1} = 1 \text{ and } A_n) = \alpha$$

Finally, since the rat can advance only if shocked,

$$P(X_n = 2 | X_{n-1} = 1) = P(X_n = 2 \text{ and } A_n | X_{n-1} = 1)$$

We combine all this information by applying the multiplication rule $P(A \text{ and } B) = P(B|A)P(A)$ to the probability measure determined by conditional probability given the event $\{X_{n-1} = 1\}$.

$$P(X_n = 2 \text{ and } A_n | X_{n-1} = 1) = P(X_n = 2 | X_{n-1} = 1 \text{ and } A_n)P(A_n | X_{n-1} = 1)$$
$$= \alpha p$$

The rat can go only to state 1 or state 2 from state 1. So we have found that

$$P(X_n = 2 | X_{n-1} = 1) = \alpha p$$
$$P(X_n = 1 | X_{n-1} = 1) = 1 - \alpha p$$

Exercise 1.12   9.   If a rat has probability $\alpha = 1/4$ of advancing after receiving a shock, compute the probability that it receives exactly four shocks during the experiment.

10.   Would this model fit an experiment identical to the one described except that the black and white compartments are interchanged? Discuss.

11.   The experimenter does not know when the rat reaches state 2. How do you think he decides when to stop taking more trials? Perhaps we can get an idea by computing the probability that a rat which is in state 1 avoids shock on five consecutive trials when $p = 1/2$. Can we conclude that a rat which avoided five straight shocks is in state 2?

12.   What are the conditional probabilities of going to each of states 0, 1, and 2 on trial n if we know that the rat was in state 2 after trial $n-1$?

13.   What is the probability that a rat starting in state 0 is in state 1 after the first trial and in state 2 after the second?

14.   A rat receives a shock on trial 2. Use Bayes' formula to find the probability that the rat is in state 0 at the beginning of trial 2 given this information. Do the same for state 1.

15.   The multiplication rule applied to probabilities conditional on event
C states that

$$P(A \text{ and } B|C) = P(A|B \text{ and } C)P(B|C)$$

Prove this fact using only the definition of conditional probability.
16.   Suppose that B is disjoint from A^c (that is only to say that B is a subset of A). Show that

$$P(B|C) = P(B \text{ and } A|C)$$

Where did we use that fact in this section?

1.13   RANDOM VARIABLES

Very often the members of a sample space are numerical measurements of
one or more quantities. A product control manager may count the number
of defective bearings in a sample of 25 taken from the assemblyline. A
sociologist may construct a test for racial prejudice and record the scores
of his subjects. He may simultaneously record their scores on a test designed
to detect authoritarian personalities. A management expert may record what
percentage of a company's middle and top management graduated from
business school. In all these cases it is convenient to have a shorthand notation for events such as "no more than five defective bearings were found"
or "the subject scored 86 on the prejudice scale and 92 on the authoritarianism scale." We might say, "let X be the number of defective bearings in
the sample" or, "let Y be the subject's prejudice score and Z his authoritarianism score." Then the events just mentioned can be abbreviated as
$\{X \leq 5\}$ and $\{Y = 86, Z = 92\}$. Here X, Y, and Z are *numerical outcomes of
random experiments.* That is, they stand not for fixed numbers but for numbers whose values depend on a chance experiment and might change if the
experiment were repeated. X, Y, and Z are called random variables. Random
variables play an important role in probability theory, and we will give a
formal definition shortly. But the basic concept is that random variables
represent numerical outcomes of random experiments.
    The sample space in a probability model need not consist of numbers,
points in the plane, or other mathematical objects. We may nonetheless
want to assign a number to each point in such a sample space. This assignment produces a numerical outcome of a random experiment, even though
the sample space for the experiment does not consist of numbers. So we
call the result of such an assignment a random variable, too.

Example 1.29   A possible sample space for tossing two coins (see Example 1.6) is

$$S = \{(H, H), (T, H), (H, T), (T, T)\}$$

Let the random variable X be the number of heads observed. We might say that X looks at the outcome and tells us the number of heads. The random variable X assigns numbers to the members of S as follows.

Outcome	Value of X
(H, H)	2
(H, T)	1
(T, H)	1
(T, T)	0

Be sure you see that X is not a fixed number—it is "random" in the sense that its value depends on the outcome of the experiment.

The formal name for a rule which assigns a number to every member of S is a function from S to the real numbers. And that is the formal definition of a random variable. Random variables are a mathematical convenience, as well as a notational simplification, for they enable us to work with ordinary numbers no matter what the original sample space is.

DEFINITION 1.10 A *random variable* X on a sample space S is a function defined on S and taking real numbers as its values.

EXAMPLE 1.29 Continued
If we use the usual function notation instead of the table given in Example 1.29, the random variable X is described by

$$X(H, H) = 2$$
$$X(H, T) = X(T, H) = 1$$
$$X(T, T) = 0$$

Then $\{X = 1\}$ is just another notation for the event $\{(H, T), (T, H)\}$, for this is the subset of the sample space on which X takes the value 1. In symbols,

$$\{X = 1\} = \{s \text{ in } S \text{ for which } X(s) = 1\}$$

That makes clear for what event (subset of S) $\{X = 1\}$ is an abbreviation. We will always use the abbreviation.

Suppose that we have on S the probability measure P which assigns probability $1/4$ to each of the four individual outcomes in S. Since $\{X = 1\}$ is an event, it has a probability. We write this probability as $P(X = 1)$. In this case

$$P(X = 1) = P(\{(H, T), (T, H)\}) = 1/2$$

Once again we have an abbreviated notation which is simple to use and easy to understand.

Any random variable has associated with it an assignment of probabilities to its possible values. For example, if X is the number of successes in n independent trials with probability p of success on each trial, we know that

$$P(X = k) = \binom{n}{k}p^k(1 - p)^{n-k}$$

for $k = 0, 1, 2, \ldots, n$. We call X a *binomial random variable*.

When several measurements are made simultaneously on the outcome of a random experiment, each can be represented by a separate random variable. We have already seen this in the example of recording scores on tests for both racial prejudice and authoritarian personality. In such cases a probability measure on the sample space assigns probabilities to any combination of values of the random variables.

Example 1.30   A large retail chain keeps punched card records on its white-collar workers. Each worker is coded as 0 for male or 1 for female. Elsewhere on the card each employee is coded by job classification, 1 for sales, 2 for buyer, and 3 for executive. Let X be the sex classification and Y the job classification. The proportions of workers having each pair of classifications is given by the following table.

		X	
		0	1
Y	1	0.28	0.42
	2	0.08	0.12
	3	0.04	0.06

For example, $\{X = 1, Y = 3\}$ is the event "the worker is a female executive."

We might ask if the chances of a worker's being an executive are affected by the knowledge that the worker is female. Formally, we ask if the events $\{X = 1\}$ and $\{Y = 3\}$ are independent. Since $\{X = 1\}$ is the union of the disjoint events $\{X = 1, Y = 1\}$, $\{X = 1, Y = 2\}$, and $\{X = 1, Y = 3\}$,

$$P(X = 1) = 0.42 + 0.12 + 0.06 = 0.6$$

Similarly

$$P(Y = 3) = 0.04 + 0.06 = 0.1$$

So

$$P(X = 1)P(Y = 3) = (0.1)(0.6) = 0.06$$

and since this is the same as $P(X = 1, Y = 3)$, these events *are* independent by Theorem 1.4.

You can easily check that in Example 1.30 *any* event $\{X = x\}$ is independent of *any* event $\{Y = y\}$. In these circumstances it is reasonable to say that the random variables X and Y are independent. This means that knowing the sex classification of a worker gives no information about the job classification—there is no discrimination by sex in job types. [*Comment*: It is very unlikely that actual data would show exact independence of X and Y. But how close $P(X = x)P(Y = y)$ is to $P(X = x, Y = y)$ for $x = 0$, 1 and $y = 1$, 2, 3 does serve as an indicator of the presence or absence of discrimination by sex in job assignment.] In defining independence of a collection of random variables, we follow the lead of Example 1.30. We should recognize that the result is similar in spirit to Definition 1.7 of independence of a collection of events.

DEFINITION 1.11 Random variables $X_1, X_2, \ldots, X_n$ defined on a sample space S are *independent* if for all possible values $x_1, x_2, \ldots, x_n$

$$P(X_1 = x_1, X_2 = x_2, \ldots, X_n = x_n) = P(X_1 = x_1)P(X_2 = x_2) \cdots P(X_n = x_n)$$

As is the case with independence of events, independence of random variables is often assumed when setting up a probability model. Thus if our experiment is to roll a fair die twice, with X the outcome on the first roll and Y the outcome on the second, we would usually *assume* X and Y to be independent. This seems reasonable, since the die has no memory. On the other hand, independence or lack of it between random variables is often a question to be answered by actually gathering data. This is certainly true, for example, when X and Y are scores on tests designed to measure prejudice and authoritarianism.

Exercise 1.13

1.   An economist has constructed an elaborate mathematical model of the American economy to be used in predicting the effects of government economic policy. In this model, which of the following do you think he might treat as random variables? Explain your answers.

a.   The gross national product (GNP) for 1970
b.   The GNP for the coming calendar year
c.   The probability that next year's GNP will exceed $1200 million
d.   The dollar value of United States exports in the coming calendar year

2.   Describe in words each of the following events from Example 1.30.

a. $\{X = 1\}$ *b.* $\{Y = 1\}$ *c.* $\{X = 1, Y = 1\}$

3.   Refer to the data of Example 1.4. The census bureau maintains a data bank in which marital status is coded as 0 for single, 1 for married, and 2 for widowed or divorced. Let X be the marital status code of a working woman.

a.   Make a table of the individual outcomes for Example 1.4 and the value of X assigned to each.

b. Find $P(X = x)$ for all possible values of x.

4.   For Example 1.30 find $P(X = x)$ and $P(Y = y)$ for all possible values of X and Y. Write the numbers $P(X = x)$ along the bottom margin of the table in Example 1.30, and the numbers $P(Y = y)$ along the right margin. Explain why this arrangement makes it easy to see that X and Y are independent.

5.   Here are data on the proportions of families below the poverty level in 1969 falling into the four combinations of classifications by race and by sex of the family head.

	White	Nonwhite
Male head	0.37	0.10
Female head	0.40	0.13

a. What is the probability that a family below the poverty level has a white head? A male head?

b. Let X be the race code for a poverty-level family and Y the sex code. (Assign values of X and Y to each classification as you choose.) Are the random variables X and Y independent?

c. Do you think the dependence between X and Y is significant? For example, would you be willing to base a program to help needy families on the statement that poor families with nonwhite heads tend to have female heads more frequently than those with white heads?

6.   Suppose we roll a fair die twice independently. Explain why the probability model assigns probability $\frac{1}{36}$ to each of the 36 possible outcomes {i on first roll, j on second roll}.

Let X be the sum of the faces showing and Y their product. Make a table showing what number each of these random variables assigns to each outcome.

Show that X and Y are *not* independent. Explain why this is intuitively reasonable.

1.14 EXPECTED VALUES

When a chance experiment has numerical outcomes, we often want to speak of the "average value" of the outcome. Since a numerical outcome of a chance experiment is a random variable, we must ask what we mean by the average value of a random variable. An example will illuminate this question.

Example 1.31   A lottery sells 4500 tickets. Three tickets are then drawn from a drum. The first wins a $3500 car, and each of the other two wins a $500 color television set. What is the "average value" of a ticket in this lottery?

The question is vague, as are most questions phrased in informal language. Since 4497 of the 4500 tickets win nothing, we could well say that the value of an average ticket is 0. Another informal answer is to say that the 4500 tickets receive among them $4500 worth of prizes, and since all tickets have the same chance of winning, the average value of a ticket is $1. It is this second answer that we want to formalize.

Suppose that you buy a single ticket. Let X be the amount that you win. If all tickets are equally likely to be drawn from the drum, the assignment of probabilities to the possible values of the random variable X is as follows.

x	$P(X = x)$
3500	$1/4500$
500	$2/4500$
0	$4497/4500$

There are only three possible values of X, and the "average value" of this random variable (the average amount won by a ticket) is clearly *not* the ordinary arithmetic mean of these three numbers

$$\frac{3500 + 500 + 0}{3} = 1333.3$$

What is wrong is that the arithmetic mean takes no account of the fact that 3500 is very unlikely to occur and 0 is very likely to occur. The arithmetic mean gives all possible values of X equal weight, with the ridiculous result that the average amount won is $1333. We must instead *weight each possible value of X by its probability of occurring*. This means taking the average value of X to be

$$3500\frac{1}{4500} + 500\frac{2}{4500} + 0\frac{4497}{4500} = \frac{4500}{4500} = 1$$

This agrees with the informal answer we gave earlier.

DEFINITION 1.12 Let X be a random variable with possible values x_1, $x_2, \ldots , x_n$. The *expected value* or *mean* of X is the number

$$E(X) = x_1 P(X = x_1) + x_2 P(X = x_2) + \cdots + x_n P(X = x_n)$$

Several comments about the expected value $E(X)$ of X should be made immediately.

a. $E(X)$ is an ordinary real number, *not* a random variable.
b. $E(X)$ need not be one of the possible values of X, as Example 1.31 shows.

A simple tabular arrangement for computing expected values is illustrated in Example 1.32.

Example 1.32 Let us choose a family in such a way that all American families are equally likely to be chosen. Let X be the number of persons in the chosen family. Then $E(X)$ is the "average family size." The first two columns below give an approximate assignment of probabilities to values of X. The third column contains the products $x_i P(X = x_i)$ whose sum is $E(X)$.

x	$P(X = x)$	$xP(X = x)$
2	0.34	0.68
3	0.21	0.63
4	0.19	0.76
5	0.13	0.65
6	0.07	0.42
7	0.04	0.28
8	0.02	0.16
		3.58

Thus $E(X) = 3.58$, so that the mean size of American families is about 3.6 persons.

The expected value of a random variable often answers practical questions about the distribution of the random variable, as Example 1.32 shows. Life insurance provides another example. An insurance company knows from past experience (written down as mortality tables) the probability that you will die in each coming year, and sets its premiums high enough to cover the expected value of its payout when you purchase a life insurance policy. Here the expected payout tells the company what it will have to pay the beneficiaries of persons like you "on the average."

Example 1.33 John agrees to insure Ivan's life for 5 years. Ivan agrees to pay John $100 at the beginning of each year, and John must pay out $10,000 if Ivan dies. Since Ivan is a healthy man of 21, John finds from a mortality table the proportions of such men who will die during each of the next 5 years. These are as follows:

Year	Probability of Death
1	0.00183
2	0.00186
3	0.00189
4	0.00191
5	0.00193

Let X be John's profit from this agreement, with negative values of X denoting losses. If Ivan dies during the first year, John has collected $100 and must pay out $10,000. So he loses $9900, and $X = -9900$ in this event. Similar reasoning allows us to complete the calculation of $E(X)$ in the following table.

Year of Ivan's Death	x	$P(X = x)$	$xP(X = x)$
1	−9900	0.00183	−18.12
2	−9800	0.00186	−18.23
3	−9700	0.00189	−18.33
4	−9600	0.00191	−18.34
5	−9500	0.00193	−18.34
6 or later	500	0.99058	495.29
			403.93

So John expects to make a substantial profit. But there is a small chance (about 0.01) that he will incur a large loss.

The usefulness of expected values is greatly increased by another interpretation of $E(X)$. We have seen that $E(X)$ is defined as an average of the possible values of X weighted by their probabilities. It is a remarkable fact that $E(X)$ is also the long-term average of the values actually taken by X when the experiment is repeated again and again. More specifically, *if a random experiment with numerical outcome X is repeated many times independently, the average of the actually observed values of X must approach the expected value E(X)*. This fact is called the *law of large numbers*; it is the precise statement of the intuitive "law of averages." We will prove a version of this theorem in the optional Sec. 1.18. The law of large numbers is clearest when applied to games of chance. It tells you that if we play such a game very many times, our average winnings will almost certainly be very close to our expected winnings in a single game. The basic idea is that two quite different "averages" are equal.

The law of large numbers is quite subtle; two comments will start you thinking. First, we can now return to the connection between probability and the long-term relative frequency with which an event occurs. Suppose A is any event with a probability of $P(A)$. Define a random variable Z by setting

$$Z = \begin{cases} 1 & \text{if } A \text{ occurs} \\ 0 & \text{if } A \text{ does not occur} \end{cases}$$

From Definition 1.12, the expected value of Z is

$$E(Z) = 1 \cdot P(A) + 0 \cdot [1 - P(A)] = P(A)$$

If the experiment is repeated n times, and we observe values $z_1, z_2, \ldots, z_n$ of Z, then since each z_i is 1 or 0, the average of the actually observed values is

$$\frac{1}{n}(z_1 + z_2 + \cdots + z_n) = \frac{\text{number of times } A \text{ occurs}}{\text{number of repetitions of the experiment}}$$

This is just the relative frequency of the event A. The law of large numbers tells us that this relative frequency must approach the probability $P(A)$ as the number of repetitions n is increased.

We have come full circle—the definition of a probability measure was intended to reflect reasonable properties of relative frequencies; we now know that starting from our definition of a probability measure, we can prove that relative frequencies do indeed approach probabilities in the long run. This is a further vindication of the "abstract" approach to probability.

Second, it is *not* true that the "law of averages" says that after 10 straight coin tosses have come up heads, a tail is more likely. That is silly, for it assumes that the coin remembers what happened on the previous tosses. The coin has no memory, of course, and we express this mathematically by saying that the tosses are independent. The law of large numbers *does* say that (if the coin is fair) the proportion of heads observed in the long run will be about $1/2$. Does that mean that the coin must compensate for those initial 10 heads? Not at all. If, after 10 straight heads, the next 10,000 tosses are evenly divided between heads and tails, the proportion of heads is then

$$\frac{5010}{10,010} = 0.5005$$

which is very close to $1/2$. This specific outcome is itself unlikely, but it indicates how the law of large numbers works. The 10 straight heads are not compensated for, but swamped by the results of later tosses.

Let us conclude this section with a simple property of expected value which will be used repeatedly in later chapters. It should be the case that any reasonable average of a set of numbers must fall between the largest and the smallest of those numbers. Lemma 1.2 says that this is true for the expected value.

LEMMA 1.2   Suppose X is a random variable with possible values x_1, $x_2, \ldots, x_n$. If m is the smallest and M the largest of the numbers x_i, then

$$m \leq E(X) \leq M$$

PROOF   By Definition 1.12

$$E(X) = x_1 P(X = x_1) + x_2 P(X = x_2) + \cdots + x_n P(X = x_n)$$

All the probabilities $P(X = x_i)$ are positive, and each possible value x_i is less than or equal to M. So each product $x_i P(X = x_i)$ is less than or equal to $M P(X = x_i)$. Therefore

$$E(X) \leq M[P(X = x_1) + \cdots P(X = x_n)]$$

Since the x_i include all possible values of X, the sum of their probabilities is 1. So we have $E(X) \leq M$. The inequality $E(X) \geq m$ is proved by a similar argument.

Exercise 1.14

1.   Teachers in the Monon Central School are allowed 7 days of paid sick leave each year. Records show the following distribution for the number of days of paid sick leave taken by teachers in this school per year.

Days taken	0	1	2	3	4	5	6	7
Percentage of teachers	20	25	15	14	8	10	3	5

What is the expected number of days of sick leave a teacher will take in a year?

2.   A fair die is rolled once. What is the expected value of the outcome? What is the probability that the die takes a value exactly equal to its expected value?

3.   If the class did Exercise 1.2.3, use the distributions that were constructed to find the expected word length in Henry James' novels and in *Popular Science.*

4.   A rural county has 1000 workers, 10 of whom own plantations. The distribution of annual income of workers in that county is as follows:

Dollars	600	1000	1500	2500	10,000	15,000
Proportion of workers	0.10	0.15	0.40	0.34	0.005	0.005

What is the mean annual income of a worker in the county? Does the mean describe the income of a typical worker in this county?

5.   Two varieties of chickens, which produce about the same number of eggs per day, produce the following proportions of eggs of different weights:

Weight, oz	0.6	0.8	1.0	1.2	1.4	1.6	1.8
Variety 1	0.00	0.10	0.20	0.40	0.20	0.10	0.00
Variety 2	0.10	0.15	0.15	0.20	0.15	0.15	0.10

Show that the expected weight is the same for both varieties. Are there any other features of the distribution of weights which might help in choosing between these varieties?

6.   A gambler is betting on tosses of a fair coin. After five heads in a row have appeared, he bets heavily on tails. A probabilist says that is silly, and explains independence and the law of large numbers to him. Shortly afterward, the gambler is dealt four black cards in a poker game. He remembers what the probabilist said and concludes that the next card is equally likely to be red or black. Explain to the gambler the difference between these two situations.

7.   Prove the other half of Lemma 1.2: $E(X) \geq m$.

8.   Use Table A of binomial probabilities to compute the mean of a binomial random variable with $n = 5$ and $p = 0.2$.

9.   Suppose that all the possible values of a random variable X are nonnegative.

a.   Show that $E(X) \geq 0$ must hold.

b.   Show that in this case $E(X) = 0$ only if $P(X = 0) = 1$.

10.   *Summation Notation.* It is very convenient in working with expected values to have an abbreviated notation for sums. An abbreviation for the sum

$$x_1 + x_2 + \cdots + x_n$$

of n numbers $x_1, x_2, \ldots, x_n$ is

$$\sum_{i=1}^{n} x_i$$

Here i is called the *index of summation*. Note that the values of i below and above the capital sigma tell us that the sum starts with $i = 1$ and takes in all values of i up to and including $i = n$. For example

$$\sum_{i=1}^{4} i^2 = 1^2 + 2^2 + 3^2 + 4^2$$
$$= 1 + 4 + 9 + 16 = 30$$

and Definition 1.12 of expected value can now be written as

$$E(X) = \sum_{i=1}^{n} x_i P(X = x_i)$$

Since Σ is only an abbreviation for ordinary addition, use of this notation does not require any new mathematics, only a little familiarization. The following "rules" are just facts about addition of numbers.

a.   For any constant a, $\displaystyle\sum_{i=1}^{n} ax_i = a\sum_{i=1}^{n} x_i$

b. $\displaystyle\sum_{i=1}^{n} (x_i + y_i) = \sum_{i=1}^{n} x_i + \sum_{i=1}^{n} y_i$

Here are some problems on summation notation.

a.   Suppose $x_1 = 1$, $x_2 = -2$, $x_3 = 4$ and $y_1 = -2$, $y_2 = 3$, $y_3 = 4$. Evaluate each of the following expressions.

i. $\displaystyle\sum_{i=1}^{3} x_i$ ii. $\displaystyle\sum_{i=1}^{3} y_i$

iii. $\displaystyle\sum_{i=1}^{3} 5x_i$ *iv.* $\displaystyle\sum_{i=1}^{3} x_i y_i$

v. $\displaystyle\left(\sum_{i=1}^{3} x_i\right)\left(\sum_{i=1}^{3} y_i\right)$ *vi.* $\displaystyle\sum_{i=1}^{3} (x_i + y_i)$

vii. $\displaystyle\sum_{i=1}^{3} x_i + \sum_{i=1}^{3} y_i$

b.   Evaluate $\displaystyle\sum_{j=1}^{5} j$ and $\displaystyle\sum_{j=1}^{5} 4j^2$.

c.   Express $2x_1{}^2 + 2x_2{}^2 + 2x_3{}^2 + 2x_4{}^2 + 2x_5{}^2$ in summation notation.
d.   Write the partition equation (Theorem 1.8) using summation notation.

We often use even a further abbreviation when the quantities being summed over are clear. If X is a random variable with possible values $x_1, x_2, \ldots, x_n$, we will sometimes write

$$E(X) = \sum_x xP(X = x)$$

It is clear that this means "sum over all possible values x of X." This is especially convenient when several random variables are involved. Thus

$$\sum_{x, y} xyP(X = x, Y = y)$$

means to form the product xy for every possible value x of X and y of Y, multiply each product by the corresponding probability $P(X = x, Y = y)$, and add up all these terms.

1.15   MORE ON EXPECTED VALUE

We are often interested in the expected value of the sum of several random variables. The theorem of this section shows us how to manipulate expected values of sums and of constant multiples of random variables. The result is very simple, but very useful.

Example 1.34   An assemblyline worker must maneuver an aluminum casting into position, then operate an automated drill press. An efficiency expert gathers data on the time required for each of these operations. These times are random variables X and Y, respectively. The data allow us to compute expected values, and we find

$$E(X) = 12 \text{ sec}$$
$$E(Y) = 21 \text{ sec}$$

These are the mean times required for these operations. The total time required for both operations is $X + Y$. This is a new random variable. What should its expected value $E(X + Y)$ be? Surely the average time required to perform both operations should be the *sum* of the average times required to perform the two individual operations. So

$$E(X + Y) = 12 + 21 = 33 \text{ sec}$$

Theorem 1.10 says that the intuitive argument used in Example 1.34 gives the correct result. Theorem 1.10 is "obviously" reasonable. The proof is surprisingly hard, so we consider it optional, and delay it until the next section.

THEOREM 1.10   If X and Y are random variables and a, b are numerical constants, then

a. $E(aX + b) = aE(X) + b$
b. $E(X + Y) = E(X) + E(Y)$

Example 1.35   Gain Radio Corporation sells VHF transceivers to the Armed Forces and police-band receivers to the civilian market. Next year's sales depend on many factors, such as success in competing for new military contracts. Gain follows the modern practice of giving probability estimates of future sales rather than a single guess. The informed estimates of officers of the transceiver operation for next year are

Unit sales	100	600	800	1000
Probability	0.2	0.4	0.3	0.1

The corresponding subjective probabilities arrived at by officers of the civilian receiver operation are

Unit sales	2000	2500	3000	4000
Probability	0.4	0.4	0.1	0.1

Gain makes a profit of $150 on a transceiver and a profit of $20 on a receiver. What is the company's expected total profit from these lines next year?

Let X be the number of transceivers sold and Y the number of receivers. The total profit is the random variable

$$T = 150X + 20Y$$

Combining both parts of Theorem 1.10 shows that the expected profit is

$$E(T) = 150E(X) + 20E(Y)$$

The rest is routine: Check that $E(X) = 600$ and $E(Y) = 2500$, so that

$$E(T) = (150)(600) + (20)(2500)$$
$$= \$140,000$$

Notice that in Example 1.35 we did *not* have to know the joint probabilities $P(X = x, Y = y)$. Theorem 1.10 requires only that we be able to compute $E(X)$ and $E(Y)$ individually. This is very important in cases such as Example 1.35 where probability distributions for X and Y are obtained from different sources, so that joint probabilities are not available.

Notice also that by applying Theorem 1.10 repeatedly it follows that the expected value of a sum of any number of random variables is the sum of their expected values. We will see that this fact and a clever idea enable us to find the expected value of binomial random variables.

Example 1.36   The probability that a newborn child will be male is about $p = 0.52$. A hospital records the sex of 100 consecutive children delivered there. What should be the expected number of males?

The answer is almost obvious—if 52 percent of newborn children are males, we should "expect" 52 males in 100 births. More formally, since it is reasonable that the sex of successive children delivered at the hospital is independent, the number of males is a binomial random variable X with $n = 100$ and $p = 0.52$. We have therefore said that $E(X) = np$.

Suppose in general that X is the number of successes in n independent trials each having probability p of success.

To show that the binomial random variable X has expected value np directly from Definition 1.12 would require us to prove that

$$\sum_{k=0}^{n} k \binom{n}{k} p^k (1-p)^{n-k} = np$$

That is unpleasant. We can find $E(X)$ with much less work by studying the individual independent trials and using Theorem 1.10. Define random variables $Z_1, Z_2, \ldots, Z_n$ by

$$Z_i = \begin{cases} 1 & \text{if the } i\text{th trial is a success} \\ 0 & \text{if the } i\text{th trial is a failure} \end{cases}$$

Then

$$X = Z_1 + Z_2 + \cdots + Z_n$$

since the sum is just the number of Z_i's which are 1 and is therefore the number of successes observed. So

$$E(X) = E(Z_1) + E(Z_2) + \cdots + E(Z_n)$$

By Definition 1.12 and the fact that each trial has probability p of succeeding,

$$E(Z_j) = 1 \cdot p + 0 \cdot (1 - p) = p$$

Therefore

$$E(X) = p + p + \cdots + p = np$$

We have shown that *if X is a binomial random variable representing the number of successes in n independent trials each having probability p of success, then*

$$E(X) = np$$

Exercise 1.15 1. Laboratory data show that the time X required for a certain chemical reaction has mean 40 min, and the time Y required for a second reaction has mean 25 min. An industrial process will run these reactions in sequence.

a. What is the expected total time required for the two reactions?
b. There is a fixed period of 5 min while the product of the first reaction is prepared for the start of the second reaction. What is the expected total time required for the entire process?

2. An expensive x-ray tube has a lifetime in service of Y mo (months), with probability distribution given below. When the tube wears out, it is temporarily replaced with an inexpensive tube while a permanent replacement is ordered. The temporary replacement has a lifetime of Z mo with probability distribution also given below.

k, mo	1	2	3	4	5	6	7
$P(Y = k)$	0.1	0.1	0.2	0.2	0.2	0.1	0.1
$P(Z = k)$	0.3	0.4	0.2	0.1			

What is the expected total lifetime in service of the two tubes?
3. The management responsible for the assemblyline of Example 1.34 is considering asking the worker to make a measurement after drilling as a check for gross errors. The efficiency expert observes workers doing this and obtains the following relative frequency data for the time required to make the measurement.

Time, sec	5	6	7	8	9
Relative frequency	0.1	0.3	0.3	0.2	0.1

What is the expected total time required to do all three operations?
4. If in fact 70 percent of the voting population opposes legalizing possession of marijuana, what is the expected number of persons who will say no if 60 persons are asked whether they favor legalizing possession?

5. Suppose that 95 percent of the vases produced by a mechanized process are free from defects. If an inspector checks 10 vases, what is the expected number of defective vases he will find?

6. A basketball player has made 80 percent of his foul shots over a long period of time. If he takes 20 foul shots in a game, what is the expected number of foul shots he makes?

7. Here are two fine points connected with Exercise 1.15.6.

a. What is the probability that the player makes exactly his expected number of shots? Is it correct to say that he will usually make his expected number?

*b. Successive foul shots are probably not independent, due to hot streaks and other factors. Does this change the result of Exercise 1.15.6?

8. Suppose X is the number of successes in n independent trials and Y the number of successes in the next m independent trials, with probability p of success on each trial.

a. Why should $X + Y$ have the binomial distribution with $n + m$ and p?

b. Use your knowledge of binomial expected values to check Theorem 1.10b in this case.

*9. The expected value of a random variable X which has positive probability of taking each of infinitely many possible values is still defined as

$$P(X) = \sum_x xP(X = x)$$

In this case the sum is over infinitely many values x. For example the mean of a geometric random variable (see Sec. 1.9) is

$$E(X) = \sum_{k=1}^{\infty} kpq^{k-1}$$
$$= p + 2pq + 3pq^2 + 4pq^3 + \cdots$$
$$= p(1 + 2q + 3q^2 + 4q^3 + \cdots)$$

where $q = 1 - p$. To find this sum, first find the sum

$$1 + 2q + 3q^2 + 4q^3 + \cdots$$

by writing the numbers to be summed in an array as follows

$$
\begin{array}{cccc}
1 & & & \\
q & q & & \\
q^2 & q^2 & q^2 & \\
q^3 & q^3 & q^3 & q^3 \\
\vdots & \vdots & \vdots & \vdots
\end{array}
$$

(Notice that the array contains a 1, two q's, three q^2's, and so on, as it ought to.) Sum down each column of the array, using your knowledge of the sum of a geometric series. Then sum the column sums to get the total sum.

Use your result to find the expected number of trials the Exache people (Example 1.23) will need to find a group which justifies their advertising claim.

*1.16 VARIANCE

The mean $E(X)$ of a random variable X is a measure of the average value of X. It is also important to have a measure of the spread or dispersion of the values of the random variable. If we are comparing two brands of $1/2$-in. ball bearings, for example, the diameters of the individual bearings are random variables. If the two brands both have a mean diameter of $1/2$ in., we will prefer the brand which has the smaller variation in diameter. Our judgment is therefore based on the spread or dispersion of the values of these random variables, as well as on their average values. Variability of a random variable is also a matter of concern in such instances as assessing the potency of a drug, predicting sales during the coming fiscal year, and measuring the time required for a worker to perform a specified task.

There are many possible measures of the spread or dispersion of the values of X. We might, for example, simply take the difference between the largest and smallest possible values of X. This measure of spread has the disadvantage that it does not take into account whether X lies near its extreme values with probability near 1 or has a probability of almost 1 of falling near some central value. Exercise 1.16.3 illustrates how this measure of spread fails to distinguish between these cases. We will use the expected value of the square of the difference of X from its mean as a measure of spread. This is the variance of X. If the probability distribution of X concentrates most of the probability near the mean (small dispersion), the variance will tend to be small. On the other hand, if values of X which are far from the mean are assigned substantial probability, the variance will increase. So we have a reasonable measure of the spread or dispersion of X. Since the variance is the average of the *square* of differences of values of X from the mean, it has units which are the square of the units of X. So we often use the square root of the variance, called the *standard deviation*, as a measure of spread. The standard deviation has the same units as does X. For example, if X is the time in seconds a worker requires to do a task, the variance of X has unit "square seconds" (sec^2), but the standard deviation is measured in seconds. Here are the formal definitions of these two measures of spread.

DEFINITION 1.13   Suppose that X is a random variable having mean m. The *variance* of X is

$$\text{Var } X = E[(X - m)^2]$$

and the *standard deviation* of X is the positive square root of its variance.

Note that for any value x of X, $(x - m)^2 \geq 0$. So by Exercise 1.14.9 the variance is always nonnegative. This means that we can always take its square root. The same exercise shows that the only random variables with variance 0 are those which take only one value: $P[(X - m)^2 = 0] = 1$, so that $P(X = m) = 1$. Such random variables are "constants in disguise." All other random variables have strictly positive variance.

Before we can compute any variances, we must meet a theoretical issue which we ignored in the last two sections. For any random variable X, $(X - m)^2$ is another random variable. Definition 1.12 says that

$$E[(X - m)^2] = \sum_z zP[(X - m)^2 = z]$$

It is very awkward to have to find $P[(X - m)^2 = z]$ when we know $P(X = x)$. It seems that the "average" of $(X - m)^2$ could be found instead by

$$\sum_x (x - m)^2 P(X = x)$$

These two numbers are always the same. Showing that is a bit of a nuisance, and we will not do it. Exercise 1.16.4 asks you to verify this fact in a particular example. This last formula makes it easy to compute variances.

Example 1.37   Return to the distribution of weights of eggs of two varieties of chickens given in Exercise 1.14.5. If X is the weight of an egg produced by a hen of variety 1, $E(X) = 1.2$ oz. Table 1.1 shows a systematic method of finding the variance using the fact that

$$\text{Var } X = \sum_x (x - m)^2 P(X = x)$$

We see that Var $X = 0.048$ oz². A similar computation shows that the variance of the weight of eggs produced by the second variety is 0.132 oz². It should be clear from studying the distributions that the second variety has much greater spread of weight, and the variances reflect this.

x	$P(X = x)$	$x - 1.2$	$(x - 1.2)^2$	$(x - 1.2)^2 P(X = x)$
0.6	0.00	−0.6	0.36	0.000
0.8	0.10	−0.4	0.16	0.016
1.0	0.20	−0.2	0.04	0.008
1.2	0.50	0.0	0.00	0.000
1.4	0.20	0.2	0.04	0.008
1.6	0.10	0.4	0.16	0.016
1.8	0.00	0.6	0.36	0.000
				0.048

Table 1.1   Computing the variance of a random variable

We have seen that applying Definition 1.12 is not the easiest way to compute $E((X - m)^2)$. Notice that $(X - m)^2$ is a function of the random variable X. Any function of a random variable X is another random variable, and Definition 1.12 is awkward in almost all such cases. As another example, the sum $X + Y$ of two random variables X and Y is a function of both of them. It turns out that for any function $g(X, Y)$ of X and Y the expected value as given by Definition 1.12, namely

$$\sum_z zP[g(X, Y) = z]$$

is the same as the simpler expression

$$\sum_{x, y} g(x, y)P(X = x, Y = y)$$

For example, if $g(x, y) = x + y$ (the sum), we can say that

$$E(X + Y) = \sum_{x, y} (x + y)P(X = x, Y = y)$$

If $g(x, y) = xy$ (the product), then

$$E(XY) = \sum_{x, y} xyP(X = x, Y = y)$$

In both cases the sum is over all possible values x of X and y of Y. We will not prove this fact. Try to see that both expressions for the expected value of $g(X, Y)$ are averages of the possible values $g(x, y)$ weighted by their probability of occurrence, and thus ought to be equal. Let us first use this fact to prove Theorem 1.10.

PROOF OF THEOREM 1.10 We will leave the proof of Theorem 1.10a as an exercise (Exercise 1.16.13) and prove only Theorem 1.10b. We saw just above that

$$E(X + Y) = \sum_{x, y} (x + y)P(X = x, Y = y)$$

$$= \sum_{x, y} xP(X = x, Y = y) + \sum_{x, y} yP(X = x, Y = y)$$

Now the events $\{Y = y\}$ form a partition of S as y ranges over all possible values of Y. So for any fixed x the event $\{X = x\}$ is the union of the disjoint events $\{X = x, Y = y\}$ over all possible values y of Y, and therefore

$$P(X = x) = \sum_y P(X = x, Y = y)$$

From this we see that if we sum first over all y for each fixed x,

$$\sum_{x, y} xP(X = x, Y = x) = \sum_x xP(X = x) = E(X)$$

and similarly, summing first over all x for each fixed y,

$$\sum_{x,y} yP(X=x, Y=y) = \sum_{y} yP(Y=y) = E(Y)$$

That completes the proof of Theorem 1.10b.

Our next task is to discover the properties of variances corresponding to those of means given by Theorem 1.10. These properties are somewhat less obvious than those of means, but we can guess some of them. First, adding a constant to a random variable X does not change the spread of X about its mean, so the variance is unchanged. Multiplying X by a constant multiplies Var X by the *square* of that constant, since Var X is a mean of squares. The variance of a sum of random variables is *not* always the sum of their variances (see Exercise 1.16.5). But if the random variables are independent, their variances *do* add.

THEOREM 1.11 Suppose X and Y are random variables and a and b are numerical constants. Then

a. Var $(aX + b) = a^2$ Var X
b. If X and Y are independent, then

$$\text{Var } (X + Y) = \text{Var } X + \text{Var } Y$$

PROOF
a. By Theorem 1.10 the mean of $aX + b$ is $am + b$, where m is the mean of X. So Definition 1.13 says that

$$\begin{aligned}
\text{Var } (aX + b) &= E[(aX + b - am - b)^2] \\
&= E[(aX - am)^2] \\
&= E[a^2(X - m)^2] \\
&= a^2 E[(X - m)^2] \\
&= a^2 \text{ Var } X
\end{aligned}$$

b. Suppose that $E(X) = m$ and $E(Y) = r$. Then by Theorem 1.10

$$E(X + Y) = m + r$$

and therefore

$$\begin{aligned}
\text{Var } (X + Y) &= E[(X + Y - m - r)^2] \\
&= E[(X - m + Y - r)^2] \\
&= E[(X - m)^2 + (Y - r)^2 + 2(X - m)(Y - r)] \\
&= E[(X - m)^2] + E[(Y - r)^2] + 2E[(X - m)(Y - r)]
\end{aligned}$$

So we see that

$$\text{Var } (X + Y) = \text{Var } X + \text{Var } Y$$

exactly when

$$E[(X - m)(Y - r)] = 0$$

This is true if X and Y are independent.

To see this, first use the fact that we now know that

$$E[(X - m)(Y - r)] = \sum_{x,y} (x - m)(y - r)P(X = x, Y = y)$$

Since X and Y are independent, this is equal to

$$\sum_{x,y} (x - m)P(X = x)(y - r)P(Y = y)$$

The summands are products of a factor involving only x and a factor involving only y. If we sum over y first (say), the factor involving only x is a constant as far as y is concerned and can be brought across the summation sign. The result is

$$E[(X - m)(Y - r)] = \sum_{x} (x - m)P(X = x) \cdot \sum_{y} (y - r)P(Y = y)$$
$$= E(X - m) \cdot E(Y - r)$$

Since $E(X - m) = E(X) - m = m - m = 0$ by Theorem 1.10, this completes the proof.

Theorem 1.11 allows us to derive some further results about variances. For a binomial random variable X with distribution described by n and p, the method used in the last section to show that $E(X) = np$ can also show that

$$\text{Var } X = np(1 - p)$$

You are asked to provide the details in Exercise 1.16.8.

As a second application of Theorem 1.11, suppose that a random experiment (which may have any number of possible outcomes, not simply success or failure) is repeated n times independently. The outcomes are represented by the values of n random variables $X_1, X_2, \ldots, X_n$ which are independent and have the same probability distribution. For example, X_i might be the monthly housing expenditure of the ith of n randomly chosen families, the weight gain of the ith of n infants fed a special diet, or the score on a standard reading test of the ith of n students who have followed a new curriculum. In all these cases we would like to know the common mean m of the random variables $X_1, X_2, \ldots, X_n$. But in practice we do not know the complete probability distribution of housing expenditures or weight gains or reading scores. We do have the n observations X_i, however, and it seems reasonable to use their arithmetic mean

$$\bar{X} = \frac{1}{n}(X_1 + X_2 + \cdots + X_n)$$

as an estimate of the unknown m.

Note that $\bar{X}$ (unlike m) is a random variable. Theorems 1.10 and 1.11

enable us to show that *if $E(X_i) = m$ and Var $X_i = v$ and the X_i are independent, then*

$$E(\bar{X}) = m$$

$$\text{Var } \bar{X} = \frac{v}{n}$$

This says that $\bar{X}$ estimates m correctly "on the average" and is more tightly concentrated about m as the number of observations increases. Applying Theorem 1.10,

$$E(\bar{X}) = E\left[\frac{1}{n}(X_1 + X_2 + \cdots + X_n)\right]$$

$$= \frac{1}{n}[E(X_1) + E(X_2) + \cdots + E(X_n)]$$

$$= \frac{1}{n}(m + m + \cdots + m)$$

$$= \frac{1}{n}(nm) = m$$

Applying Theorem 1.11,

$$\text{Var } \bar{X} = \text{Var }\left[\frac{1}{n}(X_1 + X_2 + \cdots + X_n)\right]$$

$$= \frac{1}{n^2}(\text{Var } X_1 + \text{Var } X_2 + \cdots + \text{Var } X_n)$$

$$= \frac{1}{n^2}(v + v + \cdots + v)$$

$$= \frac{1}{n^2}(nv) = \frac{v}{n}$$

Example 1.38   An economist wants to investigate homeowner expenditures for maintenance and repair in a suburban community. He cannot afford to survey every homeowner, so he chooses n homeowners at random and records their past year's expenditures $X_1, X_2, \ldots, X_n$. Each X_i is a random variable with distribution given by the overall distribution of repair expenditures in the community, and the X_i are independent. Suppose that in fact the mean expenditure is $400 with variance 2500 ("square dollars"), or more conveniently standard deviation $50.

If the economist asks only one homeowner, his datum X_1 has

$$E(X_1) = 400 \qquad \text{Var } X_1 = 2500$$

and the large variance makes this of little use in describing the community as a whole. If he asks 100 homeowners and uses the arithmetic mean $\bar{X}$ of their expenditures, then

$$E(\bar{X}) = 400 \qquad \text{Var } \bar{X} = {}^{2500}\!/_{100} = 25$$

Thus his estimate $\bar{X}$ has much less variability (as measured by its variance) than does X_1. We will make this idea more precise in Sec. 1.18.

Exercise 1.16   1.   Compute the variance of the distribution of incomes given in Exercise 1.14.4.
2.   Compute the variance of the time required for a worker to make the measurement described in Exercise 1.15.3.
3.   Here are probability distributions for two random variables with the same largest and smallest possible values. The first is clearly more spread out than the second in the sense that extreme values are much more likely to occur. Compute the variance of both random variables.

z	-10	-5	0	5	10
$P(X = z)$	0.3	0.2	0.0	0.2	0.3
$P(Y = z)$	0.1	0.1	0.6	0.1	0.1

4.   Find the variance of the random variable Y in Exercise 1.16.3 by

a.   finding the mean $m = E(Y)$
b.   finding all possible values z of $(Y - m)^2$ and the probability $P[(Y - m)^2 = z]$ [Each of these probabilities is the sum of some of the probabilities $P(Y = y)$.]
c.   computing

$$\sum_z zP[(Y - m)^2 = z]$$

The answer should be the same as that obtained in Exercise 1.16.3 by computing

$$\sum_y (y - m)^2 P(Y = y)$$

5.   A research group consists of six men and two women. To avoid argument, it is decided to choose two members of the group to present a paper by random selection. All eight names are placed in a hat, and one is drawn. Let the random variable $X = 1$ if a man's name is drawn, and $X = 0$ if it is a woman's. So $P(X = 1) = 6/8 = 3/4$ since all names are equally likely to be drawn. A second name is drawn, without replacing the first. Set $y = 1$ if the second name is that of a man, $Y = 0$ if it is that of a woman. We can easily find conditional probabilities involving Y. For example,

$$P(Y = 1 | X = 1) = 5/7$$

since if the first name drawn is male, five of the remaining seven names are male. Note that X and Y are *not* independent.

a.   Find $P(Y = 0)$ and $P(Y = 1)$ using the partition equation.
b.   Find the mean and variance of X and Y.

c. Let $Z = X + Y$ be the number of men chosen. Find $P(Z = z)$ for $z = 0$, 1, and 2.

d. Find the mean and variance of Z. Check that Var $(X + Y)$ is *not* equal to Var $X +$ Var Y. Is the standard deviation of $X + Y$ equal to the sum of the standard deviations of X and Y?

6. Use Table A of binomial probabilities and the formula

$$\text{Var } X = \sum_x (x - m)^2 P(X = x)$$

to compute the variance of a binomial random variable X with $n = 5$ and $p = 0.2$. Compare your answer with the exact result from the formula

$$\text{Var } X = np(1 - p)$$

7. An assemblyline worker fails to complete his assigned task on 10 percent of the units that pass him. If 400 units pass him during the day, what is the expected number of times he fails? What is the variance of the number of times he fails? What is the standard deviation?

8. Review the demonstration in Sec. 1.15 that a binomial random variable X has mean np. In the representation

$$X = Z_1 + Z_2 + \cdots + Z_n$$

used there, the random variables Z_i are independent because the trials are independent. Compute the variance of Z_i, and use Theorem 1.11 to obtain the variance of X.

9. The stanford-binet IQ tests are designed so that the distribution of scores over the entire nation has mean 100 and standard deviation 16. If 25 children are chosen at random from across the nation, what is the mean and standard deviation of the arithmetic mean of their 25 IQ scores?

10. Use Theorem 1.10 to show that if $m = E(X)$

$$E[(X - m)^2] = E(X^2) - m^2$$

This formula provides another method of computing variances.

11. You might suppose that another possible measure of the spread or dispersion of a random variable X having mean m is

$$E(X - m)$$

Why is this not a useful idea?

12. A reasonable measure of spread for a random variable X having mean m is $E(|X - m|)$. This is rarely used because the absolute value $|X - m|$ is harder to work with than $(X - m)^2$. Compute $E(|X - m|)$ for the random variable X of Example 1.37. Compare the result with the standard deviation of X.

13. Prove Theorem 1.10a.

1.17   APPLICATIONS

The probability models given in this section emphasize the use of expected values. The first uses the tools of probability theory to compare and evaluate three sales promotions strategies. It is a simplified version of techniques which are becoming standard corporate practice. The second application is a probability model for choice behavior when an individual is forced to choose among several courses of action with uncertain consequences.

Application 1.3   Management Decision Making

An operation within a large chemical company synthesizes a particular compound. The markets for this product are of four types:

1.   other operations within the company
2.   domestic manufacturers of consumer products
3.   domestic military contractors
4.   the overseas market

Management is considering three strategies for additional sales promotion during the coming year:

A: a direct-mail advertising campaign

B: a specified increase in the operation's sales force

C: offering additional consulting and other user services

Since budget limitations imply that only one of these strategies can be adopted, management must choose among them. Each type of promotion has differing appeal to each of the four markets. By conferring with the personnel most familiar with each market, management can assess the probable impact of each strategy on sales to the various markets.

The exact sales volume in each market if a particular sales strategy is employed depends on a number of unpredictable conditions, such as the general economic situation and the actions of competitors. It is therefore very difficult to give a single "best estimate" of sales volume. The uncertainty, as well as the informed opinion of the officers most familiar with each market, can be represented by asking them to give a probability distribution over possible volumes of sales. These subjective probability distributions are given in Table 1.2.

How can this information be used to decide which sales strategy to adopt? One method for making such decisions is to use the strategy having the greatest expected total sales. We will let X_1, X_2, X_3, and X_4 stand for the sales volumes in the four markets. If Z is the total sales volume, then

$$Z = X_1 + X_2 + X_3 + X_4$$

Market		Strategy A			Strategy B			Strategy C		
1	Sales volume	200	300	400	200	300	400	200	300	400
	Probability	0.5	0.3	0.2	0.7	0.2	0.1	0.4	0.5	0.1
2	Sales volume	100	150		100	150	200	100	150	200
	Probability	0.5	0.5		0.2	0.2	0.6	0.3	0.4	0.3
3	Sales volume	100	150	250	100	150	250	100	150	250
	Probability	0.3	0.5	0.2	0.2	0.3	0.5	0.4	0.5	0.1
4	Sales volume	100	200		100	200	300	100	200	300
	Probability	0.8	0.2		0.7	0.2	0.1	0.5	0.3	0.2

Table 1.2   Subjective probability distributions for the sales promotion problem. Sales volumes are in thousands of dollars.

and the expected total sales volume is therefore (Theorem 1.10)

$$E(Z) = E(X_1) + E(X_2) + E(X_3) + E(X_4)$$

The expected value of each X_i when a specified promotion strategy is used can be found from Table 1.2. For example, the expected sales to other operations within the company when strategy 1 is used is (in thousands of dollars)

$$E(X_1) = (200)(0.5) + (300)(0.3) + (400)(0.2) = 270$$

The results of these calculations are given in Table 1.3 together with the resulting expected total sales volume. It appears that strategy B is preferable by this criterion.

Market	Strategy A	Strategy B	Strategy C
1	270	240	270
2	125	170	150
3	155	190	140
4	120	140	170
Total	670	740	730

Table 1.3   Expected sales volumes in thousands of dollars

Expected sales is not the only decision-making criterion that management could choose to use in this problem. A cautious management will be aware that the strategy with greatest expected sales may run a greater risk of very low sales than a strategy with somewhat smaller expected sales. Corporate policy may be to avoid such risks. An alternative decision-making criterion which might be appealing is as follows:

1.   Find the probability distribution of total sales for each sales strategy.

2. Compute for each sales strategy the sales volume which will be exceeded with probability 0.90 when that strategy is used.
3. Use the strategy for which this "90 percent sure" sales volume is highest.

Step 1 of this procedure requires a good deal of calculation. We must compute the probability of each possible total sales volume under each strategy. As an example, we will find the probability of having total sales of exactly $600,000 when strategy A is used. Studying Table 1.2, we see that total sales of $600,000 under strategy A can result from any of three different combinations of possible sales volumes to the four markets. (For example, if the company has sales of $200,000 to each of markets 1 and 4 and sales of $100,000 to each of markets 2 and 3, the total sales are $600,000.) All three possibilities are displayed in Table 1.4.

Market				Probability
1	2	3	4	
200	100	100	200	0.015
200	150	150	100	0.100
300	100	100	100	0.036

Table 1.4 Possible combinations of sales to individual markets yielding $600,000 in total sales under strategy A

We will now make an assumption that was unnecessary when we were dealing only with expected values: The amounts of sales to different markets are independent random variables. We can then fill in the probability column of Table 1.4. For example, the probability of achieving the sales volumes listed in the first row is, from Table 1.2,

$$(0.5)(0.5)(0.3)(0.2) = 0.015$$

The probability of achieving exactly $600,000 in sales is then

$$0.015 + 0.100 + 0.036 = 0.151$$

To complete step 1 of this second decision-making procedure requires that a calculation like that above be done for each possible total sales volume under each of the three sales strategies. A computer programed to do this is a great help in such repetitive calculations, and allows us to study more complicated decision problems as well. You can complete the work of this example by doing Computer Project 1 at the end of this chapter.

Exercise 1.17 1. Verify the values of $E(X_2)$, $E(X_3)$, and $E(X_4)$ under sales strategy A given in the first column of Table 1.3.

2.   What is the probability that sales of exactly $700,000 will be achieved using sales strategy A? (Make the same independence assumption made in the text.)

3.   A very cautious executive suggests:

a.   Compute the minimum possible sales under each strategy.
b.   Use the strategy for which this minimum is largest.

Why is this decision-making procedure of no help in this problem?

4.   Suppose that the costs of the proposed sales strategies differ in such a way that each dollar of sales

yields $0.10 profit under strategy A

yields $0.07 profit under strategy B

yields $0.08 profit under strategy C

Which sales strategy has the greatest expected profit?

Application 1.4   A Model for Choice Behavior

If a person is forced to choose among several courses of action, each of which has uncertain consequences, how does he behave? A first step toward describing such "choice behavior" is to make the possible choices and their consequences explicit in an artificial experimental setting. One such experiment is to give the experimental subject an opportunity to choose which of several games of chance he wishes to play. The possible outcomes of each game of chance are various amounts of money. The subject is told these and the probability of winning each amount, and is rewarded by being allowed to play the game of his choice and keep his winnings. We will call each game of chance a "wager." Formally, a wager is a set of amounts of money and probabilities of winning each amount.
     For example, the experimenter could offer the subject a choice between

wager 1:   Toss a fair coin, win $2 if heads and $0 if tails.

wager 2:   Roll a fair die, win $4 if 6 comes up and $0 otherwise.

In more formal notation, the set of all possible amounts of money the subject can win in this case is

$$S = \{0,2,4\}$$

The two wagers are described by two sets of probabilities for the possible outcomes in S

$$w_1 = (\tfrac{1}{2}, \tfrac{1}{2}, 0)$$
$$w_2 = (\tfrac{5}{6}, 0, \tfrac{1}{6})$$

Several probability models for the behavior of subjects in experiments of this type are discussed by G. M. Becker, M. H. DeGroot, and J. Marschak in "Stochastic models for choice behavior." (*Behavioral Science* **8**, 1963, pp.

41–55). We will use one of the simplest and most plausible. This model assumes that a subject choosing a wager will take into account only the expected value of his winnings. It would be unrealistic to assume that the subject will always choose the wager with the higher expected winnings. Unless he has studied probability, he will not consciously compute expected values. What our model says instead is that *the subject chooses each wager with probability proportional to its expected winnings*. Thus the wager with the highest expected winnings is most likely—but not certain—to be chosen.

In our specific example, wager 1 has expected winnings

$$E_1 = 0 \cdot \tfrac{1}{2} + 2 \cdot \tfrac{1}{2} = 1$$

and wager 2 has expected winnings

$$E_2 = 0 \cdot \tfrac{5}{6} + 4 \cdot \tfrac{1}{6} = \tfrac{2}{3}$$

Since the subject must choose one of these, the model implies that he chooses wager 1 with probability

$$\frac{E_1}{E_1 + E_2} = \frac{1}{1 + \tfrac{2}{3}} = \frac{3}{5}$$

and wager 2 with probability

$$\frac{E_2}{E_1 + E_2} = \frac{\tfrac{2}{3}}{1 + \tfrac{2}{3}} = \frac{2}{5}$$

(These probabilities add to 1 and are proportional to the expected winnings E_i.)

Now neither this nor any other model is given from heaven. It is a plausible model, but it must be checked against actual experimental results. How can we do this? The "obvious" answer is to present this choice to a number of experimental subjects and compare the relative frequencies of their choices with the probabilities predicted by the model. But the choice behavior of different individuals may vary greatly. We would really like to study the individual choice behavior of a single subject by asking him to repeatedly choose from sets of wagers. But we cannot repeatedly offer him the same two wagers to choose between—he is likely to always make the same choice, so that these are not separate and independent trials.

Becker, DeGroot, and Marschak found an ingenious partial solution to the dilemma of how to test this model by experiment. They added a third wager w_3 to the two already given. The wager w_3 has the same set S of possible winnings as do w_1 and w_2. The probability of winning any amount using w_3 is the *average* of the probabilities of winning that amount using w_1 and w_2. In our case the new wager w_3 assigns to each amount in $S = \{0, 2, 4\}$ the average of the probabilities assigned by w_1 and w_2. Therefore

$$w_3 = (\tfrac{2}{3}, \tfrac{1}{4}, \tfrac{1}{12})$$

It is always the case that the "average wager" w_3 does have probabilities which add to 1 since those of w_1 and w_2 do. The expected winnings for w_3 are

$$E_3 = 0 \cdot \frac{2}{3} + 2 \cdot \frac{1}{4} + 4 \cdot \frac{1}{12} = \frac{5}{6}$$

The experimental subject is now offered a choice from among the *three* wagers w_1, w_2, and w_3. The model predicts that

$$P(w_1 \text{ is chosen}) = \frac{E_1}{E_1 + E_2 + E_3} = \frac{1}{1 + \frac{2}{3} + \frac{5}{6}} = \frac{6}{15} = \frac{2}{5}$$

$$P(w_2 \text{ is chosen}) = \frac{E_2}{E_1 + E_2 + E_3} = \frac{\frac{2}{3}}{1 + \frac{2}{3} + \frac{5}{6}} = \frac{4}{15}$$

$$P(w_3 \text{ is chosen}) = \frac{E_3}{E_1 + E_2 + E_3} = \frac{\frac{5}{6}}{1 + \frac{2}{3} + \frac{5}{6}} = \frac{5}{15} = \frac{1}{3}$$

We want to call attention to two facts in this example: E_3 is the average of E_1 and E_2, and the probability of choosing the average wager w_3 is the reciprocal of the total number of choices available (three in this case). *These facts are always true, no matter what S is and no matter what the original two wagers w_1 and w_2 are.*

This allows us to perform repeated *different* choice experiments on the same subject, as long as each trial asks the subject to choose from among three wagers, one of which is the average of the other two. The model predicts that the subject will choose the average wager (a different wager on each trial) with probability $\frac{1}{3}$. Becker, DeGroot, and Marschak performed such an experiment with 62 college students as subjects. Each subject went through 25 trials, using a different set of three wagers (one the average of the other two) on each trial. [Details of the experiment and its results are reported by these authors in "An experimental study of some stochastic models for wagers" (*Behavioral Science* **8**, 1963, pp. 199–202). If the model is correct, the number of times a subject chooses the average wager is a binomial random variable with $n = 25$ and $p = \frac{1}{3}$.

The problem of deciding whether the observed behavior of a subject supports or discredits the model belongs to statistics (which studies the process of making inferences about probability models on the basis of observed data) rather than to probability. In this case, the model appeared to fit the behavior of 38 of the 62 subjects and not to fit the behavior of the remaining 24. We want to comment that this is only tentative evidence that the model does in fact describe the choice behavior of 38 of the 62 subjects. The reason for this warning is that only one rather indirect prediction of the model was experimentally verified.

We will conclude by proving the critical facts about w_3 which made this experimental work possible. Suppose that

$$S = \{a_1, a_2, a_3\}$$

(in fact the number of different possible winnings has no effect), and that the two original wagers are

$$w_1 = (p_1, p_2, p_3)$$

and

$$w_2 = (q_1, q_2, q_3)$$

Since p_i is the probability of winning amount a_i under wager 1, the numbers p_i (and similarly the numbers q_i) are nonnegative, and their sum is 1. Then the "average wager" which completes the choices available to the subject is

$$w_3 = \left(\frac{p_1 + q_1}{2}, \frac{p_2 + q_2}{2}, \frac{p_3 + q_3}{2} \right)$$

It is easy (Exercise 1.17.8) to see that the elements of w_3 are nonnegative and that their sum is 1.

First, *the expected winnings E_3 is the average of E_1 and E_2.* For,

$$E_1 = a_1 p_1 + a_2 p_2 + a_3 p_3$$
$$E_2 = a_1 q_1 + a_2 q_2 + a_3 q_3$$

and

$$E_3 = a_1 \frac{p_1 + q_1}{2} + a_2 \frac{p_2 + q_2}{2} + a_3 \frac{p_3 + q_3}{3}$$

$$= \tfrac{1}{2} E_1 + \tfrac{1}{2} E_2$$

Second, *the probability that w_3 is chosen is $\tfrac{1}{3}$.* The model specifies that this is

$$
\begin{aligned}
P(w_3 \text{ is chosen}) &= \frac{E_3}{E_1 + E_2 + E_3} \\
&= \frac{\tfrac{1}{2} E_1 + \tfrac{1}{2} E_2}{E_1 + E_2 + \tfrac{1}{2} E_1 + \tfrac{1}{2} E_2} \qquad \text{using the first result} \\
&= \frac{\tfrac{1}{2}(E_1 + E_2)}{\tfrac{3}{2}(E_1 + E_2)} = \frac{1}{3}
\end{aligned}
$$

Exercise 1.17 5. The experimenter offers his subject a choice of the following wagers:

w_1: $3 if an odd number comes up in rolling a fair die, $0 otherwise

w_2: $4 if two consecutive odd numbers come up on independent rolls, $1 otherwise

a. Find S and write w_1 and w_2 as probability distributions on S.
b. Find the expected winnings E_1 and E_2 under w_1 and w_2.
c. Find the average wager w_3 and its expected winnings E_3. Check that E_3 is the average of E_1 and E_2.

d. Find the probability that the subject chooses each of w_1, w_2, and w_3 if the model is correct.

6.   A subject is offered a choice between

w_1: $10,000 with probability 1

w_2: $1,000,000 with probability 0.01 and $0 with probability 0.99

E_1 and E_2 are equal. Do you think most people are equally likely to choose w_1 and w_2, and why? Comment on when the assumption that probabilities of choice are proportional to expected gains is and is not reasonable.

The following exercises use the notation introduced at the end of the discussion in the text.

7.   Why are E_1 and E_2 both nonnegative if all possible winnings a_i are nonnegative? Why is this fact needed in what we did?

8.   Show that the components of w_3 are all nonnegative and that their sum is 1.

9.   Suppose each subject were offered a choice of *four* wagers, one of which was the average of the other three. Can you show that the model predicts that he will choose the average wager with probability $1/4$?

*1.18   SOME THEORY—THE LAW OF LARGE NUMBERS

In Sec. 1.14 we introduced the law of large numbers as providing an additional interpretation of the mean of a random variable. The law of large numbers was roughly stated as saying that the average of the actually observed outcomes of a sequence of independent repetitions of a random experiment must approach the mean outcome as more and more repetitions are made. This section is devoted to clarifying and proving that statement.

The mathematical model for independent repetitions of a random experiment consists of a sequence of independent random variables X_1, X_2, . . . all having the same probability distribution. The "average outcome" of the first n repetitions is the arithmetic mean

$$\bar{X}_n = \frac{1}{n}\sum_{i=1}^{n} X_i$$

We will prove that $\bar{X}_n$ approaches the common mean m of the random variables X_i in this sense: For any positive number c, however small, the probability that $\bar{X}_n$ falls below $m - c$ or above $m + c$ approaches 0 as the number n of observations increases. This means that the probability distribution of $\bar{X}_n$ concentrates more and more probability in any small interval about the mean m.

Example 1.39 A pollster wishes to estimate the proportion of voters who will vote for Burr for President in the upcoming election. Call this proportion p. If the pollster asks voters randomly selected from the entire population whom they plan to vote for, he observes independent random variables $X_1, X_2, \ldots$ where

$$X_i = \begin{cases} 1 & \text{if } i\text{th voter plans to vote for Burr} \\ 0 & \text{if not} \end{cases}$$

If the pollster has responses from a sample of n voters, the sample mean

$$\bar{X}_n = \frac{1}{n}\sum_{i=1}^{n} X_i$$

is just the proportion of these voters who plan to vote for Burr. The pollster uses $\bar{X}_n$ to estimate the unknown value p.

We expect this estimate to become more accurate as n increases. Let us illustrate this for the particular case in which $p = 0.6$ (that is, 60 percent of all voters plan to vote for Burr). We will find the pollster's chance of estimating p to within 0.05, or

$$P(0.55 \le \bar{X}_n \le 0.65)$$

If he polls $n = 10$ voters,

$$P(0.55 \le \bar{X}_{10} \le 0.65) = P\left(5.5 \le \sum_{i=1}^{10} X_i \le 6.5\right)$$

Now $\sum_{i=1}^{10} X_i$ is a binomial random variable with $n = 10$ and $p = 0.6$. So the last probability is just the probability that such a binomial random variable *equals* 6 (only integer values are possible). From Table A this is 0.251. So 10 observations give the pollster only 1 chance in 4 of meeting his goal.

If he polls $n = 20$ voters,

$$P(0.55 \le \bar{X}_{20} \le 0.65) = P\left(11 \le \sum_{i=1}^{20} X_i \le 13\right)$$
$$= P\left(\sum_{i=1}^{20} X_i = 11, 12, \text{ or } 13\right)$$
$$= 0.160 + 0.180 + 0.166$$
$$= 0.506$$

(We used the fact that $\sum_{i=1}^{20} X_i$ is a binomial random variable with $n = 20$ and $p = 0.6$.) So his chances of meeting his desired accuracy have increased to one in two.

If he polls $n = 100$ voters, similar calculations (using larger tables) show that

$$P(0.55 \leq \bar{X}_{100} \leq 0.65) = 0.692$$

This illustrates the pattern: For any fixed distance c, the probability that $\bar{X}_n$ will fall within c of the true mean m approaches 1 as n increases

To prove the law of large numbers in the form we have stated, we need an upper bound on the probability that a random variable takes a value farther than c from its mean. Here is such a result. For convenience we use absolute value notation. The event $\{|X - m| \geq c\}$ is the same as $\{X \leq m - c$ or $X \geq m + c\}$, and simply says that X takes a value at least distance c from m.

THEOREM 1.12   Chebyshev's Inequality   If X is a random variable having mean m and variance v, then for any positive number c

$$P(|X - m| \geq c) \leq \frac{v}{c^2}$$

PROOF   Suppose that the possible values of X are $x_1, x_2, \ldots$. Then the variance v is given by

$$v = \sum_i (x_i - m)^2 P(X = x_i)$$

Since each term of this sum is nonnegative, v is at least as great as the sum over any subset of the x_i. In particular, if we only sum over those values x_i which are at least at distance c from the mean m, we see that

$$v \geq \sum_{\{x_i:\, |x_i - m| \geq c\}} (x_i - m)^2 P(X = x_i)$$

(The expression under Σ specifies over which values of x_i we are summing.)

Now if $|x_i - m| \geq c$, surely $(x_i - m)^2 \geq c^2$. So if we replace $(x_i - m)^2$ in the last sum by c^2, we have a smaller sum. This means that

$$v \geq c^2 \sum_{\{x_i:\, |x_i - m| \geq c\}} P(X = x_i)$$

But the sum of $P(X = x_i)$ over all x_i falling in some event is equal to the probability of that event. The term on the right above is therefore just

$$c^2 P(|X - m| \geq c)$$

and we have shown that

$$v \geq c^2 P(|X - m| \geq c)$$

which becomes the inequality in the statement of the theorem when both sides are divided by c^2.

Notice that Chebyshev's inequality applies to any random variable having mean m and variance v. It gives an upper bound for the amount of probability which can be given to all values at a distance c or greater from the mean. The upper bound is expressed in terms of the variance v and shows that as v decreases less and less probability can be assigned to values at distance c or more from m. This reinforces our conception of the variance as a measure of the spread or dispersion of the probability distribution of X.

It is important to keep in mind that Chebyshev's inequality does not allow us to compute approximate values of

$$P(|X - m| \geq c)$$

It gives only an upper bound, which may be very much larger than the actual probability. Chebyshev's inequality is therefore of very little practical value. But when applied to $\bar{X}_n$ it enables us to prove the law of large numbers.

Example 1.40   Consider the distribution of weight X of eggs produced by hens of variety 2 in Exercise 1.14.5. We have seen (Example 1.37) that

$$m = E(X) = 1.2$$
$$v = \text{Var } X = 0.132$$

So choosing $c = 0.2$, Chebyshev's inequality says that

$$P(|X - 1.2| \geq 0.2) \leq \frac{0.132}{0.04} = 3.3$$

Since any probability is less than or equal to 1, this inequality gives no information whatsoever. The actual probability computed from the distribution is

$$P(|X - 1.2| \geq 0.2) = 1 - P(|X - 1.2| < 0.2)$$
$$= 1 - P(X = 1.2)$$
$$= 1 - 0.2 = 0.8$$

(Here we made use of the fact that all possible values of X except 1.2 itself are at least 0.2 away from 1.2.)

Example 1.41   Suppose that X is the number of heads in 10,000 tosses of a fair coin. Since X is a binomial random variable with $n = 10,000$ and $p = \frac{1}{2}$, we know that

$$E(X) = np = 5000$$
$$\text{Var } X = np(1 - p) = 2500$$

The *proportion* of heads is $Y = X/n$. By Theorems 1.10 and 1.11 the mean and variance of Y are

$$m = E(Y) = \frac{E(X)}{n} = \frac{1}{2}$$

$$v = \text{Var } Y = \frac{\text{Var } X}{n^2} = 0.000025$$

So the probability that the proportion of heads observed differs from $1/2$ by at least 0.01 is no larger than

$$P(|Y - 1/2| \geq 0.01) \leq \frac{v}{(0.01)^2} = \frac{0.000025}{0.0001} = \frac{1}{4}$$

The upper bound in this case is less than 1 and is therefore informative, but the true probability is less than 0.05.

THEOREM 1.13   Law of Large Numbers   Let X_1, X_2, . . . be independent random variables, each having mean m and variance v. Let

$$\bar{X}_n = \frac{1}{n} \sum_{i=1}^{n} X_i$$

be the average of the first n observations X_1, X_2, . . . , X_n. Then for any positive number c,

$$P(|\bar{X}_n - m| \geq c)$$

approaches 0 as n increases.

PROOF   We will apply Chebyshev's inequality to the random variable $\bar{X}_n$. We know (see Sec. 1.16) that

$$E(\bar{X}_n) = m$$

$$\text{Var } \bar{X}_n = \frac{v}{n}$$

Therefore

$$P(|\bar{X}_n - m| \geq c) \leq \frac{v}{nc^2}$$

For any fixed number c, as n increases $v/(nc^2)$ approaches 0. Since $P(|\bar{X}_n - m| \geq c)$ is nonnegative and no larger than $v/(nc^2)$ it also approaches 0.

Exercise 1.18   1.   Use Chebyshev's inequality and the result of Exercise 1.16.1 to obtain an upper bound on the probability that the income of an individual in the rural county of Exercise 1.14.4 differs by at least $800 from the mean income. Compare your upper bound with the actual value of this probability.
2.   We know from Example 1.39 that

$$P(|\bar{X}_{100} - 0.6| > 0.05) = 1 - 0.692 = 0.308$$

Apply Chebyshev's inequality to obtain an upper bound on

$$P(|\bar{X}_{100} - 0.6| \geq 0.05)$$

Notice that this upper bound is not a good approximation to the true probability.

3.   Study the proof of the law of large numbers. Apply Chebyshev's inequality to Example 1.39 just as it was applied in the proof to find the proper expression involving n to complete the inequality

$$P(|\bar{X}_n - 0.6| \geq 0.05) \leq ?$$

4.   Use the result of Exercise 1.18.3 to find an n such that the pollster can guarantee

$$P(|\bar{X}_n - 0.6| \geq 0.05) \leq 0.10$$

by interviewing n voters. (This n is very much larger than is actually needed since the true probability is much smaller than the Chebyshev bound.)

5.   Suppose that the random variable X in the statement of Theorem 1.12 has standard deviation σ. Show that Chebyshev's inequality can be restated as: for any $k > 0$

$$P(|X - m| \geq k\sigma) \leq \frac{1}{k^2}$$

1.19   FOOD FOR THOUGHT—CONTINUOUS PROBABILITY MODELS

The approach to probability which we have taken stresses that the probability of an event is a measure of the likelihood that the event will occur. A probability measure as defined in Sec. 1.4 is simply a measure of likelihood of occurrence having a few simple properties. This great generality carries over into our discussion of conditional probability and independence in the following sections, since the theorems proved there are true for any probability measure. Yet in all our applications of this general theory—in all the specific probability models we have studied—only one method of actually assigning probabilities to events has been used. We have consistently assigned probabilities to individual outcomes of a random experiment and derived the probabilities of events from these. This process often involved simply adding probabilities of individual outcomes, but also made use at times of assumptions of independence and other more subtle concepts. In this section we will briefly describe a completely different method of assigning probabilities to events—yet a method which still produces a probability measure, so that all the tools developed in this chapter apply in the new setting.

Consider first an example in which the approach of assigning probabilities to individual outcomes is inadequate. We want a probability model for the experiment of selecting a number at random from the interval of numbers between 0 and 1. Now "choosing at random" suggests that the choice is to be made so that all possible outcomes are equally likely. But that is impossible—since there are infinitely many numbers between 0 and 1, we cannot give all of them the same positive probability and still assign total probability 1. For no matter how small the probability of an individual number might be, adding up enough of them would exceed 1.

A way out of this dilemma is to assign probabilities directly to certain events (sets of numbers) rather than to individual numbers. Let us begin by assigning probabilities to intervals of numbers between 0 and 1. If A is the interval

$$A = \{x : a < x < b\}$$

for $0 < a < b < 1$, we take $P(A)$ to be the length of A,

$$P(A) = b - a$$

We claim first that this is a reasonable model for the experiment of choosing a number at random, and second that P so defined is a probability measure.

The first claim can of course only rest on persuasion, not on proof. Notice that under P all intervals of the same length contained in the unit interval have the same probability. So the probabilities of choosing a number between 0.01 and 0.02, between 0.49 and 0.50, and between 0.98 and 0.99 are all equal to 0.01. Since the number chosen has the same chance of falling in any interval of the same length, we have made the vague description "select a number at random from the unit interval" precise in a reasonable way.

To see that P is a probability measure, we must first improve the definition of P. We have defined P only for intervals, not for other sets of numbers. A better way of defining the same assignment of probabilities is this: let $f(x)$ be the function

$$f(x) = \begin{cases} 1 & 0 < x < 1 \\ 0 & x \le 0 \text{ or } x \ge 1 \end{cases}$$

and define $P(A)$ for any set of numbers A to be the *area under the graph of $f(x)$ and above the set A.* If A is the interval from a to b ($0 < a < b < 1$), then $P(A)$ is the area of a rectangle with base $b - a$ and height 1, so that $P(A) = b - a$ as before. See Fig. 1.11.

Since areas are positive or zero, all probabilities are nonnegative. The sample space S may be taken to contain all real numbers or only those falling between 0 and 1. In any case the interval between 0 and 1 receives probability 1 [the total area under the graph of $f(x)$], and no probability is assigned outside that interval, so that $P(S) = 1$. Finally, if A and B are disjoint events, it is geometrically obvious (see Fig. 1.11) that the area under the graph of $f(x)$ over the union of A and B is the sum of the individual areas over A and B. So P satisfies the three conditions of Definition 1.4 of a probability

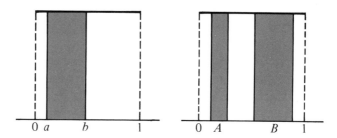

Figure 1.11   Probabilities as areas under the graph of $f(x)$

measure. The probability distribution just described is called the *uniform distribution* on the unit interval.

This example illustrates a general method of assigning probabilities when the sample space S is the real line and events are therefore sets of real numbers. Take a function $f(x)$ such that $f(x) \geq 0$ for all real numbers x and such that the total area under the graph of $f(x)$ is 1. Then the probability of any event A is taken to be the area under the graph of $f(x)$ and above A. Notice that $f(x)$ is not a probability. Rather, it is "probability per unit length" or "probability density" at the point x. Where $f(x)$ is large, an interval will receive larger probability than an interval of equal length located where $f(x)$ is small (see Fig. 1.12).

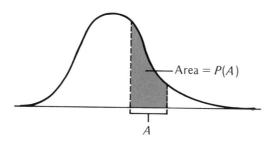

Figure 1.12   Probabilities as areas. The total area under the graph is 1.

This is the usual way of assigning probabilities when the outcome of a random experiment may be any number in an entire interval of real numbers. Since the function $f(x)$ can be very complicated, finding areas under its graph may be difficult. But the most important point is that all probability measures, including those defined in this complicated manner, obey the rules which we have learned in this chapter. The exercises at the end of this section ask you to apply some of your knowledge in this new setting.

There are two features of "probabilities as areas" which are worth thinking about. For simplicity, we will return to the uniform distribution on the unit interval, illustrated in Fig. 1.11. First, you may have wondered if we have succeeded in assigning probabilities to all sets of numbers, since we have to be able to find the area above A under the graph of $f(x)$ in order to find $P(A)$. To settle this we would have to investigate the concept of "area" very closely. It turns out that there are very complicated sets for which we cannot compute the area in question. So the uniform distribution does not assign probabilities to *all* sets of numbers in the unit interval. Strictly speaking, we must therefore amend the definition of a probability measure to allow measures which do not assign probabilities to all subsets of the sample space. Fortunately, we can ignore this problem for all practical purposes, for we can find the probability of intervals and of events which are built up from intervals by using the basic operations of set theory. That covers all events of practical interest, such as "the number chosen is less than $1/2$," "the number chosen has a 3 in the second decimal place," and so on.

Second, notice that the probability of any specified number being chosen is zero! The probability that exactly $1/4$ will be the number chosen, for example, is the length of the interval from $1/4$ to $1/4$, or the area above the single point $1/4$. That is 0 in either case. You can see that this is a general feature of this way of assigning probabilities—intervals of outcomes have positive probability, but all individual outcomes have probability 0. To see that it is not unreasonable that a specific number, say $1/4$, has probability 0, do the following "thought experiment." Suppose a blindfolded man lunges at a stretched string 1 m (meter) long with a pair of scissors. What is the probability that he cuts precisely at the 25-cm mark? Not within a millimeter of that point or within a micron of it, but exactly, precisely at the 25-cm mark. Zero is not an unreasonable probability to assign that event.

Exercise 1.19   1.   Let X be a random variable having the uniform distribution on the unit interval.

a.   Find $P(X < 1/2)$.
b.   Find $P(X \leq 1/2)$.
c.   Express "X has a 3 in the second decimal place" as a union of intervals, and find the probability of this event.

Here is a simple function which is positive between -1 and 1 and 0 elsewhere, along with its graph.

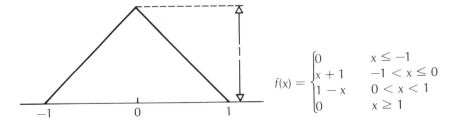

$$f(x) = \begin{cases} 0 & x \leq -1 \\ x + 1 & -1 < x \leq 0 \\ 1 - x & 0 < x < 1 \\ 0 & x \geq 1 \end{cases}$$

2. Show that the total area under the graph of this function is 1. Let X be a random variable such that the probability that X takes a value in a set A of real numbers is the area under this graph and above the set A.
3. Find $P(X < 0)$, $P(X < 1/2)$, and $P(X > 1/4)$. (*Hint*: Draw pictures.)
4. Find the probability that X takes a value having a 3 in the first decimal place.
5. Find $P(X > 1/2 \mid X > 0)$.
6. Find $P(X^2 > 1/4)$.

COMPUTER PROJECTS
Project 1: A Program for Management Decision Making

You should review the management decision-making example in Sec. 1.17, for you are now going to implement a more realistic version of it using the computer to relieve us of the many calculations required.

The input in Sec. 1.17 consisted of:

1. the number of separate markets for the product
2. the number of sales promotions strategies being compared
3. the number of possible sales volumes which could result in each market under each strategy
4. a specific set of possible sales volumes in each case and a subjective probability distribution over these possible values

The program should be able to handle all integer values of items 1, 2, and 3 up to some specified largest possible values. These largest possible values depend on the computer memory available to you; your instructor may help you set them.

To make the procedure of Sec. 1.17 more realistic, we add three complications to the example given there. First, the operation has a fixed production capacity (say \$700,000). Second, sales volume above capacity is subcontracted, and yields a fixed percentage (say 60 percent) as sales volume actually credited to the operation, due to the subcontracter's share. Third, production capacity not used has a scrap value, a fixed percentage (say 10 percent) of unused capacity. Thus in this example the credited sales volume C is different from the gross sales volume S. If sales exceed \$700,000, the credited volume C is \$700,000 plus 60 percent of the excess. If sales are less than \$700,000, then C is the actual volume S plus 10 percent of the unused capacity. In units of \$1000 this is

$$C = \begin{cases} S + (0.10)(700 - S) & \text{if } S \le 700 \\ 700 + (0.60)(S - 700) & \text{if } S > 700 \end{cases}$$

Your program should therefore include as additional inputs

5. the production capacity
6. the percentage of sales above capacity credited to the operation
7. the percentage of unused capacity credited to the operation

These additional constraints greatly complicate the computation of expected credited sales $E(C)$ under each sales strategy. For since C depends on the *total* sales S, we now must find the probability distribution of S rather than just the expected value $E(S)$.

Your program must do the following for each strategy:

1. Find all possible values of S by finding all possible sums of sales in the different markets.
2. Assuming markets to be independent, find the probability of each combination of possible sales volumes in the different markets, and from this the probability of each possible value of S.
3. Find the value of C corresponding to each value of S and the probability of each possible value of C. Printout these results.
4. Compute and printout the expected value of C.

Your main problem is to organize items 1 and 2 efficiently. If you accomplish items 1, 2, 3, and 4, it is relatively easy to add features to your program.

OPTIONAL One feature which you can add is an implementation of the second decision-making procedure discussed in Sec. 1.17. Do the following for each strategy:

1. Arrange the possible values of C in order from smallest to largest. (Your original printout will be neater if you did this as part of item 3 above.)
2. For each possible value c of C, compute $P(C \geq c)$.
3. Find the largest value c of C such that $P(C \geq c)$ is at least 0.90. Printout c and $P(C \geq c)$.

Project 2:   An Introduction to Simulation

We have seen that if a random experiment is repeated many times independently, the proportion of the repetitions on which any event A occurs should be close to its probability $P(A)$. It is therefore possible to try to determine $P(A)$ experimentally by actually repeating the experiment very many times. When the probability model is so complicated that $P(A)$ is hard to compute, experimental determination seems attractive. Unfortunately, the time and cost required to run many trials of a complex experiment are usually out of the question. If, however, we have a probability model for the experiment, it is often possible to program an electronic computer to *simulate* the experiment. It then becomes feasible to repeat the experiment hundreds or thousands of times and observe the results. Computer simulation of random experiments is widely used in scientific and industrial research. Your project here is to learn the basic ideas of simulation and write a program to simulate the learning model of Sec. 1.12.

In order to simulate any random experiment, we must be able to generate values of random variables with specified distributions. Almost all computing centers have available a routine which produces values of a sequence of independent random variables having the uniform distribution on the unit interval $0 < x < 1$. We will base our simulation on the use of this routine. You must now read Sec. 1.19 to acquaint yourself with the uniform distribution, if you have not already done so; and you must also learn how to call and use your computer system's "random number generator." A discrete random variable can be simulated by dividing the unit interval into subintervals with lengths equal to the probabilities of the possible outcomes and observing which of the intervals the generated random number falls in. To simulate rolling a fair die, for example, we record a 1 if the random number generated falls between 0 and $\frac{1}{6}$, a 2 if it falls between $\frac{1}{6}$ and $\frac{2}{6}$, a 3 if it falls between $\frac{2}{6}$ and $\frac{3}{6}$, and so on.

Your program to simulate the model for "avoidance conditioning" of rats should have as input

1. the probability α of learning when shocked
2. the probability p of avoiding a shock while in state 1
3. the number of times the experiment is to be repeated (This will depend on the computer time available to you.)

The computations proceed as follows. Begin in state 0. Generate a random variable to see if the rat advances to state 1 (probability α of occurrence) or remains in state 0. If the latter, repeat this step. If the rat advances, generate a random variable to see if it avoids shock (probability p) or is shocked (probability $1 - p$) on the next trial. If it avoids shock, repeat this step. If the rat is shocked, generate a random variable to see if it advances to state 2 or remains in state 1. Continue until the rat reaches state 2, which completes one repetition of the experiment.

You should store

1. the number of trials the rat spends in each of states 0 and 1 and the total number of trials required to reach state 2
2. the number of shocks received in each of states 0 and 1 and the total number of shocks received

As you repeat the simulated experiment many times, you should keep track of the number of times each of the random variables mentioned in items 1 and 2 takes each of its possible values. When the simulation is completed, divide these frequencies by the number of repetitions of the experiment to obtain the relative frequencies of each value. Print these out. Finally, compute the probability of each possible total number of shocks from the formula for $P(D_k)$ given in Sec. 1.12 and print these out for comparison with the observed relative frequencies. (*Warning*: These random variables have infinitely many possible values, so that you cannot actually consider "each possible value." You decide how to deal with this complication.)

When your program is debugged, simulate the learning model when $\alpha = 0.4$ and $p = 0.5$ (values which seem to fit live rats fairly well).

2   MARKOV CHAINS

2.1   INTRODUCTION

In many instances a single random variable is not adequate to provide a mathematical model for a random phenomenon we wish to study. Often data are obtained from a *sequence* of random outcomes such as a patient's pulse rate recorded each hour, the monthly consumer price index, or the daily smog index. Each such reading is a random variable, and the list of those readings is a sequence of random variables. We will study probability models for sequences of random variables in this chapter.

We used sequences of random variables experimentally while gaining experience with relative frequencies and the properties of probability. The random variables in those sequences were assumed to be mutually independent. In contrast to sequences of independent random variables are sequences of random variables which record developing or evolving processes for which the value of one observation determines the likelihood of the various possible values for the next observation. To be worthwhile, a model for such a sequence of observations must use random variables which reflect this fact.

A probability model for a sequence of random observations entails a different random variable for each observation. To be more explicit, we consider a random variable X_1 which records the outcome of the first observation, X_2 which records the second outcome, and so on. Thus X_1, X_2, X_3, . . . , X_{10} record the first 10 outcomes of a developing process. We call the set of all values for the random variables the *state space*, and each individual value is called a *state* of this process. It is customary to view the subscripts as the times of observations.

113

Example 2.1   The We Have All Types employment agency evaluates the efficiency of its operations weekly. The weekly efficiency ratings form a sequence of random outcomes recording the evolution of a random phenomenon over time.

If the set of possible ratings, the state space, is {inefficient, satisfactory, efficient}, all possible sequences of outcomes in the first two weekly ratings are given by Fig. 2.1.

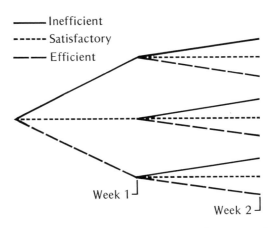

Figure 2.1   All possible sequences of outcomes in a 2-wk period

Various assumptions about dependence or independence of the random variables $X_1, X_2, \ldots, X_n$ can be made. We are going to consider one particular type of dependence relation which distinguishes a *Markov chain* as a probability model for a random sequence. This dependence relation is intuitively described in terms of conditional probabilities when we say, "The conditional probability of the state at any time is determined by the state at the preceding time and is unaffected by any additional information about earlier outcomes."

Example 2.2   Suppose that the weekly efficiency ratings from Example 2.1 form a Markov chain. Then the conditional probability for a rating efficient on the third week, given the rating satisfactory on the second week, does not depend on the rating observed the first week. Thus

$$P(X_3 = \text{efficient}|X_2 = \text{satisfactory},\ X_1 = \text{efficient})$$
$$= P(X_3 = \text{efficient}|X_2 = \text{satisfactory},\ X_1 = \text{inefficient})$$

and so on, no matter what the value of X_1 may be. This means that the probabilities above are all equal to

$$P(X_3 = \text{efficient}|X_2 = \text{satisfactory})$$

which also does not depend on the value of X_1.

Note that we do intend for these probabilities to depend on the value for X_2. Thus we may well have

$$P(X_3 = \text{efficient}|X_2 = \text{satisfactory}) = \frac{1}{2}$$

and

$$P(X_3 = \text{efficient}|X_2 = \text{efficient}) = \frac{2}{3}$$

for example.

The dependence relation stated in Example 2.2 is called *Markov's principle*. The principle was put forth by A. A. Markov, a pioneer in the modern study of probability, who first proposed his idea in 1906. The adequacy of the Markov principle in describing the dependence relation for many naturally occurring random sequences has been proved again and again by experience. The success of Markov chains can be gauged by the many and diverse applications found in research literature. Some of these applications are to social mobility, learning theory, market trends, and molecular movements. All have the unifying factor that they are based on an analysis of the same type of probability model, the Markov chain.

A formal definition of a Markov chain will now be given. For notational convenience we represent each state of the Markov chain by a letter from the alphabet.

DEFINITION 2.1 The random variables X_1, X_2, . . . , X_n, . . . are called a *Markov chain* if they satisfy Markov's principle

$$P(X_n = j|X_{n-1} = k, X_{n-2} = l, \ldots, X_2 = t, X_1 = u) = P(X_n = j|X_{n-1} = k)$$

whatever be the states $\{j, k, l, \ldots, t, u\}$ and the time n, and if

$$P(X_n = j|X_{n-1} = k) = P(X_2 = j|X_1 = k)$$

whatever be the states j, k and the time n.

The intuitive idea discussed earlier is made precise in Definition 2.1. The probability (on the left) that X_n will have the value j given the states at each preceding time is equal to the probability (on the right) that X_n will have the value j given only that the state at time $n - 1$ was k. The last condition of Definition 2.1 states that the probability distribution of X_n given that the state at time $n - 1$ is k will be the same as the probability distribution of X_2 given the state at time 1 is k.

Since the probability $P(X_n = j|X_{n-1} = k)$ is the same whatever the value of n might be, the notation is simplified by writing

$$P(X_n = j|X_{n-1} = k) = p_{k,j}$$

The conditional probability $p_{k,j}$ is called a *transition probability*, and we say "the probability of a transition from state k to state j is $p_{k,j}$."

Example 2.3   The We Have All Types agency has observed that following a week rated inefficient the proportion of weeks rated inefficient is one-half while one-third of them are rated satisfactory. Following a week rated satisfactory, each of the three ratings is equally likely to be given.

Here the transition probabilities from state i (inefficient) are $p_{i,i} = 1/2$, $p_{i,s} = 1/3$, and, therefore, $p_{i,e} = 1/6$, while the transition probabilities from state s (satisfactory) are all equal, $p_{s,i} = p_{s,s} = p_{s,e} = 1/3$.

Exercise 2.1   1.   Would you expect a Markov chain to provide a reasonable model for each of the following? If not, why not?

    a.   daily fluctuation of DJIA
    b.   employee absenteeism
    c.   population migrations recorded by decennial census data
    d.   your score on weekly quizzes given in this course
    e.   weekly UPI ratings of the football team

2.   Suggest a state space for each of Exercise 2.1.1a through e. (You need not think a Markov chain is appropriate as a model to have a state space.)

3.   Refer to Example 2.3. Find the desired probabilities.

    a. $P(X_{19} = s | X_{18} = i)$ *b.* $P(X_4 = e | X_3 = s)$ *c.* $P(X_6 = s | X_5 = s)$

Exercises 2.1.4 and 2.1.5 refer to Example 2.2; s represents satisfactory and i, e similarly represent the states inefficient, efficient.

4.   Write out in words the meaning of each transition probability

    a. $p_{s,i}$ *b.* $p_{e,e}$ *c.* $p_{e,i}$

5.   Describe an event for which these are the transition probabilities.

    a. $p_{i,s}$ *b.* $p_{s,i}$ *c.* $p_{i,i}$

6.   Suppose that $X_1, X_2, \ldots$, are *independent* random variables. Is this sequence a Markov chain?

2.2   TRANSITION PROBABILITIES

As we learn about Markov chains, it will become apparent that the transition probabilities are used to compute all probabilities for events determined by the Markov chain. In this section we will use transition probabilities to compute the probabilities of events determined by X_1 and X_2. These computations introduce you to the use of transition probabilities.

It is common to imagine a random variable labeled X_0 which denotes the "starting state" for the Markov chain. We often call X_0 the *initial state* and assign X_0 a probability distribution which reflects our knowledge of the starting state. In some instances this initial state is known with certainty so we assign probability 1 to that outcome, while in other instances X_0 will have a probability distribution over all the states. The distribution for X_0 is called the *initial distribution*.

Example 2.4 If the We Have All Types agency wishes to analyze a Markov chain model for future weekly efficiency ratings using last week's rating as the starting state, then, if the rating last week was state i (inefficient), the appropriate probability distribution for X_0 is $P(X_0 = i) = 1$, and all other states are assigned probability zero.

If, however, the model is intended for next year's ratings to begin in January, then the appropriate probability distribution for X_0 is our best knowledge about the probability distribution for the rating of the last week of this year. Past experience should indicate the distribution for X_0 in this case.

We now assume that the transition probabilities and an initial probability distribution have been given to us. From the multiplication rule of page 28 we are easily able to compute

$$P(X_1 = j, \ X_0 = k) = P(X_1 = j \mid X_0 = k)P(X_0 = k) = p_{k,j}P(X_0 = k)$$

This equation determines the probability of the event $\{X_1 = j, \ X_0 = k\}$ in terms of a transition probability and the initial probability for the state k.

The more complicated event $\{X_0 = i, \ X_1 = j, \ X_2 = k\}$ can be analyzed in a similar manner. In addition to the repeated use of the multiplication rule (see Exercise 1.6.12) we will use Markov's principle in the third line of this computation.

$$
\begin{aligned}
P(X_0 = i, \ X_1 = j, \ X_2 = k) &= P(X_0 = i, \ X_1 = j)P(X_2 = k \mid X_0 = i, \ X_1 = j) \\
&= P(X_0 = i)P(X_1 = j \mid X_0 = i)P(X_2 = k \mid X_0 = i, \ X_1 = j) \\
&= P(X_0 = i)P(X_1 = j \mid X_0 = i)P(X_2 = k \mid X_1 = j) \\
&= P(X_0 = i)p_{i,j}p_{j,k}
\end{aligned}
$$

Example 2.5 For the We Have All Types Agency suppose that the rating last week was i and that the ratings this week and next are given by X_1 and X_2. Since last week's rating is known to be i, we write $P(X_0 = i) = 1$. Then if $p_{i,s} = \frac{1}{3}$ and $p_{s,e} = \frac{1}{2}$, the probability of the event $\{X_0 = i, \ X_1 = s, \ X_2 = e\}$ can be computed, and we find

$$P(X_0 = i, \ X_1 = s, \ X_2 = e) = P(X_0 = i)p_{i,s}p_{s,e} = 1 \cdot \tfrac{1}{3} \cdot \tfrac{1}{2} = \tfrac{1}{6}$$

These calculations are typical, and they illustrate a relatively simple method for computing probabilities of all events determined by the outcomes X_0, X_1, X_2. In Fig. 2.2 we have used the method of Fig. 2.1 in listing all events determined by X_1 and X_2 for which $X_0 = i$. Symbols i, s, e represent states; transition probabilities are written next to the lines joining the various states. On the right of this diagram is written the probability for each of the possible sequences of states. The symbol p_i represents $P(X_0 = i)$ which in general is not equal to 1.

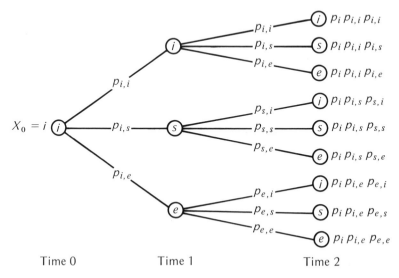

Figure 2.2   Probabilities for all events determined by X_1 and X_2 when $X_0 = i$ and $P(X_0 = i) = p_i$

From Fig. 2.2 we are able to determine the probabilities of other events. If the probability of the event $\{X_2 = e, X_0 = i\}$ is desired, for example, we can add the probabilities of the events

$$\{X_2 = e, X_1 = i, X_0 = i\} \qquad \{X_2 = e, X_1 = s, X_0 = i\}$$
$$\{X_2 = e, X_1 = e, X_0 = i\}$$

since these events are disjoint and their union is the desired event $\{X_2 = e, X_0 = i\}$. Thus we find

$$P(X_2 = e, X_0 = i) = p_i p_{i,i} p_{i,e} + p_i p_{i,s} p_{s,e} + p_i p_{i,e} p_{e,e}$$

Example 2.6   Salesmen for the Shining Wax Company make two sales pitches per day to industrial accounts. The probability of no sale on an average day is 0.2 while the probabilities of one sale or two are equal. On a day following no sale it is observed that the probability of no sale the next day decreases to 0.1 and the probability of two sales is twice that of one sale, while on a day following two sales the probability of two sales decreases to 0.2 and no sale is equally as likely as one sale. Following a day with one sale, the sales have the same probabilities as an average day.

A Markov chain model for the number of sales might use the probability distribution of sales for an average day as the initial distribution; hence $P(X_0 = 0) = 0.2$ and $P(X_0 = 1) = P(X_0 = 2) = 0.4$. The transition probabilities expressed above will be listed in tabular form with the "given" number of sales labeling each row.

| | | Sales next day | | |
		0	1	2
	0	0.1	0.3	0.6
Sales one day	1	0.2	0.4	0.4
	2	0.4	0.4	0.2

The transition probabilities from this table can be used to compute probabilities of events determined by the Markov chain. For example, the event that the sales the initial day are 2 and the sales the second day are also 2 is

$$P(X_0 = 2, X_2 = 2) = P(X_0 = 2, X_1 = 0, X_2 = 2) + P(X_0 = 2, X_1 = 1, X_2 = 2)$$
$$+ P(X_0 = 2, X_1 = 2, X_2 = 2)$$
$$= P(X_0 = 2)p_{2,0}p_{0,2} + P(X_0 = 2)p_{2,1}p_{1,2}$$
$$+ P(X_0 = 2)p_{2,2}p_{2,2}$$
$$= (0.4)(0.4)(0.6) + (0.4)(0.4)(0.4) + (0.4)(0.2)(0.2)$$
$$= 0.096 + 0.064 + 0.016 = 0.176$$

Exercise 2.2

1. Express the indicated probability in symbols using transition probabilities. Figure 2.2 may be helpful. Assume $P(X_0 = i) = p_i = 1$ and that the state space is $\{i,s,e\}$.

a. $P(X_1 = e)$ c. $P(X_2 = s)$
b. $P(X_1 = e, X_2 = s)$ d. $P(X_1 = e \text{ or } X_2 = s)$

2. If the state space is $\{i,s,e\}$, $p_i = 1$, $p_{i,e} = 1/4$, $p_{i,i} = 1/4$, $p_{e,s} = 1/2$, $p_{s,i} = 1/2$, $p_{i,s} = 1/2$, and $p_{s,s} = 1/4$, find the probabilities below.

a. $P(X_1 = e)$ c. $P(X_2 = s)$
b. $P(X_1 = e, X_2 = s)$ d. $P(X_1 = e \text{ or } X_2 = s)$

Exercises 2.2.3 through 2.2.6 refer to Example 2.6.

3. Find $P(X_0 = 0, X_1 = 1, X_2 = 2)$.
4. Find $P(X_0 = 0, X_2 = 2)$.
5. Find $P(X_0 = 2 \text{ or } X_1 = 2)$.
6. Find $P(X_0 + X_1 + X_2 = 1)$.

7.   Draw a figure which extends Fig. 2.2 so as to describe all possible outcomes for the first 3 weeks. (Can you guess how many possible outcomes there are for the first 4 weeks?)

2.3   VECTORS

Quantities of data are regularly recorded and handled in a routine manner. Often these data consist of a set of numbers, each number a measure of a different entity. For example, if each student in a schoolroom were to list the numbers of pencils, erasers, paper clips, and pieces of stationery in his desk drawer, we would have many sets of numbers to record. There are systematic methods for working with such sets of numbers. For example, writing the numbers in a specific order avoids the necessity of individually identifying each number. Such an ordered list of numbers is called a *vector*.

Example 2.7   The inventory lists of pencils, erasers, paper clips, and stationery owned by a student can be recorded by writing the quantity of each item in the order listed above. Thus Bob reports [6, 1, 12, 10] and Sally lists [2, 0, 50, 16]. Note that Sally was clever to realize the list [2, 50, 16] is not sufficient to convey the information that she has no erasers. Only by inclusion of zero in her list are *all* the numbers identified.

DEFINITION 2.2   A *vector* is an ordered list of numbers. The numbers in the list are called *components* of the vector; the first number is referred to as the first component, and so on. When the vector is a list of n numbers, we call it an *n*-component vector or, sometimes, an *n-dimensional vector* depending on the context. A vector **v** with components $v_1, v_2, \ldots, v_n$ is written $\mathbf{v} = [v_1, v_2, \ldots, v_n]$ and denoted by the boldface type **v**. Remember that the components $v_1, v_2, \ldots, v_n$ are ordinary numbers.

Example 2.8   Points in the plane are customarily denoted by vectors. The vector [3, 1] refers to the point with x (horizontal) coordinate 3 and y (vertical) coordinate 1. In this context we usually call the vector [3, 1] a two-dimensional vector. Notice the amount of written detail which is omitted by the use of the vector notation.

Our primary interest in vectors is as a computational device. Thus we will not dwell on other aspects of vector usage but will begin to explore vector arithmetic. To discover how to add two vectors, we consider the inventory example again.

Example 2.9 Two inventory vectors, say [3, 0, 100, 8] and [9, 1, 50, 1], if added should have a clear meaning—the combined list of objects in the two inventories. Thus we write

$$[3, 0, 100, 8] + [9, 1, 50, 1] = [12, 1, 150, 9]$$

This suggests a meaningful definition for addition of vectors which we now give.

DEFINITION 2.3 We add two n-component vectors by adding their corresponding components. If $\mathbf{v} = [v_1, v_2, \ldots, v_n]$ and $\mathbf{u} = [u_1, u_2, \ldots, u_n]$, then $\mathbf{v} + \mathbf{u} = [v_1 + u_1, v_2 + u_2, \ldots, v_n + u_n]$. The sum is also an n-component vector.

The result of subtracting one vector from another can best be determined by considering inventory vectors again. The expression $[8, 2, 25, 6] - [3, 0, 5, 2]$ if thought of as the removal of [3, 0, 5, 2] from the inventory [8, 2, 25, 6] results in the vector

$$[8, 2, 25, 6] - [3, 0, 5, 2] = [5, 2, 20, 4]$$

Symbolically then the difference of two vectors is

$$\mathbf{u} - \mathbf{v} = [u_1 - v_1, u_2 - v_2, \ldots, u_n - v_n]$$

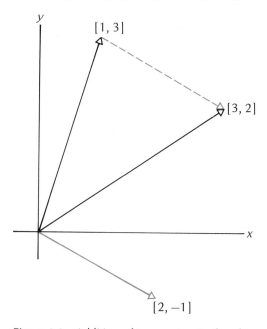

Figure 2.3   Addition of two vectors in the plane

With this definition of addition, each time we use a vector representation we must decide how to interpret the sum of two vectors. For example, if our vectors represent points in the plane, what meaning does the sum have? Adding $[1, 3]$ and $[2, -1]$ results in $[3, 2]$. If we interpret $[3, 2]$ as representing another point in the plane, how is this point geometrically related to the points $[1, 3]$ and $[2, -1]$?

Figure 2.3 locates the three points $[1, 3]$, $[2, -1]$, and $[3, 2]$. Notice that the blue arrow pointing at $[2, -1]$ and the blue arrow pointing at $[3, 2]$ both point in the same direction and have the same length. Adding $[2, -1]$ to $[1, 3]$ can therefore be interpreted as starting at $[1, 3]$ and going in the same direction as and the same distance as the vector $[2, -1]$. This interpretation is valid for points in space, as well as for points in the plane as we have drawn them.

Addition of two vectors need not always have a meaningful interpretation. Our next example illustrates an important application of vectors and also demonstrates that the sum of two vectors may not have an apparent meaning.

Example 2.10   A random variable X has possible outcomes x_1, x_2, x_3. If $P(X = x_1) = 1/2$, $P(X = x_2) = 1/3$, and $P(X = x_3) = 1/6$, then the vector $[1/2, 1/3, 1/6]$ can be used to record that information. The vector $[1/2, 1/3, 1/6]$ is called a *probability vector*. Any vector whose components determine a probability distribution ($v_i \geq 0$ and $\sum_{i=1}^{n} v_i = 1$) is called a *probability vector*.

The two probability vectors $[1/2, 1/3, 1/6]$ and $[1/3, 1/3, 1/3]$ when added together produce the vector $[5/6, 2/3, 1/2]$ which is not a probability vector (its components have a sum of 2) and does not have an apparent interpretation.

It is useful to have available the notion of "proportionality" for vectors. When we say that one vector is "twice as large as" another vector, we are saying that the vectors are proportional to each other. Once again an appeal to your intuition for inventory vectors should be ample to justify an agreement that if **u** is twice as large as **v** then

$$u_1 = 2v_1, \qquad u_2 = 2v_2, \ldots, \qquad u_n = 2v_n$$

DEFINITION 2.4   A vector $\mathbf{u} = [u_1, u_2, \ldots, u_n]$ is said to be *proportional* to a vector $\mathbf{v} = [v_1, v_2, \ldots, v_n]$ if $u_i = cv_i$ for $i = 1, 2, \ldots, n$. The number c determines the proportion. We write $\mathbf{u} = c\mathbf{v}$ to denote this proportionality.

The statement $\mathbf{u} = 2\mathbf{v}$ lends itself to arithmetic manipulations very nicely since obviously it is also true that $\mathbf{v} = 1/2\mathbf{u}$ and if $\mathbf{v} = 2\mathbf{w}$, then $\mathbf{u} = 2\mathbf{v} = 4\mathbf{w}$ and so forth. Geometrically, $\mathbf{u} = 2\mathbf{v}$ denotes a vector $\mathbf{u}$ in the same direction as **v** with a length twice that of **v**.

Combining the operations of vector addition and proportional vectors leads to an interesting and useful geometric fact. When two vectors $\mathbf{x}_1$ and $\mathbf{x}_2$ represent points in space, then any vector $t\mathbf{x}_1 + (1 - t)\mathbf{x}_2$, where t is any number, is a point on the line through the points $\mathbf{x}_1$ and $\mathbf{x}_2$; and if $0 \le t \le 1$, then this point lies on the line segment joining $\mathbf{x}_1$ and $\mathbf{x}_2$.

Notice that $t\mathbf{x}_1 + (1 - t)\mathbf{x}_2$ can be written in another way; by regrouping the terms we have

$$t\mathbf{x}_1 + (1 - t)\mathbf{x}_2 = \mathbf{x}_2 + t(\mathbf{x}_1 - \mathbf{x}_2)$$

Notice also that if $\mathbf{x}_1 - \mathbf{x}_2$ is added to $\mathbf{x}_2$, the result is $\mathbf{x}_1$. Now think of $t(\mathbf{x}_1 - \mathbf{x}_2)$ as a vector in the same direction as $\mathbf{x}_1 - \mathbf{x}_2$ but having a different length (namely t times the length of $\mathbf{x}_1 - \mathbf{x}_2$) so that $\mathbf{x}_2 + t(\mathbf{x}_1 - \mathbf{x}_2)$ is the point reached by starting at $\mathbf{x}_2$ and going in the same direction and length as $t(\mathbf{x}_1 - \mathbf{x}_2)$ (see Fig. 2.4).

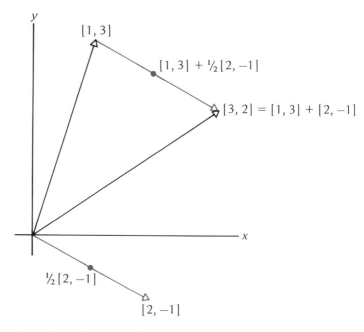

Figure 2.4   Proportional vectors and vector sums in the plane

Example 2.11   Refer to Fig. 2.4. Note the points $[1, 3]$ and $[3, 2]$ in Fig. 2.4. Now $[3, 2] - [1, 3] = [2, -1]$ and $[1, 3] + [2, -1] = [3, 2]$ as shown in Fig. 2.4. Consider next $[1, 3] + \frac{1}{2}([3, 2] - [1, 3]) = [1, 3] + \frac{1}{2}[2, -1] = [2, \frac{5}{2}]$. Since $\frac{1}{2}[2, -1]$ is "half as large as" $[2, -1]$, the resulting point should be (and is) halfway between $[1, 3]$ and $[3, 2]$.

One final computational operation for vectors which we will use is the multiplication of two vectors. Our multiplication will result in a number rather than another vector and is usually called the *dot product* of the vectors because this product of vectors **u** and **v** is written **u·v**.

Before we motivate the definition of this product, it should be mentioned that **u** and **v** may be vectors with different meanings. To emphasize this, we will write the first vector as a "row" of numbers as we have been doing but will write the other "factor" of the product as a "column" of numbers. Thus if **v** has the components $v_1, v_2, \ldots, v_n$, we write

$$\mathbf{v} = \begin{bmatrix} v_1 \\ v_2 \\ \vdots \\ v_n \end{bmatrix}$$

v is called a *column vector* whereas the vector **u**, the row of numbers, is called a *row vector*.

Example 2.12   We return to the inventory example once again. Suppose pencils cost 5¢ each, erasers 15¢ each, paper clips 0.1¢ each, and stationery 1¢ per sheet. We can write this information as a vector, the price vector, and write it as a column vector **p**. Stated in dollars, this is

$$\mathbf{p} = \begin{bmatrix} 0.05 \\ 0.15 \\ 0.001 \\ 0.01 \end{bmatrix}$$

The value of the inventory **v** = [3, 0, 100, 8] is found by the computation

$$3(0.05) + 0(0.15) + 100(0.001) + 8(0.01) = \$0.33$$

This computation is called the dot product of **v** and **p** or

$$\mathbf{v \cdot p} = \$0.33$$

DEFINITION 2.5   The dot product of an *n*-component row vector $\mathbf{u} = [u_1, u_2, \ldots, u_n]$ and an *n*-component column vector

$$\mathbf{v} = \begin{bmatrix} v_1 \\ v_2 \\ \vdots \\ v_n \end{bmatrix}$$

is the number

$$\mathbf{u \cdot v} = u_1 v_1 + u_2 v_2 + \cdots + u_n v_n = \sum_{j=1}^{n} u_j v_j$$

The row vector **u** is always written to the left and the column vector **v** to the right in the dot product **u·v**.

Example 2.13   The row vector $\mathbf{p} = [p_1,\ p_2,\ p_3]$ gives the probability distribution of the random variable X. Let the possible values of X be given by the column vector

$$\mathbf{x} = \begin{bmatrix} x_1 \\ x_2 \\ x_3 \end{bmatrix}$$

Then

$$\mathbf{p \cdot x} = p_1 x_1 + p_2 x_2 + p_3 x_3 = E(X)$$

Thus **p·x** has a meaning and produces a familiar number, the expected value of the random variable X.

Example 2.14   In economics a vector **c** is used to represent a commodity bundle and **p** a column vector of prices. Then **c·p** is the cost of the commodity bundle **c**. Our Example 2.12 illustrates this practice.

If **c** is unspecified while **p** is known, then for any fixed expenditure e we can write the equation

$$\mathbf{c \cdot p} = e \qquad \text{that is,} \qquad c_1 p_1 + c_2 p_2 + \cdots + c_n p_n = e$$

This equation represents all commodity bundles which are available for a fixed expenditure e under the set of prices **p**. The equation $\mathbf{c \cdot p} = e$ is called the *budget equation*.

Exercise 2.3   1.   Find the resulting vectors.

a. $[-1, 2, 3, 0] + [5, 3, -1, -2]$ d. $[2, 0, 5, 3] + 2[1, 2, 0, 4]$
b. $[2, a, 7] + [b, 6, -3]$ e. $[2, 2, 1, 0] - [1, 2, -1, -2]$
c. $[4, 5, -3] + \frac{1}{3}[-12, 6, 0]$ f. $[2, -3, 1] - [2, -3, 1]$

2.   Find the dot products.

a. $[2, -3, 5] \cdot \begin{bmatrix} -2 \\ -1 \\ 6 \end{bmatrix}$ c. $[3, 2] \cdot \begin{bmatrix} x \\ y \end{bmatrix}$

b. $[2, a, 7] \cdot \begin{bmatrix} -2a \\ 4 \\ 2 \end{bmatrix}$ d. $[1, -1, 2] \cdot \begin{bmatrix} x \\ y \\ z \end{bmatrix}$

3.   Find the vector which gives the direction and distance from the point [6, 0] to the point [5, 4].

4.   Verify that the point $[1, 2] + \frac{1}{2}([5, 4] - [1, 2])$ is the midpoint of the line segment joining the points [1, 2] and [5, 4].

5.   On a piece of graph paper (you can make your own) locate and label the points:

a. [−1, 2]   e. $[−1, 2] + \frac{1}{4}[2, 4]$

b. [1, 6]   f. $[−1, 2] + \frac{3}{4}[2, 4]$

c. $[−1, 2] + \frac{1}{2}[2, 4]$ g. $[−1, 2] + 2[2, 4]$

d. $[−1, 2] + [2, 4]$ h. $[−1, 2] − [2, 4]$

6.   Graph the equation

$$[e, b] \cdot \begin{bmatrix} 0.45 \\ 0.78 \end{bmatrix} = 3.50$$

in the two-commodity space of eggs and butter. The variable e is measured in units of dozens of eggs and b is in units of pounds of butter. Prices are quoted in dollars. The graph of the equation is the budget line; interpret the meaning of any point on that line.

7.   Write a vector equation equivalent to the given equation in each case.

a. $3x − 2y = 4$ c. $c_1 p_1 + c_2 p_2 + c_3 p_3 = 5$

b. $3x_1 − 2x_3 = 4$ d. $p_{1,1} p_{1,3} + p_{1,2} p_{2,3} + p_{1,3} p_{3,3} = P$

2.4 MATRICES

When we wish to list several n-component vectors, we sometimes write them in tabular form. Such a tabular array of vectors is called a *matrix*.

Example 2.15   The table of transition probabilities from Example 2.6 is given below. Each row of that table is a probability vector, and the complete table is a matrix, an array of transition probability vectors. Such matrices are employed throughout the remaining sections of this chapter.

		Sales next day		
		0	1	2
	0	0.1	0.3	0.6
Sales one day	1	0.2	0.4	0.4
	2	0.4	0.4	0.2

Table 2.1   A matrix of transition probability vectors

Example 2.16 Two students report inventory vectors [3, 0, 100, 8] and [9, 1, 75, 1]. A matrix which records these inventories is shown in Table 2.2.

	Pencils	Erasers	Paper clips	Stationery
Student 1	3	0	100	8
Student 2	9	1	75	1

Table 2.2 An example of a matrix

DEFINITION 2.6 An $n \times m$ *matrix* consists of n rows and m columns of numbers, which are called the *elements* of the matrix. Each of the n rows is an m-component row vector, and each of the m columns is an n-component column vector. Any element of the matrix can be referred to by designating the number of the row and the number of the column in which it appears. The i,jth element appears at the intersection of the ith row with the jth column.

We will always denote matrices by boldface capital letters. If

$$\mathbf{A} = \begin{bmatrix} a_{11} & a_{12} & \cdots & a_{1m} \\ a_{21} & a_{22} & \cdots & a_{2m} \\ \cdots\cdots\cdots\cdots\cdots \\ a_{n1} & a_{n2} & \cdots & a_{nm} \end{bmatrix}$$

then $a_{11}, a_{12}, \ldots, a_{1m}, a_{21}, \ldots, a_{nm}$ are the nm elements of $\mathbf{A}$. Note that a_{ij} are numbers. In the matrix of Table 2.2, the 2,3 element is 75 while the 1,2 element is 0.

A computation which we will use extensively in this and later chapters is given by the product of a vector and a matrix. The *matrix-vector product* is defined in terms of vector dot products. Because a dot product is defined only for a row vector (on the left) times a column vector (on the right), we are very careful in our definition to not violate that restriction. If you thoroughly understand this multiplication, you will find much of the later material of this text easier to follow.

DEFINITION 2.7 If $\mathbf{A}$ is an $n \times m$ matrix, and $\mathbf{v}$ is an m-component column vector, then $\mathbf{Av}$ is an n-component column vector. The ith component of $\mathbf{Av}$ is the dot product of the ith row of $\mathbf{A}$ with $\mathbf{v}$

$$\sum_{j=1}^{m} a_{ij}v_j = a_{i1}v_1 + a_{i2}v_2 + \cdots + a_{im}v_m \qquad i = 1, 2, \ldots, n$$

If $\mathbf{u}$ is an n-component row vector, then $\mathbf{uA}$ is an m-component row vector. The jth component of $\mathbf{uA}$ is the dot product of the row vector $\mathbf{u}$ and the jth column of $\mathbf{A}$

$$\sum_{i=1}^{n} u_i a_{ij} = u_1 a_{1j} + u_2 a_{2j} + \cdots + u_n a_{nj} \qquad j = 1, 2, \ldots, m$$

To compute $\mathbf{Av}$, we view each row of $\mathbf{A}$ as an m-component row vector, so that the ith component of $\mathbf{Av}$ is the dot product of the ith row of $\mathbf{A}$ with $\mathbf{v}$. The n rows of $\mathbf{A}$ then account for the n components of the resulting product. In like fashion $\mathbf{uA}$ has jth element obtained from the dot product of $\mathbf{u}$ with the jth column of $\mathbf{A}$.

Example 2.17   Consider the 2×3 matrix

$$\mathbf{A} = \begin{bmatrix} 5 & 2 & 8 \\ 0 & 6 & 5 \end{bmatrix}$$

and the three-component vector

$$\mathbf{v} = \begin{bmatrix} 5 \\ 8 \\ 2 \end{bmatrix}$$

A product can only be formed if this vector is written to the right of $\mathbf{A}$. The product results in a two-component vector, one component for each row of $\mathbf{A}$.

The first component of the product $\mathbf{Av}$ is

$$[5, 2, 8] \begin{bmatrix} 5 \\ 8 \\ 2 \end{bmatrix} = 5 \cdot 5 + 2 \cdot 8 + 8 \cdot 2 = 57$$

the second component is $0 \cdot 5 + 6 \cdot 8 + 5 \cdot 2 = 58$ so that

$$\mathbf{Av} = \begin{bmatrix} 57 \\ 58 \end{bmatrix}$$

Example 2.18   Consider the matrix

$$\mathbf{A} = \begin{bmatrix} 5 & 2 & 8 \\ 0 & 6 & 5 \end{bmatrix}$$

and the vector $\mathbf{u} = [1, 1]$. The vector $\mathbf{uA}$ is a three-component row vector whose components are the sums of the column elements of $\mathbf{A}$. If $\mathbf{A}$ is a matrix of inventory vectors, then $\mathbf{uA} = [5, 8, 13]$ represents the sum of the two inventories itemized by $\mathbf{A}$, and $\mathbf{uA}$ is another inventory vector.

Systems of linear equations can be expressed by matrix-vector products. We used dot products to express single equations in the last section, here the matrix-vector product expresses several such equations.

Example 2.19   Consider the equation

$$\begin{bmatrix} 7 & -2 & 3 \\ 0 & 6 & 4 \\ -2 & -3 & 9 \end{bmatrix} \begin{bmatrix} x \\ y \\ z \end{bmatrix} = \begin{bmatrix} -1 \\ 2 \\ 7 \end{bmatrix}$$

Multiplying the left-hand side and equating components of the resulting vector on the left to corresponding components on the right, we obtain the system of equations

$$7x - 2y + 3z = -1$$
$$0x + 6y + 4z = 2$$
$$-2x - 3y + 9z = 7$$

The matrix above is called the *matrix of coefficients* for the system of equations.

Exercise 2.4   1.   Multiply

$$\mathbf{M} = \begin{bmatrix} 1 & -2 & 6 \\ 5 & 0 & 3 \end{bmatrix}$$

by

a. $\begin{bmatrix} 1 \\ 2 \\ -1 \end{bmatrix}$ b. $[1, -1]$ c. $\begin{bmatrix} 0 \\ 1 \\ 0 \end{bmatrix}$ d. $[1, 0]$

2.   Four commodities stored in three warehouses provide us with the inventories $[50, 100, 6, 0]$ in Newark, $[25, 50, 0, 17]$ in Chicago, and $[75, 20, 60, 80]$ in Los Angeles.

a. Write a matrix of these inventory figures.
b. Compute the sum of these inventories by a vector-matrix product.
c. The value of the four commodities is given by the vector

$$\begin{bmatrix} 10 \\ 2 \\ 80 \\ 3 \end{bmatrix}$$

Find the value of each inventory by using a vector-matrix product.

3.   Write the set of equations expressed by the vector-matrix product

$$\begin{bmatrix} 6 & 3 \\ 2 & -1 \\ 4 & 7 \end{bmatrix} \begin{bmatrix} x \\ y \end{bmatrix} = \begin{bmatrix} 0 \\ 7 \\ 3 \end{bmatrix}$$

4.   Write a vector-matrix product to represent the system of equations.

$$y + 2z = -1$$
$$x + 8z = 7$$
$$x - y + 3z = 4$$

5.   If pencils cost 5¢, erasers 7¢, paper clips 0.1¢ and stationery 1¢, the vector

$$\begin{bmatrix} 0.05 \\ 0.07 \\ 0.001 \\ 0.01 \end{bmatrix}$$

is a cost vector. Give an interpretation of the product

$$\begin{bmatrix} 3 & 0 & 100 & 8 \\ 9 & 1 & 75 & 1 \end{bmatrix} \begin{bmatrix} 0.05 \\ 0.07 \\ 0.001 \\ 0.01 \end{bmatrix}$$

2.5 TRANSITION PROBABILITIES—COMPUTATIONS

Probability vectors enter extensively into the analysis of probability distributions for a Markov chain. In this section we introduce two types of probability vectors which are important for such analysis. The first of these gives the probability distribution of X_n (the state of the Markov chain at time n), and the second gives the transition probabilities (the conditional probabilities of the next state given the current state).

A specific order must be designated for the state space to allow the use of vector notation. The order of the states can be arbitrarily chosen but must remain the same throughout the analysis once it has been specified. Recall that any set of nonnegative numbers which have a sum equal to 1 determines a probability distribution (refer to Theorem 1.3 for that fact). We now introduce a vector notation for the probability distribution of X_n.

DEFINITION 2.8 Let the ordered list of states of a Markov chain be x_1, $x_2, \ldots, x_k$. The *state distribution at* n is the probability vector

$$\mathbf{p}^{(n)} = [p_1^{(n)}, p_2^{(n)}, \ldots, p_k^{(n)}]$$

which gives the probability distribution of X_n, so that $p_i^{(n)} = P(X_n = x_i)$ for $i = 1, 2, \ldots, k$.

Notice that we are defining many probability vectors—one for each value of n—and that we use a superscript n to identify the probability vector $\mathbf{p}^{(n)}$ with the random variable X_n.

Example 2.20 A Markov chain model for the We Have All Types Agency given in Example 2.1 had state space $\{i,s,e\}$. If $P(X_0 = i) = 1/3$, $P(X_0 = s) = 1/2$, and $P(X_0 = e) = 1/6$, and the order i, s, e is specified for the state space, then the probability distribution for the initial state X_0 is given by the state distribution at 0, $\mathbf{p}^{(0)} = [1/3, 1/2, 1/6]$.

We have previously used transition probabilities in Sec. 2.2. Recall that a transition probability is a conditional probability for the next state given the current state. In this and subsequent sections when the states are $x_1, x_2, \ldots, x_k$, we will write

$$p_{i,j} = P(X_1 = x_j | X_0 = x_i)$$

for the probability that the next state after x_i is x_j. In cases when the states are not numbered, as with the states i, s, e, we write $p_{i,e}$ for the probability that the next state after i is e.

We saw in Chapter 1 (Theorem 1.6) that the conditional probabilities determined by a fixed given event determine a probability measure. The transition probabilities $p_{i,1}, p_{i,2}, \ldots, p_{i,k}$ are all determined for the fixed given event $\{X_0 = x_i\}$; thus the vector $[p_{i,1}, p_{i,2}, \ldots, p_{i,k}]$ determines the probability distribution for the next state when the state x_i has been observed, this probability vector is called a *transition vector*. Since each state x_i, $i = 1$, $2, \ldots, k$, determines a transition vector, it is convenient to give the set of all such vectors as a matrix which we now define.

DEFINITION 2.9 The *transition probability matrix* $\mathbf{P}$ of a Markov chain with states $x_1, x_2, \ldots, x_k$ is a $k \times k$ matrix with ith row equal to the transition vector for state x_i: $[p_{i,1}, p_{i,2}, \ldots, p_{i,k}]$. The i,jth element of $\mathbf{P}$ is the transition probability $p_{i,j} = P(X_1 = x_j | X_0 = x_i)$.

Each row of a transition probability matrix is a probability vector (all elements are nonnegative and have a sum equal to 1). A typical transition probability matrix is given in Table 2.3.

$$\begin{bmatrix} 1/2 & 3/8 & 1/8 \\ 1/16 & 3/4 & 3/16 \\ 1/8 & 5/8 & 1/4 \end{bmatrix}$$

Table 2.3 A transition probability matrix

Let us suppose a Markov chain with states $x_1 = i$, $x_2 = s$, $x_3 = e$ has the transition probability matrix given by Table 2.3. If we wish to find the probability that state e will be the next state after state s has been observed, we proceed as follows: Locate the row which corresponds to state s (the second row), this row gives the probability distribution for the next state; follow along that row to the column corresponding to state e (the third column) where we read $^3/_{16}$—the probability of observing state e next after state s has occurred is $^3/_{16}$.

Several computations will help us to learn this terminology better. We also wish to illustrate the usefulness of the vector and matrix notation. To this end we continue to discuss a Markov chain with states i, s, e, given in that order. We will assume that $\mathbf{p}^{(0)}$, the state distribution at 0, is known. We will show how to determine $\mathbf{p}^{(1)}$, using Fig. 2.5 to illustrate the discussion.

We first consider $p_i^{(1)} = P(X_1 = i)$. Figure 2.5 shows that the event $\{X_1 = i\}$ can be partitioned by the three events $\{X_0 = i\}$, $\{X_0 = s\}$, and $\{X_0 = e\}$. The partition equation (Theorem 1.8) says that

$$p_i^{(1)} = P(X_0 = i)P(X_1 = i \mid X_0 = i) + P(X_0 = s)P(X_1 = i \mid X_0 = s)$$
$$+ P(X_0 = e)P(X_1 = i \mid X_0 = e)$$
$$= p_i^{(0)}p_{i,i} + p_s^{(0)}p_{s,i} + p_e^{(0)}p_{e,i}$$

This expression is equal to the dot product of $\mathbf{p}^{(0)}$ with the column of $\mathbf{P}$ corresponding to the state i (the first column).

We can verify that to find $p_s^{(1)}$ we compute the dot product of $\mathbf{p}^{(0)}$ with the column of $\mathbf{P}$ corresponding to the state s (the second column) and that $p_e^{(1)}$ is the dot product of $\mathbf{p}^{(0)}$ with the third column of $\mathbf{P}$. This shows that the

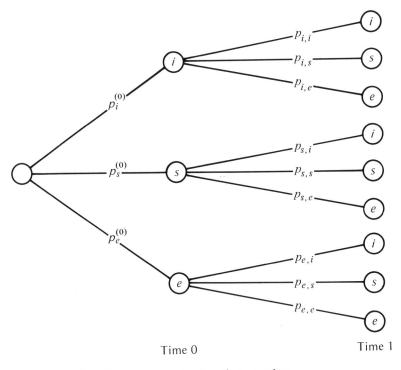

Time 0   Time 1

Figure 2.5   Possible outcomes for X_1 with X_0 random

probability vector $\mathbf{p}^{(1)}$ is given by the equation

$$\mathbf{p}^{(1)} = \mathbf{p}^{(0)}\mathbf{P}$$

Example 2.21 Consider the vector $\mathbf{p}^{(0)}$ and matrix $\mathbf{P}$ given in Example 2.20 and Table 2.3 respectively. Our computations show that

$$p_i^{(1)} = p(X_1 = i) = \mathbf{p}^{(0)} \cdot \begin{bmatrix} p_{i,i} \\ p_{s,i} \\ p_{e,i} \end{bmatrix} = [1/3,\ 1/2,\ 1/6] \cdot \begin{bmatrix} 1/2 \\ 1/16 \\ 1/8 \end{bmatrix}$$

$$= 1/6 + 1/32 + 1/48 = 7/32$$

$$p_s^{(1)} = P(X_1 = s) = \mathbf{p}^{(0)} \cdot \begin{bmatrix} p_{i,s} \\ p_{s,s} \\ p_{e,s} \end{bmatrix} = [1/3,\ 1/2,\ 1/6] \cdot \begin{bmatrix} 3/8 \\ 3/4 \\ 5/8 \end{bmatrix}$$

$$= 1/8 + 3/8 + 5/48 = 29/48$$

$$p_e^{(1)} = P(X_1 = e) = \mathbf{p}^{(0)} \cdot \begin{bmatrix} p_{i,e} \\ p_{s,e} \\ p_{e,e} \end{bmatrix} = [1/3,\ 1/2,\ 1/6] \cdot \begin{bmatrix} 1/8 \\ 3/16 \\ 1/4 \end{bmatrix}$$

$$= 1/24 + 3/32 + 1/24 = 17/96$$

Hence

$$\mathbf{p}^{(1)} = \mathbf{p}^{(0)}\mathbf{P} = [7/32,\ 29/48,\ 17/96]$$

Note that this *is* a probability vector.

We next compute $\mathbf{p}^{(2)}$, the state distribution at 2, in the same manner, using Fig. 2.6 as an illustration of the partitioning of events. Again the partition equation (Theorem 1.8) is the key to the analysis.

We note that the event $\{X_2 = i\}$ is partitioned by the three events $\{X_1 = i\}$, $\{X_1 = s\}$, and $\{X_1 = e\}$. This time we will first consider the event $\{X_1 = s, X_2 = i\}$ to show that the probability of this event can be computed by taking the product of the probabilities along the corresponding path shown in Fig. 2.6 (the path marked by an asterisk at its right endpoint).

$$P(X_1 = s,\ X_2 = i) = P(X_1 = s)P(X_2 = i|X_1 = s) = p_s^{(1)}p_{s,i}$$

This analysis applies to each path which ends with state i, leading to the results $P(X_1 = s, X_2 = i) = p_s^{(1)}p_{s,i}$ and $P(X_1 = e, X_2 = i) = p_e^{(1)}p_{e,i}$. The partition equation combines these products to give

$$p_i^{(2)} = p_i^{(1)}p_{i,i} + p_s^{(1)}p_{s,i} + p_e^{(1)}p_{e,i}$$

This sum equals the dot product

$$\mathbf{p}^{(1)} \cdot \begin{bmatrix} p_{i,i} \\ p_{s,i} \\ p_{e,i} \end{bmatrix}$$

The corresponding dot products for $p_s^{(2)}$ and $p_e^{(2)}$ are

$$p_s^{(2)} = \mathbf{p}^{(1)} \cdot \begin{bmatrix} p_{i,s} \\ p_{s,s} \\ p_{e,s} \end{bmatrix} \quad \text{and} \quad p_e^{(2)} = \mathbf{p}^{(1)} \cdot \begin{bmatrix} p_{i,e} \\ p_{s,e} \\ p_{e,e} \end{bmatrix}$$

Thus we conclude that $\mathbf{p}^{(2)}$ satisfies the equation

$$\mathbf{p}^{(2)} = \mathbf{p}^{(1)}\mathbf{P}$$

Notice the similarity of this equation to the equation $\mathbf{p}^{(1)} = \mathbf{p}^{(0)}\mathbf{P}$. Comparison of Figs. 2.5 and 2.6, as well as the two arguments used in finding $\mathbf{p}^{(1)}$ and $\mathbf{p}^{(2)}$, reveals the only difference to be the shift in time, considering transitions between times 0 and 1 in the first case and between times 1 and 2 in the second case.

Substituting $\mathbf{p}^{(0)}\mathbf{P}$ for $\mathbf{p}^{(1)}$ in the equation for $\mathbf{p}^{(2)}$ gives the result

$$\mathbf{p}^{(2)} = (\mathbf{p}^{(0)}\mathbf{P})\mathbf{P}$$

(In words this is "to find $\mathbf{p}^{(2)}$, form the vector $\mathbf{p}^{(0)}\mathbf{P}$ then multiply this vector on the right by the matrix $\mathbf{P}$.")

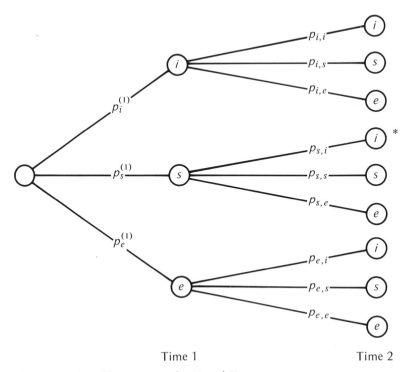

Time 1   Time 2

Figure 2.6   Possible outcomes for X_1 and X_2

Example 2.22 We refer again to the transition matrix and probability vector computed in Example 2.21. We found there that $\mathbf{p}^{(1)} = \mathbf{p}^{(0)}\mathbf{P} = [7/32, 29/48, 17/96]$, and we now conclude that the state distribution at 2, $\mathbf{p}^{(2)}$, equals the vector-matrix product

$$[7/32, 29/48, 17/96] \begin{bmatrix} 1/2 & 3/8 & 1/8 \\ 1/16 & 3/4 & 3/16 \\ 1/8 & 5/8 & 1/4 \end{bmatrix}$$

This computation produces the probability vector

$$\mathbf{p}^{(2)} = [7/64 + 29/768 + 17/768, \; 21/256 + 87/192 + 85/768,$$
$$7/256 + 87/768 + 17/384]$$

$$= [65/384, \; 248/384, \; 71/384]$$

Exercise 2.5 *In Exercises 2.5.1 to 2.5.3, refer to the transition probability matrix*

$$\mathbf{P} = \begin{bmatrix} 1/2 & 3/8 & 1/8 \\ 1/4 & 1/2 & 1/4 \\ 1/3 & 2/3 & 0 \end{bmatrix}$$

1. (a) For $\mathbf{p}^{(0)} = [1, 0, 0]$ find $\mathbf{p}^{(1)}$ and $\mathbf{p}^{(2)}$. (b) For $\mathbf{p}^{(0)} = [0, 1, 0]$ find $\mathbf{p}^{(1)}$ and $\mathbf{p}^{(2)}$. (c) For the states x_1, x_2, x_3 give a verbal interpretation of the answers to (a) and (b).

2. Let $\mathbf{p}^{(0)} = [1, 0, 0]$ and determine

a. $P(X_2 = x_1, X_1 = x_3)$ d. $P(X_2 = x_1, X_1 = x_1)$
b. $P(X_2 = x_2, X_1 = x_3)$ e. $P(X_2 = x_1, X_1 = x_2)$
c. $P(X_2 = x_3, X_1 = x_3)$ f. $P(X_2 = x_1)$

3. Determine the following probabilities when $\mathbf{p}^{(0)} = [1/2, 1/2, 0]$.

a. $P(X_1 = x_1, X_2 = x_3)$ d. $P(X_1 = x_2, X_2 \neq x_2)$
b. $P(X_1 = x_3, X_2 = x_1)$ e. $P(X_2 \neq x_2 | X_1 = x_2)$
c. $P(X_1 = x_3 \text{ or } X_2 = x_3)$

4. The General Giant Corporation seeks to appease three minority groups, identified only as S, T, and U, by appointing a personnel manager responsible for hiring workers from these minority groups. It is observed that the hiring practices of this personnel manager are dependent only on the minority group from which the last worker was hired. Thus the probability that the next worker hired is a member of group S depends only on which of the three groups the last worker hired identified with. Observations of the group identity of workers hired in the past gives the transition matrix

$$\begin{bmatrix} 0 & 1/3 & 2/3 \\ 1/3 & 0 & 2/3 \\ 1/4 & 1/2 & 1/4 \end{bmatrix}$$

with the order S, T, U for the states. The last worker hired belonged to group T.

a. Find the probability that the next two workers hired belong to the group U.
b. Find the probability that the next two workers hired belong to the group S.
c. Find the probability that at least one of the next two workers hired belongs to the group T.

5.   The game of matching pennies can be modeled as a Markov chain. The game amounts to two players, each having at least one penny; each player selects a penny and flips it. By prearrangement one player wins if the pennies match (both are heads or both are tails) and loses if they do not match. Winning means you get to keep your opponent's penny, and losing means you lose your penny. The game continues until one player becomes penniless.

The model for this game consists of X_0, the number of pennies you own at the beginning of the game, X_1 the number of pennies you own after the first play of the game (obviously $X_1 = X_0 + 1$ or $X_1 = X_0 - 1$). For a game in which you begin with one penny and your opponent has two, the state space is $\{0,1,2,3\}$. If the pennies are fair coins, $P(\text{heads}) = 1/2$, then the transition matrix is

$$\begin{bmatrix} 1 & 0 & 0 & 0 \\ 1/2 & 0 & 1/2 & 0 \\ 0 & 1/2 & 0 & 1/2 \\ 0 & 0 & 0 & 1 \end{bmatrix}$$

The Markov chain we have defined does not stop when this game stops; rather, if you lose at time n, the outcomes X_n, X_{n+1}, X_{n+2}, . . . will all be 0, and if you win at time n, then X_n, X_{n+1}, X_{n+2}, . . . will all be 3.

a. Find the state distribution at 1 and 2 (when you begin with one penny).
b. Do you think this is a fair game; that is, do you have an even chance to win? (*Hint*: Compute the state distribution at 1 and 2 when you begin with two pennies, then compare this with Exercise 2.5.5a.)

2.6   *n*-STEP TRANSITION PROBABILITIES

In this section we will continue to develop the use of matrices and vectors for computational purposes. Our emphasis will be concentrated on the role of the transition probability matrix for computations of the type done in Sec. 2.5.

We found in Sec. 2.5 that the state distribution at 2, $\mathbf{p}^{(2)}$, was equal to the product $\mathbf{p}^{(1)}\mathbf{P}$ and that $\mathbf{p}^{(1)}$ was equal to the product $\mathbf{p}^{(0)}\mathbf{P}$. Thus we wrote

$$\mathbf{p}^{(2)} = (\mathbf{p}^{(0)}\mathbf{P})\mathbf{P}$$

Removing the parentheses from the expression on the right results in our writing $\mathbf{p}^{(0)}\mathbf{PP}$ which leads us to consider the "product" $\mathbf{PP}$. We will show that a product for matrices can be defined so that $\mathbf{PP} = \mathbf{P}^2$ is a matrix for which the equation $\mathbf{p}^{(2)} = \mathbf{p}^{(0)}\mathbf{P}^2$ holds.

DEFINITION 2.10   If $\mathbf{A}$ is a $m \times k$ matrix, and $\mathbf{B}$ is a $k \times n$ matrix, then $\mathbf{AB}$ is a $m \times n$ matrix whose i, jth element (the element in row i and column j) is the dot product of the ith row of $\mathbf{A}$ with the jth column of $\mathbf{B}$. In symbols the i,jth element of the product $\mathbf{AB}$ is

$$a_{i_1}b_{1j} + a_{i_2}b_{2j} + \cdots + a_{ik}b_{kj} = \sum_{l=1}^{k} a_{il}b_{lj}$$

It is essential that $\mathbf{A}$ has the same number of columns as $\mathbf{B}$ has rows, otherwise this product is *not defined*.

Example 2.23   The product $\mathbf{AB}$ of the matrices

$$\mathbf{A} = \begin{bmatrix} 3 & 2 \\ 4 & 1 \end{bmatrix} \quad \text{and} \quad \mathbf{B} = \begin{bmatrix} 0 & -2 & 6 \\ 1 & -1 & 2 \end{bmatrix}$$

is a 2×3 matrix. The first row of $\mathbf{AB}$ is the vector $[2, -8, 22]$ obtained from the dot products of the first row of $\mathbf{A}$ with each of the three successive columns of $\mathbf{B}$

$$[3, 2] \cdot \begin{bmatrix} 0 \\ 1 \end{bmatrix} = 2 \qquad [3, 2] \cdot \begin{bmatrix} -2 \\ -1 \end{bmatrix} = -8$$

and

$$[3, 2] \cdot \begin{bmatrix} 6 \\ 2 \end{bmatrix} = 22$$

Computing the second row of $\mathbf{AB}$, we use the second row of $\mathbf{A}$ with the successive columns of $\mathbf{B}$. This results in the matrix

$$\mathbf{AB} = \begin{bmatrix} 2 & -8 & 22 \\ 1 & -9 & 26 \end{bmatrix}$$

Notice that the product $\mathbf{BA}$ cannot be defined as $\mathbf{B}$ has three columns and $\mathbf{A}$ has only two rows.

Example 2.24   We refer again to the transition matrix from Example 2.21. By finding the product $\mathbf{PP}$ and then multiplying by $\mathbf{p}^{(0)}$, we will verify that $\mathbf{p}^{(2)}$ as found in Example 2.22 is equal to $\mathbf{p}^{(0)}(\mathbf{PP}) = \mathbf{p}^{(0)}\mathbf{P}^2$.

$$\mathbf{PP} = \begin{bmatrix} 1/2 & 3/8 & 1/8 \\ 1/16 & 3/4 & 3/16 \\ 1/8 & 5/8 & 1/4 \end{bmatrix} \begin{bmatrix} 1/2 & 3/8 & 1/8 \\ 1/16 & 3/4 & 3/16 \\ 1/8 & 5/8 & 1/4 \end{bmatrix}$$

$$= \begin{bmatrix} 39/128 & 35/64 & 21/128 \\ 13/128 & 45/64 & 25/128 \\ 17/128 & 43/64 & 25/128 \end{bmatrix}$$

and

$$\mathbf{p}^{(0)}(\mathbf{PP}) = \begin{bmatrix} 1/3, & 1/2, & 1/6 \end{bmatrix} \begin{bmatrix} 39/128 & 35/64 & 21/128 \\ 13/128 & 45/64 & 25/128 \\ 17/128 & 43/64 & 25/128 \end{bmatrix}$$

$$= \begin{bmatrix} 65/384, & 284/384, & 71/384 \end{bmatrix}$$

which agrees with our computations in Examples 2.21 and 2.22.

Exercises 2.6.9 to 2.6.11 explore several properties of matrix multiplication. The first of those exercises gives an example which shows that **AB** is *not* always equal to **BA** (matrix multiplication is not commutative). The other exercises show that matrix multiplication *is* associative [i.e., $(\mathbf{AB})\mathbf{C} = \mathbf{A}(\mathbf{BC})$ whenever the product is defined], so that the product $\mathbf{P}^2\mathbf{P}$ always equals $\mathbf{PP}^2$ and $\mathbf{P}^3$ is a meaningful symbol for that product.

We will now use matrix multiplication together with the type of argument used in Sec. 2.5 to prove the important formula for the state distribution at n stated formally below.

THEOREM 2.1   The state distribution at n, $\mathbf{p}^{(n)}$, of a Markov chain with transition matrix **P** and state distribution at 0 equal to $\mathbf{p}^{(0)}$ is the vector

$$\mathbf{p}^{(n)} = \mathbf{p}^{(0)}\mathbf{P}^n$$

PROOF   A partition of the sample space given by the events $\{X_{n-1} = x_1\}$, $\{X_{n-1} = x_2\}, \ldots, \{X_{n-1} = x_k\}$ is used in the partition equation to provide the expression for $P(X_n = x_j)$

$$P(X_n = x_j) = P(X_{n-1} = x_1)P(X_n = x_j | X_{n-1} = x_1)$$
$$+ P(X_{n-1} = x_2)P(X_n = x_j | X_{n-1} = x_2)$$
$$+ \cdots + P(X_{n-1} = x_k)P(X_n = x_j | X_{n-1} = x_k)$$

We note that $P(X_{n-1} = x_i) = p_i^{(n-1)}$ and that $P(X_n = x_j | X_{n-1} = x_i) = p_{i,j}$ so that we can write

$$P(X_n = x_j) = p_1^{(n-1)}P_{1,j} + p_2^{(n-1)}p_{2,j} + \cdots + p_i^{(n-1)}p_{i,j}$$
$$+ \cdots + p_k^{(n-1)}p_{k,j}$$

This formula is nothing more than a shift in time from the argument used to compute $\mathbf{p}^{(1)}$ and $\mathbf{p}^{(2)}$ in Sec. 2.5. We recognize this sum as a dot product of

$\mathbf{p}^{(n-1)}$ with the *j*th column of **P**. When all states are considered, $j = 1$, $2, \ldots, k$, the state distribution at *n* is found and we have the equation

$$\mathbf{p}^{(n)} = \mathbf{p}^{(n-1)}\mathbf{P}$$

Since *n* is not a particular integer, we can apply this result to find that $\mathbf{p}^{(n-1)} = \mathbf{p}^{(n-2)}\mathbf{P}$ and hence

$$\mathbf{p}^{(n)} = \mathbf{p}^{(n-1)}\mathbf{P} = (\mathbf{p}^{(n-2)}\mathbf{P})\mathbf{P} = \mathbf{p}^{(n-2)}\mathbf{P}^2$$

We continue to substitute for $\mathbf{p}^{(m)}$, $m = 1, 2, \ldots, n-1$, to obtain the equalities

$$\mathbf{p}^{(n)} = \mathbf{p}^{(n-1)}\mathbf{P} = \mathbf{p}^{(n-2)}\mathbf{p}^2 = \cdots = \mathbf{p}^{(0)}\mathbf{P}^n$$

Diagrams similar to Fig. 2.5 or Fig. 2.9 create a mental image of a process which hops along a path jumping from one state to the next. In an attempt to keep this image vivid in our minds, we refer to the steps (hops) of the Markov chain and to the elements of the matrix $\mathbf{P}^n$ as the *n-step transition probabilities*, since they tell us the probabilities of transitions in *n* steps from one state to another.

Each of the equations for $\mathbf{p}^{(n)}$ given at the end of the proof for Theorem 2.1 has a reasonable intuitive interpretation—$\mathbf{p}^{(n)} = \mathbf{p}^{(n-1)}\mathbf{P}$ says that from the state distribution at $n - 1$ ($\mathbf{p}^{(n-1)}$) the one-step transition probabilities (**P**) determine the state distribution at *n*. The second, $\mathbf{p}^{(n)} = \mathbf{p}^{(n-2)}\mathbf{P}^2$, says that from the state distribution at $n - 2(\mathbf{p}^{(n-2)})$ the two-step transition probabilities ($\mathbf{P}^2$) determine the state distribution at *n*. The last equation reflects the fact that from the initial state distribution ($\mathbf{p}^{(0)}$) the *n*-step transition probabilities determine the state distribution at *n*.

Our next theorem gives substance to the use of the term *transition probability* in describing the elements of $\mathbf{P}^n$, as it identifies them as conditional probabilities for making *n*-step transitions from a given state.

THEOREM 2.2 The *i,j*th element of $\mathbf{P}^n$ is the conditional probability

$$P(X_n = x_j | X_0 = x_i)$$

We will write $p_{i,j}^{(n)}$ for this *n*-step transition probability.

PROOF We need to show that the *i,j*th component of $\mathbf{P}^n$, $p_{i,j}^{(n)}$, equals $P(X_n = x_j | X_0 = x_i)$. This fact is easiest to demonstrate by considering the situation when $P(X_0 = x_i) = 1$ for then $P(X_n = x_j | X_0 = x_i) = P(X_n = x_j)$ (see Exercise 2.6.7).

When $P(X_0 = x_i) = 1$, we have $p_i^{(0)} = 1$ and $p_h^{(0)} = 0$ for $h \neq i$. With the vector $\mathbf{p}^{(0)}$ determined by those values, the state distribution at *n*, $\mathbf{p}^{(n)}$, is simply the *i*th row of the matrix $\mathbf{P}^n$, for Theorem 2.1 says $\mathbf{p}^{(n)} = \mathbf{p}^{(0)}\mathbf{P}^n$.

The *j*th component of the *i*th row of $\mathbf{P}^n$ is the element $p_{i,j}^{(n)}$, so that the result follows, i.e.,

$$p_{i,j}^{(n)} = P(X_n = x_j | X_0 = x_i)$$

One interesting result of this theorem is that every row of $\mathbf{P}^n$ is a probability vector, since the ith row contains the n-step transition probabilities from state i. This fact will often help you to catch errors in calculating $\mathbf{P}^n$

Example 2.25 Nanosecond Computers has found that a Markov chain provides a good model for the number of sales each day by their ace salesman. He can make only two presentations each day, so that the possible numbers of sales are 0, 1, 2. The transition matrix for the Markov chain with the order of states 0,1,2 is

$$\begin{bmatrix} 0.1 & 0.4 & 0.5 \\ 0.1 & 0.5 & 0.4 \\ 0.3 & 0.4 & 0.3 \end{bmatrix}$$

On Monday he makes two sales. That evening he speculates on the outlook for sales on Friday. The transition matrix provides a probability distribution for the number of sales on Friday as follows: Label Monday's sales as X_0; then Friday's sales are X_4. Since two sales were made on Monday, we know that $\mathbf{p}^{(0)} = [0, 0, 1]$. Thus the state distribution at 4 is $\mathbf{p}^{(0)}\mathbf{P}^4$, which is just the third row of $\mathbf{P}^4$ and equals

$$[0.1752, 0.4444, 0.3804]$$

Summary

1. The probability distribution of X_n, given by the state distribution at n, $\mathbf{p}^{(n)}$, can be computed by any of the formulas

$$\mathbf{p}^{(n)} = \mathbf{p}^{(n-1)}\mathbf{P} = \mathbf{p}^{(n-2)}\mathbf{P}^2 = \cdots = \mathbf{p}^{(0)}\mathbf{P}^n$$

2. The matrix $\mathbf{P}^n$ is the matrix of n-step transition probabilities. The ith row of $\mathbf{P}^n$ is a probability vector which gives the probability distribution of the Markov chain n transitions (steps) after the state i is observed.

Exercise 2.6 1. Compute $\mathbf{p}^{(3)}$ for a Markov chain with transition matrix

$$\mathbf{P} = \begin{bmatrix} 1/4 & 3/4 \\ 1/2 & 1/2 \end{bmatrix} \quad \text{and} \quad \mathbf{p}^{(0)} = [1/3, 2/3]$$

in two ways.

a. Find $\mathbf{P}^3$ first, then $\mathbf{p}^{(0)}\mathbf{P}^3$.
b. Use successive products, $\mathbf{p}^{(1)} = \mathbf{p}^{(0)}\mathbf{P}$, $\mathbf{p}^{(2)} = \mathbf{p}^{(1)}\mathbf{P}$, and $\mathbf{p}^{(3)} = \mathbf{p}^{(2)}\mathbf{P}$.

2.   Suppose the We Have All Types agency (see Example 2.1) uses the transition matrix

$$\mathbf{P} = \begin{bmatrix} 1/2 & 1/2 & 0 \\ 1/3 & 1/3 & 1/3 \\ 0 & 1/2 & 1/2 \end{bmatrix} \quad \text{and} \quad \mathbf{p}^{(0)} = [1, 0, 0]$$

Determine the state distributions

a. $\mathbf{p}^{(1)}$ c. $\mathbf{p}^{(3)}$
b. $\mathbf{p}^{(2)}$ d. $\mathbf{p}^{(4)}$

3.   Refer to the We Have All Types agency's Markov chain with state space $\{i,s,e\}$ (Example 2.1). Give a *verbal interpretation* for each expression

a. $p_{s,e}^{(5)}$ c. $p_{s,s}^{(3)}p_{s,e}^{(2)}$
b. $p_{s,s}^{(3)}$ d. $p_{s,i}p_{i,e}p_{e,s}p_{s,e}^{(2)}$

4.   Refer to the Markov chain for sales in Example 2.25.

a. Find the probability that Friday's sales are less than two.
b. Find the probability that exactly three sales are made on Tuesday and Wednesday combined.
c. What is the expected number of sales on Friday? (Example 2.24 determined the probability distribution for this random variable. Refer to Example 2.13 and determine the answer by a dot product.)

5.   For the transition matrix given in Exercise 2.6.2 above, suppose that $\mathbf{p}^{(0)} = [2/7, 3/7, 2/7]$.

a. Determine $\mathbf{p}^{(1)}$.
b. Determine $\mathbf{p}^{(2)}$.
c. Can you determine $\mathbf{p}^{(n)}$ for any value of n?

6.   Refer to Exercise 2.5.5 (matching pennies).

a. Supposing you start with one penny, find $\mathbf{p}^{(2)}$, $\mathbf{p}^{(4)}$, $\mathbf{p}^{(6)}$. [*Hint:* After finding $\mathbf{P}^2$, note that $\mathbf{P}^4 = (\mathbf{P}^2)^2$ and $\mathbf{P}^6 = \mathbf{P}^4\mathbf{P}^2$.]
b. Find the probability that you *lose* on or before the second play, fourth play, sixth play.
c. Find the probability that you *win* on or before the second play, fourth play, sixth play.
d. Suppose you start with two pennies. Find the probability you win on or before the second play, fourth play, sixth play. (Before doing any computations, can you reason that the answers should be the same as in Exercise 2.6.6*b*?)

7.   If an event A has probability 1, $P(A) = 1$, show that for any event B,

the conditional probability $P(B|A) = P(B)$ [*Hint*: The events A and A^c form a partition for which $P(A^c) = 0$. Use the partition equation to conclude the result.]

Note that this result establishes the fact—$P(X_n = x_j|X_0 = x_i) = P(X_n = x_j)$ when $P(X_0 = x_i) = 1$—used in the proof of Theorem 2.2.

8.   Compute the matrix products.

a. $\begin{bmatrix} 1/2 & 1/2 \\ 1/2 & 1/2 \end{bmatrix} \begin{bmatrix} 1/3 & 2/3 \\ 1/3 & 2/3 \end{bmatrix}$

b. $\begin{bmatrix} 3 & 2 \\ 1 & 4 \end{bmatrix} \begin{bmatrix} 1 & 3 & -1 \\ 2 & 6 & 1 \end{bmatrix}$

c. $\begin{bmatrix} 1 & 0 \\ 2 & 4 \\ 0 & 3 \end{bmatrix} \begin{bmatrix} 1 & 3 & -1 \\ 2 & 6 & 1 \end{bmatrix}$

Exercises on properties of matrix multiplication.

9.   Show that multiplication of matrices is not commutative by forming **AB** then **BA** for the matrices

$$\mathbf{A} = \begin{bmatrix} 1 & 2 \\ 3 & 4 \end{bmatrix} \qquad \mathbf{B} = \begin{bmatrix} 1 & 3 \\ 2 & 4 \end{bmatrix}$$

10.   Show that $(\mathbf{AB})\mathbf{C} = \mathbf{A}(\mathbf{BC})$ with **A** and **B** as in Exercise 2.6.6 and $\mathbf{C} = \begin{bmatrix} 1 & 2 & 0 \\ 0 & 1 & 1 \end{bmatrix}$. Note that on the left we want the product **AB** formed first, then the product of (**AB**) with **C**, and that on the right we want the product **BC** formed first, then the product of **A** with (**BC**).

11.   Verify the associativity of matrix multiplication in the general case; i.e., let **A** be a $k \times l$ matrix, **B** a $l \times m$ matrix, and **C** a $m \times n$ matrix. The product (**AB**)**C** is a $k \times n$ matrix; the i,jth element of **AB** is the number

$$\sum_{k=1}^{l} a_{ik}b_{kj} = (ab)_{ij},$$ so that (**AB**)**C** has the i,hth element given by the sum

$$\sum_{j=1}^{m} (ab)_{ij}c_{jh}.$$ You must verify that this sum is equal to the sum $\sum_{k=1}^{l} a_{ik}(bc)_{kh}$

in which $(bc)_{kh}$ is defined by the sum $\sum_{j=1}^{m} b_{kj}c_{jh}$.

2.7   APPLICATIONS

This section will present two applications of Markov chains; the first application concerns social mobility, and the second a personnel screening test. The discussion of social mobility is shallow from the sociologist's view and is used here to demonstrate formulation of questions about the model and interpreting results obtained from the model. The Markov chain model for a personnel screening test enables us to analyze the responses to a test which allows branching.

Application 2.1   Social Mobility

A study of social mobility in Denmark provides us with data which we use to give a Markov chain model for the mobility. We will use two basic sets of data which the study produced. The first set of data resulted from the classification of a group of fathers into one of five social states determined according to education and occupation. When these social states are identified only as 1 (lowest), 2, 3, 4, and 5 (highest), the proportion of fathers classified in each of those states respectively is given by the vector [0.22, 0.33, 0.30, 0.13, 0.02].

The sons of those fathers who were classified to provide the first set of data were studied when mature; that study produced a second set of data. In this study the social states of both the father and the son were used to determine the relative frequency of change in social state from father to son. These relative frequencies were determined by grouping all fathers of like social state and finding the proportions of their sons in each of the five states.

Table 2.4 presents the relative frequency data which were reported. The first row of this table was used in Example 1.5.

$$\mathbf{P} = \begin{bmatrix} 0.46 & 0.38 & 0.13 & 0.02 & 0.01 \\ 0.25 & 0.45 & 0.23 & 0.06 & 0.01 \\ 0.13 & 0.31 & 0.41 & 0.12 & 0.03 \\ 0.07 & 0.18 & 0.34 & 0.33 & 0.08 \\ 0.03 & 0.07 & 0.30 & 0.30 & 0.30 \end{bmatrix}$$

Table 2.4   Observed relative frequency of change in social state.

Each row of this matrix is a probability vector—the observed relative frequency of each social state among the sons of fathers whose social state was the same. The first row corresponds to fathers who were in the social state 1; the number 0.46 indicates that 46 percent of their sons were also in social state 1. Table 2.4 provides us with the transition matrix for a Markov chain whose states are 1, 2, 3, 4, 5 and which provides a model for social mobility in Denmark.

In our model we will use time zero to refer to the initial generation of fathers studied, thus X_0 is the social state of a randomly selected individual from that group. The state distribution at 0, $\mathbf{p}^{(0)}$, is provided by the first set of data considered—the proportion of fathers classified to each social state. We repeat that vector here and write

$$\mathbf{p}^{(0)} = [0.22, 0.33, 0.30, 0.13, 0.02]$$

The next generation, the sons in the study, has social state represented by X_1 in the model. The social state for each successive generation is represented in the model by the successive random variables. X_2 representing the state for the generation of grandsons, and so on.

The state distribution at 1, $\mathbf{p}^{(1)}$, is information which was observed in the original study (recall that the social state of the sons was used to determine the transition frequencies given in Table 2.4). The state distribution at 2, $\mathbf{p}^{(2)} = \mathbf{p}^{(0)}\mathbf{P}^2$, is a result *projected* by our model. We will call it a *projection* because it uses the same set of transition probabilities for transitions from X_1 to X_2 as were observed from X_0 to X_1. $\mathbf{p}^{(2)}$ is a result derived from our model; the extent to which $\mathbf{p}^{(2)}$ agrees with reality (the actual social state of grandsons) is a factor in determining how well this Markov chain model is able to describe social mobility in Denmark.

Table 2.5 gives the two-step transition matrix in which we have rounded the numbers to three decimal places for convenience.

$$\mathbf{P}^2 = \begin{bmatrix} 0.326 & 0.390 & 0.210 & 0.057 & 0.017 \\ 0.262 & 0.380 & 0.254 & 0.082 & 0.022 \\ 0.200 & 0.340 & 0.306 & 0.119 & 0.035 \\ 0.147 & 0.278 & 0.326 & 0.186 & 0.063 \\ 0.100 & 0.211 & 0.335 & 0.230 & 0.124 \end{bmatrix}$$

Table 2.5   Two-step transition probabilities for change in social state

Conclusions which are drawn from this model must be tempered by the fact that the transition probabilities are obtained from relative frequency data. Thus, while $\mathbf{P}$ is a precise statement of probability for change in social state from X_0 to X_1, to use $\mathbf{P}$ to describe changes in social state from X_1 to X_2 is to assume that social trends will not have changed from one generation to the next. This assumption is likely to be justified when most factors affecting social trends remain constant and is in doubt when many social values are changing.

From Table 2.5 and $\mathbf{p}^{(0)}$, we find the state distribution at 2,

$$\mathbf{p}^{(2)} = \mathbf{p}^{(0)}\mathbf{P}^2 = [0.239, 0.354, 0.271, 0.104, 0.032]$$

A comparison of $\mathbf{p}^{(0)}$ with $\mathbf{p}^{(2)}$ reveals an increase in the two lowest states and an increase in the highest state. The sociologist might ask about the ancestry of those individuals represented by $\mathbf{p}^{(2)}$. For example, if the highest class shows a 50 percent increase (from 0.02 to 0.032), the social state of their grandfathers is of interest. Similarly the ancestry of those individuals in the lower states would reveal which segments of the population have shifted to these lowest states during two generations.

Consider the five individual terms in the dot product of $\mathbf{p}^{(0)}$ with the last column of $\mathbf{P}^2$. The first of those terms, $p_1^{(0)}p_{1,5}^{(2)}$, equals the probability that the grandfather was in social state 1 and the grandson in social state 5; in symbols this is

$$p_1^{(0)}p_{1,5}^{(2)} = P(X_0 = 1, X_2 = 5)$$

Similar interpretations hold for the other four terms. The symbolic expressions and their values in this problem are:

$$P(X_0 = 1, X_2 = 5) = p_1^{(0)}p_{1,5}^{(2)} = (0.22)(0.017) = 0.004$$
$$P(X_0 = 2, X_2 = 5) = p_2^{(0)}p_{2,5}^{(2)} = (0.33)(0.022) = 0.007$$
$$P(X_0 = 3, X_2 = 5) = p_3^{(0)}p_{3,5}^{(2)} = (0.30)(0.035) = 0.011$$
$$P(X_0 = 4, X_2 = 5) = p_4^{(0)}p_{4,5}^{(2)} = (0.13)(0.063) = 0.008$$
$$P(X_0 = 5, X_2 = 5) = p_5^{(0)}p_{5,5}^{(2)} = (0.02)(0.124) = 0.002$$

From this we see that those individuals whose grandfathers were in state 3 are the largest group [about $1/3 - 0.011$ of the total probability $P(X_2 = 5) = 0.032$].

The dot product of $\mathbf{p}^{(0)}$ with the first column of $\mathbf{P}^2$ provides a similar analysis of the ancestry of the lowest social class. The five terms are:

$$P(X_0 = 1, X_2 = 1) = p_1^{(0)}p_{1,1}^{(2)} = (0.22)(0.326) = 0.072$$
$$P(X_0 = 2, X_2 = 1) = p_2^{(0)}p_{2,1}^{(2)} = (0.33)(0.262) = 0.086$$
$$P(X_0 = 3, X_2 = 1) = p_3^{(0)}p_{3,1}^{(2)} = (0.30)(0.200) = 0.060$$
$$P(X_0 = 4, X_2 = 1) = p_4^{(0)}p_{4,1}^{(2)} = (0.13)(0.147) = 0.019$$
$$P(X_0 = 5, X_2 = 1) = p_5^{(0)}p_{5,1}^{(2)} = (0.02)(0.100) = 0.002$$

These probabilities reveal that the grandfathers of these individuals were about equally likely to be in states 1, 2, or 3 with states 4 and 5 less likely. Notice that, while 10 percent of the second-generation decendants of individuals in state 5 are in state 1 ($p_{5,1}^{(2)} = 0.100$), this group is only 0.002 of the total population.

More information can be obtained about the two-generation descent from state 5 to state 1. For example,

$$P(X_0 = 5, X_1 = 3, X_2 = 1) = p_5^{(0)}p_{5,3}p_{3,1} = (0.02)(0.30)(0.13)$$
$$= 0.00078$$

This is a very small probability, but 0.00078 as compared to $P(X_0 = 5, X_2 = 1) = 0.002$ is about 40 percent of that probability; i.e., about 40 percent of those individuals in state 1 whose grandfathers were in state 5 had fathers in state 3 (middle class).

The essential point to focus on is the importance of translating a question in one context (say about mobility) to a question about the model and vice versa. A secondary point is the realization of limitations on interpreting results from the model due to the effects of assumptions made in forming the model (no change in mobility between X_1 and X_2 from that observed between X_0 and X_1, for instance).

Exercise 2.7   1.   A three-state social mobility transition matrix is

$$\begin{bmatrix} 0.7 & 0.2 & 0.1 \\ 0.3 & 0.5 & 0.2 \\ 0.3 & 0.2 & 0.5 \end{bmatrix}$$

Give the second- and third-generation transition matrices found from this transition matrix.

Ans. $\mathbf{P}^2 = \begin{bmatrix} 0.58 & 0.26 & 0.16 \\ 0.42 & 0.35 & 0.23 \\ 0.42 & 0.26 & 0.32 \end{bmatrix}$ $\mathbf{P}^3 = \begin{bmatrix} 0.532 & 0.278 & 0.190 \\ 0.468 & 0.305 & 0.227 \\ 0.468 & 0.278 & 0.254 \end{bmatrix}$

2.   Using the answers from Exercise 2.7.1 and the initial distribution of states, $\mathbf{p}^{(0)} = [0.2, 0.5, 0.3]$, find the proportions in each state at the first, second, and third generation.

Exercises 2.7.3 to 2.7.5 refer to the model of Exercises 2.7.1 and 2.7.2.

3.   Determine the proportion of the population for which (a) $\{X_0 = 1, X_2 = 3\}$, (b) $\{X_0 = 2, X_2 = 3\}$, and (c) $\{X_0 = 3, X_2 = 3\}$. (d) Give a verbal description of the events in 2.7.3a, b, and c.

4.   What proportion of the descendants of the middle class (state 2) are found in each of the three states at the third generation?

5.   Determine the proportion of descendants from those individuals who are in state 1 who will remain in state 1 for three generations or more (these are the so-called "hard core" of the lowest social state).

6.   There is an old adage which says "from rags to riches and back to rags in three generations." At least two formulations of this adage are plausible. Using the matrix given in Exercise 2.7.1, determine which formulation has the largest probability. Give a verbal interpretation which differentiates the two formulations.

formulation 1: $P(X_2 = 1, X_1 = 3 | X_0 = 1)$
formulation 2: $P(X_2 = 1 | X_1 = 3, X_0 = 1)$

Application 2.2   A Personnel Screening Test

Companies often use a test to screen job applicants. Such a test can be an invaluable aid to the people who must interview applicants and make decisions about hiring. Since a company can lose money both by hiring an individual who does not contribute and by not hiring an individual who could have contributed, the personnel executive who uses a screening test for his decisions will want to have information about the test's accuracy. One method of measuring the accuracy of a test is to analyze the probability that the test results are wrong. We will describe a test for which the analysis of these probabilities uses a Markov chain model.

BRANCH TESTING A branch test is constructed in such a way that each answer choice determines the next question. This is accomplished by a statement such as "go to question xx" following each answer choice. The number xx is generally different for each answer choice, so that the applicant "branches" at each question as he chooses his response. Each applicant answers only those questions he is instructed to answer; usually only one-third or one-fourth of the questions are considered by any one applicant.

The question statement and answer choices use a format similar to that given below:

29. Which statement best describes your professional compe-
     tence?
     a. My college grades are as good as those of my friends.
             Go to question 35.
     b. I have a distinct bent for genius and inventiveness.
             Go to question 28.
     c. I can usually do the tasks assigned to me.
             Go to question 31.

Choosing answer b sends you off to question 28, you ignore questions 35 and 31 and will probably never be instructed to answer them.

In such a test the questions are closely related to the applicant's responses, thus providing flexibility. Since the applicant answers a "randomly se-lected" set of questions—it cannot be determined in advance which questions will be answered—we can see that a probability model will be helpful in analyzing the test. To specify a model we will first assume a more definite structure for the test.

TEST STRUCTURE The employer is interested in the four attributes:

1. loyalty
2. financial drive
3. desire for professional success
4. decision-making ability

Each applicant will answer four questions pertaining to each of the four attributes, hence a total of 16 questions. The three answer choices for each question are, in random order, considered desirable, undesirable, and unin-formative to the employer.

TEST SCORING The applicant records a sequence of answer choices, called a "track." To the person scoring the test, each answer choice can be trans-lated as desirable, undesirable, or uninformative. An applicant is rejected with no further interview if his "track" contains too many undesirable and/or too few desirable responses. One problem which must be resolved by the personnel department is a determination of how many "too many" is, as well as how many "too few" should be. Our analysis will give some information which is helpful in making such decisions.

Results from standard tests administered in the past compared with later employee records provide the personnel department with information on how "desirable" employees responded to individual questions. These data make it possible to carefully choose each question and set of answer choices according to a specific set of criteria as described in the next paragraph.

The personnel department has provided answer choices to each question so that *among those applicants whom the company wishes to hire the relative frequency of each answer choice is as follows:*

1. The first question will have relative frequencies of 0.8 desirable, 0.1 uninformative, and 0.1 undesirable responses.
2. The answer choices to a question following a desirable response will have relative frequencies of 0.6 desirable, 0.3 uninformative, and 0.1 undesirable.
3. The answer choices to a question following an uninformative response will have relative frequencies of 0.8 desirable, 0.15 uninformative, and 0.05 undesirable.
4. The answer choices to a question following an undesirable response will have relative frequencies of 0.9 desirable, 0.09 uninformative, and 0.01 undesirable.

THE MODEL   We will use a Markov chain with three states 1, 2, and 3 with the correspondence, 1 = desirable, 2 = uninformative, and 3 = undesirable. The transition matrix **P** is determined from the conditions given in items 1, 2, 3, and 4 in the last paragraph above. The result is obviously.

$$\mathbf{P} = \begin{bmatrix} 0.6 & 0.3 & 0.1 \\ 0.8 & 0.15 & 0.05 \\ 0.9 & 0.09 & 0.01 \end{bmatrix}$$

Condition 1 above determines the frequency of responses to the first question. We will use that information to determine $\mathbf{p}^{(1)}$ rather than $\mathbf{p}^{(0)}$ since the "time" will correspond to the number of questions answered. Thus the state distribution at 1 is

$$\mathbf{p}^{(1)} = [0.8, 0.1, 0.1]$$

This Markov chain provides a model for the "track" of responses given by an applicant whom the company wishes to hire. We will use the model to compute two quantities which would help to decide if an applicant is "desirable." We will find the probability that a "desirable" applicant would give no undesirable responses. Our second computation will find the expected number of each of the three types of response.

ANALYSIS We first consider the probability that the applicant gives no undesirable responses. This asks for the probability of an event which is easy to describe in words and symbols but more difficult to compute. We can write

$$P(X_1 \neq 3, X_2 \neq 3, X_3 \neq 3, \ldots, X_{16} \neq 3)$$

but to evaluate this expression requires tedious calculation.

We will begin by looking at a simpler case of this question. The event $\{X_1 \neq 3, X_2 \neq 3\}$ is sufficiently simple that we can analyze its probability by considering all possible ways that the event can occur. Thus we write

$$P(X_1 \neq 3, X_2 \neq 3) = P(X_1 = 1, X_2 = 1) + P(X_1 = 2, X_2 = 1)$$
$$+ P(X_1 = 1, X_2 = 2) + P(X_1 = 2, X_2 = 2)$$
$$= p_1^{(1)}p_{1,1} = p_1^{(1)}p_{1,2} + p_2^{(1)}p_{2,1} + p_2^{(1)}p_{2,2}$$

The first two terms above give us the probability that $X_2 = 1$ and $X_1 \neq 3$, and the other two terms give the probability that $X_2 = 2$ and $X_1 \neq 3$. If we write the product

$$[p_1^{(1)}, p_2^{(1)}] \begin{bmatrix} p_{1,1} & p_{1,2} \\ p_{2,1} & p_{2,2} \end{bmatrix} = [p_1^{(1)}p_{1,1} + p_2^{(1)}p_{2,1}, p_1^{(1)}p_{1,2} + p_2^{(1)}p_{2,2}]$$

we see that the first component of the vector on the right equals $P(X_1 \neq 3, X_2 = 1)$ and the second component $P(X_1 \neq 3, X_2 = 2)$.

For this model the actual numbers are given by

$$[0.8, 0.1] \begin{bmatrix} 0.6 & 0.3 \\ 0.8 & 0.15 \end{bmatrix} = [0.56, 0.255]$$

Since we do not wish to consider "tracks" which use state 3, we do not need to use transition probabilities which correspond to state 3 either as the current state or as the next state. This is the basis for a proof that the probability of a 15-step transition which avoids state 3 at each step is found from the matrix

$$\begin{bmatrix} p_{1,1} & p_{1,2} \\ p_{2,1} & p_{2,2} \end{bmatrix}^{15}$$

Further, when we multiply this matrix by the vector $[p_1^{(1)}, p_2^{(1)}]$, we find the probabilities for X_{16} being state 1 and state 2 with the added proviso that state 3 did not occur along the way. We find (after much computation) that

$$[0.8, 0.1] \begin{bmatrix} 0.6 & 0.3 \\ 0.8 & 0.15 \end{bmatrix}^{15} = [0.167, 0.067]$$

We conclude from this calculation that the probability of no undesirable answer choices is $0.167 + 0.067 = 0.234$. These numbers are added because they correspond to the two ways that X_{16} can avoid state 3 after having avoided state 3 on the first 15 steps. So it is not unusual for a "desirable" applicant to have at least one undesirable response—but how many should we allow?

It may become necessary to find the exact probability distribution for the number of undesirable responses to determine how many such responses can be allowed. We will be content to determine the expected value of that random variable.

Let the number of undesirable responses be the random variable U. Then if random variables $U_1, U_2, \ldots, U_{16}$ are defined by the events $\{X_1 = 3\}$, $\{X_2 = 3\}, \ldots, \{X_{16} = 3\}$ as

$$U_i = \begin{cases} 1 & \text{if } X_i = 3 \\ 0 & \text{if } X_i \neq 3 \end{cases} \qquad i = 1, 2, \ldots, 16$$

we have $U = U_1 + U_2 + \cdots + U_{16}$.

By Theorem 1.10

$$E(U) = E(U_1) + E(U_2) + \cdots + E(U_{16})$$

and

$$E(U_i) = 0 \cdot P(U_i = 0) + 1 \cdot P(U_i = 1) = P(U_i = 1) = P(X_i = 3)$$

n	$p_1^{(n)}$	$p_2^{(n)}$	$p_3^{(n)}$
1	0.8	0.1	0.1
2	0.65	0.264	0.086
3	0.679	0.242	0.079
4	0.672	0.247	0.081
5	0.6737	0.2459	0.0804
6	0.6733	0.2462	0.0805
7	0.6734	0.2462	0.0804
8	0.6734	0.2462	0.0804
9	0.6734	0.2462	0.0804
10	0.6734	0.2462	0.0804
11	0.6734	0.2462	0.0804
12	0.6734	0.2462	0.0804
13	0.6734	0.2462	0.0804
14	0.6734	0.2462	0.0804
15	0.6734	0.2462	0.0804
16	0.6734	0.2462	0.0804
Total	10.8820	3.8071	1.3109

Table 2.6   The state distributions at n, $n = 1, 2, \ldots, 16$

Since $P(X_i = 3) = p_3^{(i)}$, we have

$$E(U) = p_3^{(1)} + p_3^{(2)} + \cdots + p_3^{(16)}$$

Clearly this same reasoning applies to the other two states as well. The expected number of desirable and the expected number of uninformative responses can be computed just as easily.

In Table 2.6 we give the state distributions at $1, 2, \ldots, 16$ and from them compute the expected values.

We now can say that only 1.31 undesirable responses on the average will be given by a "desirable" applicant and only 3.81 uninformative responses on the average. Notice that the expected values add up to the total number of questions answered. This is a check on our calculations, as well as an example of the fact that the sum of the expected values of the random variables must add to the expected value of their sum.

Exercise 2.7   7.   Find the probability that *no* undesirable responses will be given to the first three questions.
8.   What is the probability that *all* responses are desirable?
9.   What is the probability that *no* desirable responses are given to the first four questions?
10.   Find the expected number of responses of each type to the first eight questions.
11.   Using the result found in Exercise 2.7.10 and Theorem 1.10, find the expected number of responses of each type to the last eight questions.
12.   Demonstrate that $P(X_4 \neq 3, X_3 \neq 3, X_2 \neq 3, X_1 \neq 3)$ can be determined from the vector-matrix product

$$[0.8, 0.1] \begin{bmatrix} 0.6 & 0.3 \\ 0.8 & 0.15 \end{bmatrix}^3$$

2.8 LIMITING STATE DISTRIBUTIONS

The equations for the state distributions at n given in Secs. 2.5 and 2.6 are practical for computations provided n is not too large. When n is a large number, the computations became impossible to do with paper and pencil and impractical even for a computer. On the other hand, there are perfectly reasonable questions about Markov chain models which cannot be answered unless we know $\mathbf{p}^{(n)}$ for very large values of n. For instance, a sociologist studying social mobility may want to make projections for many generations into the future (see Application 2.1) or an analysis of a 100-question branching test may be needed (see Application 2.2). In the first instance, the sociologist would likely be satisfied to know that when n is very large all the state distributions are nearly equal to a specific probability vector. In the second instance, we will see that an analysis requiring each state distribution can be less lengthy than we might expect. This section presents some mathematical results which provide a method for approximating the state distributions for large values of n. As we will see, these results are somewhat theoretical, but they have useful and practical applications. We will introduce our first main idea by reprinting the state distributions for the personnel screening test of Application 2.2 in Sec. 2.7.

Example 2.26   A Markov chain model for analyzing a branching test was developed in Application 2.2. The Markov chain had the transition matrix

$$\mathbf{P} = \begin{bmatrix} 0.6 & 0.3 & 0.1 \\ 0.8 & 0.15 & 0.05 \\ 0.9 & 0.09 & 0.01 \end{bmatrix}$$

and state distribution at 1 given by $\mathbf{p}^{(1)} = [0.8, 0.1, 0.1]$. The state distributions were needed at 1, 2, . . . , 16 for the analysis given in Sec. 2.7, and those results were given in Table 2.6.

n	$p_1^{(n)}$	$p_2^{(n)}$	$p_3^{(n)}$
1	0.8	0.1	0.1
2	0.65	0.264	0.086
3	0.679	0.242	0.079
4	0.672	0.247	0.081
5	0.6737	0.2459	0.0804
6	0.6733	0.2462	0.0805
7	0.6734	0.2462	0.0804
8	0.6734	0.2462	0.0804
9	0.6734	0.2462	0.0804
10	0.6734	0.2462	0.0804
11	0.6734	0.2462	0.0804
12	0.6734	0.2462	0.0804
13	0.6734	0.2462	0.0804
14	0.6734	0.2462	0.0804
15	0.6734	0.2462	0.0804
16	0.6734	0.2462	0.0804
Total	10.8820	3.8071	1.3109

Table 2.7   (reprinted from Table 2.6)

In Example 2.26 we see that the state distributions at 1, 2, . . . approach the vector $[0.6734, 0.2462, 0.0804]$. In fact, from time 7 to time 16 the table gives the same vector. The numbers reported in Table 2.6 are rounded to four decimal places, so that if we multiply the vector $[0.6734, 0.2462, 0.0804]$ times the matrix $\mathbf{P}$ of Example 2.26, we will get a vector which is different in the fifth and higher decimal places; however, when using a certain degree of accuracy, such as four decimal places, we will continue to get the same result as soon as the same vector results twice in succession. We know that from that time on all state distributions will be the same because our basic equation for state distributions is

$$\mathbf{p}^{(n+1)} = \mathbf{p}^{(n)}\mathbf{P}$$

and so if $\mathbf{p}^{(n+1)} = \mathbf{p}^{(n)}$, we will find that

$$\mathbf{p}^{(n+2)} = \mathbf{p}^{(n+1)}\mathbf{P} = \mathbf{p}^{(n)}\mathbf{P} = \mathbf{p}^{(n+1)}$$

Thus if two consecutive state distributions are the same, we have shown that there must be three consecutive state distributions which are equal. We can continue this reasoning to show that *all* successive state distributions will be equal after the first time that two consecutive ones are equal.

When $\mathbf{p}^{(n)}$ approaches a vector $\mathbf{b}$ as n becomes large, we say that $\mathbf{p}^{(n)}$ has a *limit* and that $\mathbf{b}$ is a *limiting vector*. As we have seen in Example 2.26 where the state distributions had a limit $\mathbf{b} = [0.6734, 0.2462, 0.0804]$, the vector $\mathbf{b}$ when multiplied by the transition matrix $\mathbf{P}$ produced the vector $\mathbf{b}$ again (after rounding the components to four decimal places). A vector with this characteristic is important enough to our discussion that we give a formal definition here.

DEFINITION 2.11   A *fixed vector* of the matrix $\mathbf{P}$ is a vector $\mathbf{v}$ which satisfies the equation

$$\mathbf{v} = \mathbf{v}\mathbf{P}$$

The importance of fixed vectors to the topic of this section is pointed out by our next theorem.

THEOREM 2.3   If the state distributions of a Markov chain with transition matrix $\mathbf{P}$ have a limit $\mathbf{b}$, then $\mathbf{b}$ is a fixed vector of $\mathbf{P}$, that is

$$\mathbf{b} = \mathbf{b}\mathbf{P}$$

PROOF   (Our proof is informal but gives the essence of a logically rigorous argument.) Suppose that the state distributions have a limit $\mathbf{b}$. Then $\mathbf{p}^{(n)}$ and $\mathbf{p}^{(n+1)}$ are nearly equal to $\mathbf{b}$, and so from the equation

$$\mathbf{p}^{(n+1)} = \mathbf{p}^{(n)}\mathbf{P}$$

when we allow n to increase on both sides, both $\mathbf{p}^{(n+1)}$ and $\mathbf{p}^{(n)}$ approach the vector $\mathbf{b}$, and we must then have the equation

$$\mathbf{b} = \mathbf{b}\mathbf{P}$$

One clear application for Theorem 2.3 is that it tells us how to search for a possible limiting vector. Notice that any limiting vector is a fixed vector; thus we know what to look for. Also, an inspection of the equation $\mathbf{b} = \mathbf{bP}$ reveals that this equation determines a system of linear equations (see Example 2.19). Since $\mathbf{b}$ is not known to us, we can take its components to be the unknowns and solve the system of equations.

Before considering a specific example of such a system of equations, one additional remark needs to be made. First, it is possible to guess a solution— the vector of all zero components, $\mathbf{0}$, is a solution—$\mathbf{0} = \mathbf{0P}$. But of course $\mathbf{0}$ is not a probability vector, and we are looking only for probability vectors for our solution. Thus we always add the equation

$$b_1 + b_2 + \cdots + b_k = 1$$

to the system of equations given by $\mathbf{b} = \mathbf{bP}$ to ensure that a probability vector will be the solution—here we write $b_1, b_2, \ldots, b_k$ as the components of $\mathbf{b}$, they are the unknowns in the equations.

Example 2.27 Find a fixed vector of the transition matrix

$$\mathbf{P} = \begin{bmatrix} 1/3 & 1/3 & 1/3 \\ 0 & 0 & 1 \\ 1 & 0 & 0 \end{bmatrix}$$

We first let the fixed vector have components b_1, b_2, b_3 and require that

$$b_1 + b_2 + b_3 = 1$$

Next we write the system of equations given by $\mathbf{b} = \mathbf{bP}$:

$$\begin{aligned} 1/3\,b_1 \quad\;\; + b_3 &= b_1 \\ 1/3\,b_1 \qquad\quad &= b_2 \\ 1/3\,b_1 + b_2 \quad &= b_3 \end{aligned}$$

We see that the first equation states

$$b_3 = 2/3\,b_1$$

and that the second equation gives

$$b_2 = 1/3\,b_1$$

Next we consider the requirement that $b_1 + b_2 + b_3 = 1$ to find

$$b_1 + 1/3\,b_1 + 2/3\,b_1 = 1 \qquad \text{hence} \qquad 2b_1 = 1 \qquad \text{and} \qquad b_1 = 1/2$$

Then from above we see that $b_2 = 1/6$ and $b_3 = 1/3$. Thus the vector $[b_1, b_2, b_3] = [1/2, 1/6, 1/3]$ for which we check our arithmetic by computing

$$[1/2,\ 1/6,\ 1/3] \begin{bmatrix} 1/3 & 1/3 & 1/3 \\ 0 & 0 & 1 \\ 1 & 0 & 0 \end{bmatrix} = [1/2,\ 1/6,\ 1/3]$$

so $[1/2,\ 1/6,\ 1/3]$ is indeed a fixed vector.

Note that we have found the only possible fixed vector, since the equations we solved had only one solution.

The result found in Example 2.27 is not unusual. It is possible to give conditions for a transition matrix **P** which will ensure us that **P** has one and only one fixed vector which is a probability vector and, most important, that the state distributions will always have that fixed vector as a limit. On the other hand, it is not always the case that the state distributions have a limit if **P** does not satisfy some extra condition. Our next example verifies that some transition matrices exist for which we are not guaranteed that the state distributions have a limit.

Example 2.28   The two-state Markov chain which switches states at each step has the transition matrix

$$\mathbf{P} = \begin{bmatrix} 0 & 1 \\ 1 & 0 \end{bmatrix}$$

If the initial state distribution is $\mathbf{p}^{(0)} = [a, 1 - a]$, then at time 1 there is a switch of states and $\mathbf{p}^{(1)} = [1 - a, a]$. One more transition returns the initial state and $\mathbf{p}^{(2)} = [a, 1 - a]$. Thus the state distributions for this Markov chain alternate between $[a, 1 - a]$ at even times and $[1 - a, a]$ at odd times.

The value $a = \frac{1}{2}$ determines the one and only fixed vector since $a = 1 - a$ in that case and only in that case. Thus $\mathbf{p}^{(0)} = [\frac{1}{2}, \frac{1}{2}]$ will determine $\mathbf{p}^{(n)} = [\frac{1}{2}, \frac{1}{2}]$ for all n. When $a \neq \frac{1}{2}$, we have seen that $[a, 1 - a] \neq [1 - a, a]$; and, since these two vectors alternate as the state distributions, it is not possible for there to be a limiting vector.

In Example 2.28 there is not a limiting vector unless the vector $\mathbf{p}^{(0)}$ happens to be the fixed vector for **P**. Since it is not practical to set the state distribution equal to the fixed vector in order to get a limiting vector, we will list conditions which guarantee that there is a limiting vector no matter what initial state distribution $\mathbf{p}^{(0)}$ is appropriate. In Sec. 2.10 we will prove that there is a limit when these conditions are fulfilled. There is no obvious way to motivate the condition needed. Mathematicians have discovered this condition, and find that it is a convenient one to work with so we will pass it on to you.

DEFINITION 2.12   A Markov chain is called *regular* if there is some positive integer n so that every element of the n-step transition matrix $\mathbf{P}^n$ is positive.

Since every element of the n-step transition matrix is positive, there is a positive probability of being at each state at time n no matter which state you started in.

EXAMPLE 2.28 Continued
This Markov chain switches states at each step and has only two states, so after two steps it has returned to the starting state. This is reflected also in the two-step matrix

$$\mathbf{P}^2 = \begin{bmatrix} 1 & 0 \\ 0 & 1 \end{bmatrix}$$

We can see that this Markov chain will always have n-step transition matrix equal to $\mathbf{P}$ (if n is odd) or $\mathbf{P}^2$ (if n is even). If an even number of transitions has occurred, the original state is the current one, while after an odd number of transitions, the other state must appear. We can never have a value of n for which all elements of $\mathbf{P}^n$ are positive. A Markov chain with this transition matrix is *not* regular.

Example 2.29   Consider the transition matrix

$$\mathbf{P} = \begin{bmatrix} 1/3 & 1/3 & 1/3 \\ 0 & 0 & 1 \\ 1 & 0 & 0 \end{bmatrix}$$

of Example 2.27.

$$\mathbf{P}^2 = \begin{bmatrix} 4/9 & 1/9 & 4/9 \\ 1 & 0 & 0 \\ 1/3 & 1/3 & 1/3 \end{bmatrix} \quad \text{and} \quad \mathbf{P}^3 = \begin{bmatrix} 16/27 & 4/27 & 7/27 \\ 1/3 & 1/3 & 1/3 \\ 4/9 & 1/9 & 4/9 \end{bmatrix}$$

Since every element of $\mathbf{P}^3$ is positive, we conclude that $\mathbf{P}$ determines a regular Markov chain.

Our next theorem states the facts that we have referred to here. We will discuss the proof of this theorem in great detail in Sec. 2.10.

THEOREM 2.4   If a Markov chain is regular, then the state distributions $\mathbf{p}^{(n)}$ have the same limiting vector $\mathbf{b}$ no matter what the initial state distribution $\mathbf{p}^{(0)}$ may be.
    Furthermore, there is only one fixed probability vector for $\mathbf{P}$, which is necessarily $\mathbf{b}$.

EXAMPLE 2.29 Continued
In Example 2.27 we showed that $[1/2, 1/6, 1/3]$ was the fixed probability vector for the matrix

$$\begin{bmatrix} 1/3 & 1/3 & 1/3 \\ 0 & 0 & 1 \\ 1 & 0 & 0 \end{bmatrix}$$

In Example 2.29 we showed that **P** determines a regular Markov chain. Here we show the result of successive state distributions when two different initial state distributions are used.

We will give the successive state distributions as three-place decimals, so for comparison the fixed vector is $[0.5, 0.167, 0.333]$. Our two initial vectors will be $[1/3, 1/3, 1/3]$ and $[1/2, 0, 1/2]$.

n	$\mathbf{p}^{(0)} = [1/3, 1/3, 1/3]$ $\mathbf{p}^{(n)}$	$\mathbf{p}^{(0)} = [1/2, 0, 1/2]$ $\mathbf{p}^{(n)}$
1	$[0.444, 0.112, 0.444]$	$[0.666, 0.167, 0.167]$
2	$[0.592, 0.148, 0.260]$	$[0.389, 0.222, 0.389]$
3	$[0.458, 0.197, 0.345]$	$[0.519, 0.130, 0.351]$
4	$[0.497, 0.153, 0.350]$	$[0.524, 0.173, 0.303]$
5	$[0.516, 0.166, 0.318]$	$[0.478, 0.175, 0.347]$
6	$[0.490, 0.172, 0.338]$	$[0.507, 0.159, 0.334]$
7	$[0.501, 0.164, 0.335]$	$[0.503, 0.169, 0.328]$
8	$[0.502, 0.167, 0.331]$	$[0.496, 0.168, 0.336]$
9	$[0.498, 0.167, 0.335]$	$[0.501, 0.166, 0.333]$
10	$[0.501, 0.166, 0.333]$	$[0.500, 0.167, 0.333]$

Notice that n does not have to be very large before the state distributions are nearly equal to the fixed vector (rounded to three decimals).

Exercise 2.8

1. Let

$$\mathbf{P} = \begin{bmatrix} 1/2 & 1/2 \\ 1/2 & 1/2 \end{bmatrix}$$

and $\mathbf{p}^{(0)} = [1/3, 2/3]$.

a. Compute $\mathbf{p}^{(1)}$ and $\mathbf{p}^{(2)}$.
b. Find $\mathbf{p}^{(n)}$ for $n \geq 2$.
c. Find the fixed probability vector for **P**.

2. Let

$$\mathbf{P} = \begin{bmatrix} 1/3 & 2/3 \\ 1/3 & 2/3 \end{bmatrix}$$

and $\mathbf{p}^{(0)} = [a, 1 - a]$.

a. Compute $\mathbf{p}^{(1)}$ and $\mathbf{p}^{(2)}$.
b. Find $\mathbf{p}^{(n)}$ for $n \geq 2$.
c. Find the fixed probability vector for **P**.

3.   Find the fixed probability vector for

$$P = \begin{bmatrix} 0 & 1/3 & 2/3 \\ 1/2 & 0 & 1/2 \\ 1/3 & 1/3 & 1/3 \end{bmatrix}$$

4.   (a) Find the fixed probability vector for

$$P = \begin{bmatrix} 0 & 0 & 1 \\ 0 & 0 & 1 \\ 1/2 & 1/2 & 0 \end{bmatrix} .$$

(b) Let $p^{(0)} = [1, 0, 0]$ and find $p^{(1)}$, $p^{(2)}$, $p^{(3)}$, $p^{(4)}$. (c) Compare the vectors in 2.8.4b to the fixed vector found in 2.8.4a.

5.   Show that a Markov chain with matrix given by P in Exercise 2.8.4 is not regular. (*Hint*: Find P^2 and P^3; do you see a pattern?)

6.   Let $P = \begin{bmatrix} 0.3 & 0.7 \\ 1 & 0 \end{bmatrix}$

a.   Show that P determines a regular Markov chain.
b.   Find the fixed probability vector for P.
c.   For $p^{(0)} = [0.6, 0.4]$ compute enough state distributions so that the state distributions and limiting vector differ by at most 0.01.
d.   For $p^{(0)} = [0.5, 0.5]$ repeat the instructions for 2.8.6c.

7.   The matrix

$$P = \begin{bmatrix} 1/2 & 1/4 & 1/4 \\ 1/2 & 1/2 & 0 \\ 1/4 & 0 & 3/4 \end{bmatrix}$$

is a matrix for social mobility in Zorkaland. Find the theoretical proportion of the population which will be in social state 3 after many generations.

8.   The matrix

$$P = \begin{bmatrix} 0.6 & 0.39 & 0.01 \\ 0.8 & 0.19 & 0.01 \\ 0.9 & 0.09 & 0.01 \end{bmatrix}$$

gives a model for a branching test, the states are the types of answer choices.

a.   Find the limiting state distribution for P.
b.   What is the probability that the hundredth answer choice is type 1?

9.   Let $P = \begin{bmatrix} 1/2 & 1/2 \\ 1 & 0 \end{bmatrix}$

a.   Compute P^2.
b.   Using item a, compute $P^4 = P^2 P^2$.

c. Using item b, compute $\mathbf{P}^8 = \mathbf{P}^4\mathbf{P}^4$.
d. Do you think that $\mathbf{P}^n$ has a limit?
e. Compare the rows of $\mathbf{P}^8$ with the fixed vector for $\mathbf{P}$.

10.   Recall the basic concepts of fixed vector and limiting vector. If a Markov chain has a transition matrix $\mathbf{P}$:

Does the existence of a limiting vector for the Markov chain imply that $\mathbf{P}$ has a fixed vector?

Does the existence of a fixed vector for $\mathbf{P}$ imply that the Markov chain has a limiting vector?

11.   Using the data of Example 2.26, find the fixed vector for $\mathbf{P}$. Compare the fixed vector with the vector given in the table as a fixed vector, and note that they are different in the fourth decimal place. This is the effect of rounding off the components of the vectors before multiplying (the accumulated roundoff error explains the difference between the fixed vector found and the one given in Table 2.7).

2.9   APPLICATIONS

The first Markov chain model discussed in this section analyzes an inventory policy to determine the long-term effects of the policy. Our second Markov chain model is intended to study certain axioms of human behavior.

Application 2.3   An Inventory Model

Companies which maintain an inventory of commodities to sustain their business require careful management policies in order to provide reliable service at a minimal investment. Two counter forces are at work which management must seek to balance. On the one hand it is necessary to keep capital investments at a minimum and hence maintain a small inventory; while on the other hand a larger inventory is needed to guard against depletion and the consequent loss of sales.

We will specify an inventory control policy, then give a Markov chain model for that policy. From the model we will find the long-term probability that inventory is depleted with that policy. As a measure of average capital investment, we will determine the average number of items on hand.

When a large number of different items are kept in inventory an efficient system will use a control policy for each type of item. A typical control policy can be stated as follows:

INVENTORY CONTROL POLICY   Order three items whenever only one item remains in inventory.

Execution of this policy could be done in several ways. A card describing the item to be ordered might be sent on to the order department as soon as the second to the last item is sold, or perhaps at the end of the day one employee could be given the task of reviewing the inventory to ascertain which orders must be placed.

To build our model, we will make several assumptions about the number of sales made and about the delivery of items ordered. This sort of information would ordinarily be available through the records of sales and deliveries in the past. Our discussion now centers on a single type of item for which the inventory policy given above is followed. The assumptions we make about sales and deliveries are given below.

ASSUMPTIONS

1. On any given day there is a probability a that one item of this type is sold and a probability $1 - a$ that none are sold. Sales on any day are independent of sales on previous days given that the inventory is not depleted.

2. If the inventory is depleted and a customer arrives, it is assumed that the customer will make his purchase from a competitor (the sale is lost).

3. Delivery of an order requires a random number of days. There is probability d that delivery of an order is made on any day given that it has not been made previously.

Since the control policy specifies an order for three items when inventory has reached one item, there should never be more than four items on hand. We will take the state space for the Markov chain model to coincide with the number of items in inventory; hence the integers 0, 1, 2, 3, 4 will describe the states.

Analysis of assumptions 1, 2, and 3 determines the transition probabilities. Because of assumption 1 there are many transitions which will be assigned probability 0. From each of states 2, 3, and 4, we are only allowed to remain at the state or to decrease by 1 on any given day. Transitions from states 0 and 1 are more complicated.

In our analysis we will take the value of X_i to be the inventory at the end of day i and assume that deliveries arrive before the opening for business, so that if an order arrives the items are available for the entire day during which a sale can be made. The possible transitions which can occur are indicated by Fig. 2.7.

From state 0, a transition to state 0 means no delivery has been made, an event to which assumption 3 assigns probability $1 - d$; thus we find

$$p_{0,0} = 1 - d$$

A transition from state 0 to state 2 can occur if delivery of the three items is made; then one item is sold during the day. We will write a word descrip-

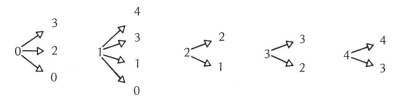

Figure 2.7   Possible one-step transitions

tion of that event using X_0 to denote the previous day with inventory 0, so that X_1 has value 2.

$$P(X_1 = 2|X_0 = 0) = P(\text{delivery is made and a sale is made}|X_0 = 0)$$
$$= P(\text{delivery is made}|X_0 = 0) \times$$
$$P(\text{a sale is made}|\text{delivery is made and } X_0 = 0)$$

What we claim in these expressions is that the transition probability $p_{0,2}$ should be given as the product of the probability of a delivery (which is d) times the probability that a sale is made given that delivery was made, which is a. The two values d and a should be apparent from assumption 3 and 1 respectively. Thus we have found

$$p_{0,2} = ad$$

We could now find $p_{0,3}$ by subtracting the quantity $p_{0,0} + p_{0,2}$ from 1. A more instructive method and one which checks our previous calculations is to argue from basic principles. For this we write the expressions

$$P(X_1 = 3|X_0 = 0) = P(\text{delivery is made and no sale is made}|X_0 = 0)$$
$$= P(\text{delivery is made}|X_0 = 0)$$
$$\times P(\text{no sale is made}|\text{delivery is made and } X_0 = 0)$$
$$= d(1 - a)$$

We will leave the determination of the remaining transition probabilities to you as Exercise 2.9.5, but present the results here as the transition matrix

$$\mathbf{P} = \begin{bmatrix} 1-d & 0 & ad & d(1-a) & 0 \\ a(1-d) & (1-a)(1-d) & 0 & ad & d(1-a) \\ 0 & a & 1-a & 0 & 0 \\ 0 & 0 & a & 1-a & 0 \\ 0 & 0 & 0 & a & 1-a \end{bmatrix}$$

We will first show that this Markov chain is regular. Formally this requires computing $\mathbf{P}^4$—informally you can see this by reasoning as follows:

> From state 0 is it possible to be at any of the states 0,1,2,3,4 after four steps? The answer is yes since, for example, we can be at state 0 by just staying there, which has a positive probability of occurring.

We can be at state 4 by going to state 3 at time 1, state 2 at time 2, state 1 at time 3, and state 4 at time 4. It is possible to get to states 1, 2, and 3 in fewer than four steps and then to stay there.

From states 1, 2, 3, 4 similar arguments show that after four transitions it is possible to be in any state no matter which state we started in.

Now that we know the Markov chain to be regular we will use Theorem 2.4 to assert the existence of a limiting state distribution which can be found by solving the equations for the fixed probability vector for $\mathbf{P}$. The equations expressed in terms of the fixed vector $\mathbf{b} = [b_0, b_1, b_2, b_3, b_4]$ determined by the vector-matrix equation $\mathbf{bP} = \mathbf{b}$ are:

$$(1 - d)b_0 + a(1 - d)b_1 = b_0$$
$$(1 - a)(1 - d)b_1 + ab_2 = b_1$$
$$adb_0 + (1 - a)b_2 + ab_3 = b_2$$
$$d(1 - a)b_0 + adb_1 + (1 - a)b_3 + ab_4 = b_3$$
$$d(1 - a)b_1 + (1 - a)b_4 = b_4$$
$$b_0 + b_1 + b_2 + b_3 + b_4 = 1$$

The last equation assures us that $\mathbf{b}$ is a probability vector.

To express the components of the fixed vector in terms of a and d after solving these equations, we introduce another symbol; namely, we let

$$c = \frac{1}{3[d + a(1 - d)]/a + a/d + a(d - 2)}$$

This allows us to write the fixed vector as

$$\mathbf{b} = \left[\frac{a(1 - d)}{d}c, \; c, \; \frac{d + a(1 - d)}{a}c, \; \frac{d + a(1 - a)(1 - d)}{a}c, \; \frac{d(1 - a)}{a}c\right]$$

From the existence of a limiting state distribution, we conclude that the probability of having a depleted inventory on any given day is approximately equal to the first component of $\mathbf{b}$, the fixed vector. This number for any values of a and d is

$$b_0 = \frac{a(1 - d)}{d[3(d/a + 1 - d) + a/d + a(d - 2)]}$$

Table 2.8 gives numerical values for b_0 when a and d have various different values. Such a table is convenient if the inventories for several types of items are controlled by the same policy, but the probabilities a and d differ for the various items. For us the table assigns a more tangible value to b_0 and illustrates the effect on the value of b_0 when the values of a and d change.

Note the intuitive appeal of the values presented by Table 2.8. For a certain chance of making a sale—the value of a—as the probability of delivery increases the chance of a depleted inventory diminishes. Thus in the first row of Table 2.8 ($a = 0.100$), as d increases from 0.100 to 0.667 the probability of a depleted inventory decreases from 0.138 to 0.002.

d \ a	0.100	0.250	0.333	0.500	0.667
0.100	0.138	0.030	0.016	0.006	0.002
0.250	0.380	0.129	0.079	0.033	0.014
0.333	0.476	0.190	0.122	0.054	0.024
0.500	0.612	0.308	0.214	0.105	0.049
0.667	0.702	0.410	0.304	0.163	0.081

Table 2.8   Values of b_0, the limiting probability for state 0

Table 2.8 presents only one side of the picture when evaluating an inventory control policy. The other considerations concern the average number of items in inventory. Knowing the average number of items in inventory allows us to determine the average amount of capital invested to maintain the inventory of this item.

To calculate the average number of items in inventory, we again use the existence of a limiting state distribution and do all computations using the limiting distribution. We are trying to find $E(X_n)$ but will content ourselves with the expected value of a random variable with the distribution given by **b**. This expected value is then given by the dot product

$$\mathbf{b} \cdot \begin{bmatrix} 0 \\ 1 \\ 2 \\ 3 \\ 4 \end{bmatrix} = 0\frac{a(1-d)}{d}c + 1(c) + 2\frac{d + a(1-d)}{a}c$$

$$+ 3\frac{d + a(1-a)(1-d)}{a}c + 4\frac{d(1-a)}{a}c$$

We will not attempt to present a simpler formula for this expected value but will give a table of values for the same choices of a and d as were used in Table 2.8.

d \ a	0.100	0.250	0.333	0.500	0.667
0.100	2.12	2.61	2.70	2.80	2.85
0.250	1.35	2.10	2.29	2.51	2.63
0.333	1.10	1.86	2.08	2.35	2.50
0.500	0.76	1.46	1.71	2.05	2.26
0.667	0.54	1.15	1.41	1.78	2.03

Table 2.9   The expected number of items in inventory

Suppose that a certain item has this inventory control policy with $a = 0.333$ and $d = 0.250$. From Table 2.8 the probability of a depleted inventory is 0.190, and from Table 2.9 the average inventory size is 1.86. Thus on the average there are 1.86 of these items in inventory, and there is a probability of 0.190 that inventory is depleted on any given day. Management knows the cost of purchasing and storing each item, as well as the cost of losing a sale when inventory is depleted. With that information they can compute an average cost per day of this inventory policy to determine if it is a policy which earns money or loses money for the firm.

We can find the expected cost of a lost sale if we know the probability of the event that a sale is lost on a given day. Again we consider the limiting state distribution. Thus we find

$$P(\text{a sale is lost on day } n) = P(X_n = 0)P(\text{a customer arrives}|X_n = 0)$$
$$= b_0 \cdot a$$

A customer's arrival is presumably independent of the size of inventory, and so we set $P(\text{a customer arrives}|X_n = 0) = a$. For the situation when $a = 0.333$ and $d = 0.250$, we find that the probability of losing a sale is $(0.190)(0.333) = 0.063$ since in this case $b_0 = 0.190$. If a lost sale costs D dollars, and the cost of purchasing and storing one item costs E dollars per day, then the expected cost per day for this control policy with a and d as above is

$$0.063D + 1.86E$$

From this computation it is possible to assess the worth of the inventory control policy.

Exercise 2.9   1.   The transition matrix for a certain inventory model is

$$\mathbf{P} = \begin{bmatrix} 0 & 0 & 0 & 1 \\ 1/6 & 1/3 & 1/6 & 1/3 \\ 1/6 & 1/6 & 2/3 & 0 \\ 0 & 1/6 & 1/6 & 2/3 \end{bmatrix}$$

a.   Find b_0, the limiting probability of zero inventory.
b.   Compute the expected inventory size.

2.   Find the transition matrix for the following inventory policy:

a.   When inventory reaches one, an order is placed for two items.
b.   When inventory reaches zero, a rush order is placed, and the two items previously ordered are delivered the next day with probability 1.
c.   On any given day one item is sold with probability $1/5$, no sale is made with probability $4/5$.
d.   Delivery of a regular order (not a rush order) is made on any day with probability $2/3$, given that delivery has not occurred earlier.

3.   For the inventory policy of Exercise 2.9.2 find the limiting probability of 0 inventory and the expected inventory size.

4.   Amend the inventory policy of Application 2.3 to read: Order three items whenever only two items remain in inventory. For $a = \frac{1}{4}$ and $d = \frac{1}{3}$ find:

a.   the transition matrix for a Markov chain model of this policy
b.   the limiting state distribution
c.   the probability of a depleted inventory on a given day
d.   the expected inventory size [In items c and d use the limiting state distribution for the computations.]

5.   Determine the remaining transition probabilities for the model of Application 2.3. In particular, the probabilities $p_{1,0}, p_{1,1}, p_{1,2}, p_{1,3}, p_{1,4}$ offer a challenge, while the transitions from states 2, 3, 4 are all similar and much easier.

Application 2.4   A Model for Human Behavior

An experiment in human behavior involves a type of "game" between two players. At each play of the game each player may respond with one of two actions, say "bluff" or "raise." Neither player is told his opponent's response but is told by the experimenter if he has won or lost that play of the game. The players are each rewarded when they win so that they have an incentive to win as often as possible.

The experimenter uses a random method of deciding if the players have won or lost. His method is:

1.   If both players "bluff," then they will both be told they lost.
2.   If one player "bluffs" and the other "raises," then the experimenter randomly decides what to tell the players. The experimenter's chance mechanism informs the bluffer, with probability a, that he has won and, with probability $1 - a$, that he has lost. On the basis of a second and independent random choice, the raiser is told with probability $1 - a$ that he has won and with probability a that he has lost.
3.   When both players "raise," the experimenter again randomly decides what to tell the players. His technique is the same as in rule 2 above except that $a = \frac{1}{2}$ is used.

Two axioms are assumed to govern the player's actions:

Axiom 1:   If a player wins on a given play, he will repeat the same action the next play of the game.

Axiom 2:   If a player's response results in a loss, then on the next play of the game he will use the alternative response with probability q.

The experimenter wishes to test the validity of these axioms. To do so, he first analyzes the probability model which encompasses these axioms and rules of play. After the analysis indicates the type of behavior to expect, the experimenter will use subjects to "play" the game. The experimenter will look for the behavior predicted by the model.

The game described can be analyzed by a Markov chain model having just three states. These states can be designated as BB, BR, and RR; they are described by the phrases, both players bluff, one player bluffs while the other player raises, and both players raise. These are the only interesting outcomes of each play (the experimenter does not care which player gives which response when we record BR).

Transitions from the state BB are relatively simple to analyze since rule 1 states that each player is told that he lost. Each other state of the Markov chain is possible at the next play of the game, and the relevant probabilities are determined from axiom 2. The next state is:

BB: if both players repeat their actions—probability $= (1 - q)^2$

BR: when only one player repeats his action—probability $= 2q(1 - q)$

RR: when both players choose alternative actions—probability $= q^2$

These probabilities determine the transition probabilities from state BB to each other state. Transition probabilities for transitions from state RR are more complicated as rule 3 comes into play, and each player is told independently, with probability $1/2$, that he won. The possible outcomes resulting from this rule are:

Both players are told that they won—probability $= 1/4$.

Exactly one player is told that he won—probability $= 1/2$.

Neither player is told that he won—probability $= 1/4$.

Axiom 1 and axiom 2 now can be used to determine the transition probabilities if we partition the sample space according to these three outcomes. We consider first the transition probability $P(X_1 = RR | X_0 = RR)$ and use the partitioning as follows:

$$P(X_1 = RR | X_0 = RR) = P(X_1 = RR \text{ and both won at } X_0 | X_0 = RR)$$

$$+ P(X_1 = RR \text{ and neither won at } X_0 | X_0 = RR)$$

Now we consider the first term on the right above, it is

$$P(X_1 = RR \text{ and both won at } X_0 | X_0 = RR)$$
$$= P(\text{both won at } X_0 | X_0 = RR)P(X_1 = RR | \text{both won at } X_0$$
$$\text{and } X_0 = RR)$$
$$= 1/4 \cdot 1 = 1/4$$

(Axiom 1 says that both will respond with R the next time; the first factor uses rule 3.)

Similarly we find

$P(X_1 = RR$ and exactly one won at $X_0|X_0 = RR)$
$= P(\text{exactly one won at } X_0|X_0 = RR)$
$\qquad \times\ P(X_1 = RR|\text{exactly one won at } X_0 \text{ and } X_0 = RR)$
$= \frac{1}{2} \cdot (1 - q)$

The third term produces the product

$P(\text{neither won at } X_0|X_0 = RR)P(X_1 = RR|\text{neither won at } X_0$
$\qquad\qquad\qquad\qquad \text{and } X_0 = RR) = \frac{1}{4} \cdot (1 - q)^2$

Thus $P(X_1 = RR|X_0 = RR) = \frac{1}{4} + \frac{1}{2}(1 - q) + \frac{1}{4}(1 - q)^2$.

To find the probability of transition from state RR to state BB, we reason that it must be the case that both players were told that they lost; hence, they both chose the alternative action (axiom 1).

$P(X_1 = BB|X_0 = RR) = P(\text{both lost at } X_0|X_0 = RR)$
$\qquad\qquad \times\ P(X_1 = BB|\text{both lost at } X_0 \text{ and } X_0 = RR)$
$\qquad = \frac{1}{4} \cdot q^2$

These methods for finding the transition probabilities are now routine; we leave the remaining computations to the student who does Exercise 2.9.6. The results are given in the matrix **P**.

$$\mathbf{P} = \begin{bmatrix} (1 - q)^2 & 2q(1 - q) & q^2 \\ aq(1 - q + aq) & 1 - q + 2aq^2 - 2a^2q^2 & q(1 - a)(1 - aq) \\ \dfrac{q^2}{4} & q - \dfrac{q^2}{2} & 1 - q + \dfrac{q^2}{4} \end{bmatrix}$$

Solving for the fixed vector of **P** is appropriate since, if neither a nor q is equal to 0 or 1, the Markov chain will be regular. To express the fixed vector in terms of a and q, we introduce a symbol by letting

$$c = 32 - 34q + 9q^2 - 8a - 8aq + 24a^2q + 6aq^2 - 12a^2q^2$$

With this value for c we find (after some work) that the components of the fixed vector $\mathbf{b} = [b_{BB}, b_{BR}, b_{RR}]$ are

$$b_{BB} = \frac{2a(1 - q + aq)(4 - 3q) + q(2 - 2a + 2a^2q - q)}{c}$$

$$b_{BR} = \frac{2(2 - q)(4 - 3q)}{c}$$

$$b_{RR} = \frac{4(2 - q)(2 - 2a + 2a^2q - q)}{c}$$

This experiment was carried out with 25 pairs of players. Each pair played

210 games and produced relative frequencies of the states BB, BR, and RR which were remarkably close to those predicted by b_{BB}, b_{BR}, and b_{RR}. This experimenter claims the axioms are substantiated by this evidence.

Note that while the experimenter controls the value of a (the probability mechanism used in rule 2) the player has his own q. For these reasons the experimenter must make a determination of the value of q from his observations, then find b_{BB}, b_{BR}, and b_{RR} to make final comparisons with the data.

Exercise 2.9   6.   Verify the transition probabilities in the matrix **P** by using the experimenters rules 1, 2, and 3 and the two axioms.

7.   Suppose that in this model we have $q = \frac{1}{2}$.

a.   If $a = \frac{1}{3}$, find the matrix **P** and its fixed vector **b**.

b.   Would seven occurrences of state BB, 126 occurrences of state BR, and 84 occurrences of state RR tend to support or to refute the axioms? (Use $a = \frac{1}{3}$.)

c.   Would 70 occurrences of each state in 210 games tend to support or to refute the axioms? (Use $a = \frac{1}{3}$.)

8.   In Exercise 2.9.7 all computations use the limiting state distributions as comparison to the observed relative frequencies. Verify that this is justified by finding the exact state distributions for $n = 1, 2, 3$, and so on until they agree to two decimal places with the limiting state distribution. (Use the initial state distribution $p_{BB}{}^{(0)} = 1$, also assume $q = \frac{1}{2}$ and $a = \frac{1}{3}$.)

*2.10   SOME THEORY—THE BASIC LIMIT THEOREM

Our purpose in this section is to develop an understanding of the condition that a Markov chain be regular by presenting a proof of Theorem 2.4. Our proof consists of several pertinent facts which, when properly related to each other, give a logical argument for the truth of Theorem 2.4. This approach is logically complex, but it has the advantage that the proof emphasizes the important facts and the condition that the Markov chain be regular is used in a very obvious way.

We state Theorem 2.4. Then we state three theorems which provide the three main parts in our proof of Theorem 2.4.

THEOREM 2.4   If a Markov chain is regular then the state distributions, $\mathbf{p}^{(n)}$, have the same limiting vector **b** no matter what the initial state distribution $\mathbf{p}^{(0)}$ may be.

Furthermore, there is only one fixed probability vector for **P**, which is necessarily **b**.

PROOF

1.   The first part of our proof is the assertion given in Theorem 2.5.

THEOREM 2.5   If a Markov chain is regular, and $\mathbf{y}^{(0)}$ is any column vector with nonnegative components, then the column vectors $\mathbf{y}^{(n)}$ defined by the equation

$$\mathbf{y}^{(n)} = \mathbf{P}^n \mathbf{y}^{(0)}$$

approach a vector $\mathbf{y}$ as n becomes large, and all the components of $\mathbf{y}$ are equal. (The components of $\mathbf{y}$ will of course depend on the choice of the vector $\mathbf{y}^{(0)}$.)

2.   The next theorem uses the conclusion of Theorem 2.5 as its hypothesis, so the conclusion of Theorem 2.6 takes us one more step toward the conclusion of Theorem 2.4.

THEOREM 2.6   If $\mathbf{y}^{(0)}$ is any column vector with nonnegative components and the column vectors $\mathbf{y}^{(n)}$, defined as in Theorem 2.5, approach a vector $\mathbf{y}$ as n becomes large and all the components of $\mathbf{y}$ are equal, then the matrices $\mathbf{P}^n$ approach a matrix $\mathbf{B}$ as n becomes large and all the rows of $\mathbf{B}$ are equal to the same vector $\mathbf{b}$.

3.   The concluding step to proving Theorem 2.4 is provided by Theorem 2.7 which takes the conclusion of Theorem 2.6 as its hypothesis and the conclusion of Theorem 2.4 as its conclusion.

THEOREM 2.7   If the matrices $\mathbf{P}^n$ approach a matrix $\mathbf{B}$ as n becomes large and all the rows of $\mathbf{B}$ are equal to the same vector $\mathbf{b}$, then if $\mathbf{p}^{(0)}$ is any probability vector, the state distributions at n, $\mathbf{p}^{(n)} = \mathbf{p}^{(0)}\mathbf{P}^n$, approach the vector $\mathbf{b}$ as n becomes large.

The logical progression from a regular Markov chain to the limiting state distribution is clear; it only remains to prove the three theorems as stated. Because of the complexity of the three proofs, we will give the proofs in the reverse order from that stated. Thus we first prove Theorem 2.7.

PROOF OF THEOREM 2.7   Suppose that the matrix $\mathbf{P}$ satisfies the hypotheses; that is, the matrices $\mathbf{P}^n$ approach a matrix $\mathbf{B}$ as n becomes large and all the rows of $\mathbf{B}$ are equal to a probability vector $\mathbf{b}$. We must show that for any initial probability vector $\mathbf{p}^{(0)}$, the state distributions at n, $\mathbf{p}^{(n)}$, approach the vector $\mathbf{b}$.

We assume that n is large enough so that all the rows of $\mathbf{P}^n$ are nearly equal to the vector $\mathbf{b}$. Then all the elements in each column are nearly equal; in the ith column, all the elements are nearly equal to b_i, the ith component of **b**.

Now the ith component of $\mathbf{p}^{(n)}$ is given by the dot product of $\mathbf{p}^{(0)}$ with the ith column of $\mathbf{P}^n$. This dot product is

$$p_1^{(0)}p_{1,i}^{(n)} + p_2^{(0)}p_{2,i}^{(n)} + \cdots + p_k^{(0)}p_{k,i}^{(n)}$$

and since each element of the ith column is nearly equal to b_i, this sum is approximately

$$p_1^{(0)}b_i + p_2^{(0)}b_i + \cdots + p_k^{(0)}b_i = b_i(p_1^{(0)} + p_2^{(0)} + \cdots + p_k^{(0)}) = b_i$$

because $p_1^{(0)} + p_2^{(0)} + \cdots + p_k^{(0)} = 1$. Thus $\mathbf{p}^{(n)}$ is approaching the vector **b** as n becomes large.

We next go to the theorem which provides the second step in the proof of Theorem 2.4.

PROOF OF THEOREM 2.6 Our assumptions are that for any column vector $\mathbf{y}^{(0)}$ the products $\mathbf{P}^n\mathbf{y}^{(0)}$ approach a vector **y** and all the components of **y** are equal.

Since $\mathbf{y}^{(0)}$ can be any column vector with nonnegative components, we are allowed to choose $\mathbf{y}^{(0)}$. We will first consider

$$\mathbf{y}^{(0)} = \begin{bmatrix} 1 \\ 0 \\ \vdots \\ 0 \end{bmatrix}$$

That is, $y_1^{(0)} = 1$ and $y_j^{(0)} = 0$ for $j = 2, 3, \ldots, k$. Now the product $\mathbf{P}^n\mathbf{y}^{(0)}$ is a column vector equal to the first column of $\mathbf{P}^n$. As n increases, the vectors $\mathbf{y}^{(n)}$ approach a vector **y** with all components equal according to the hypothesis of Theorem 2.6, thus the components in the first column of $\mathbf{P}^n$ are all approaching the same number.

Now for the ith column of $\mathbf{P}^n$ we consider the vector

$$\mathbf{y}^{(0)} = \begin{bmatrix} 0 \\ \vdots \\ 0 \\ 1 \\ 0 \\ \vdots \\ 0 \end{bmatrix}$$

for which $y_i^{(0)} = 1$ and $y_j^{(0)} = 0$ for $j \neq i$. This choice of $\mathbf{y}^{(0)}$ will result in a different vector **y**, but the argument is the same. The product $\mathbf{P}^n\mathbf{y}^{(0)}$ is the ith column of $\mathbf{P}^n$.

The column vectors $\mathbf{y}^{(n)}$ approach a (different) vector **y**, which has all components equal, and so the components of the ith column of $\mathbf{P}^n$ are all approaching the same number as n becomes large.

This completes the proof that the matrices $\mathbf{P}^n$ are approaching a matrix $\mathbf{B}$ all of whose rows are equal (the elements of each column of $\mathbf{B}$ are equal).

Example 2.30   We will now give an example of a transition matrix for which Theorem 2.6 is shown to hold and for which Theorem 2.5 is also illustrated. Consider the matrix

$$\mathbf{P} = \begin{bmatrix} 0.1 & 0.9 \\ 0.7 & 0.3 \end{bmatrix}$$

and the vector

$$\mathbf{y}^{(0)} = \begin{bmatrix} a \\ 1-a \end{bmatrix}$$

After much computation we find the several matrices $\mathbf{P}^n$ given below and then the vectors $\mathbf{y}^{(n)}$.

$$\mathbf{y}^{(1)} = \mathbf{P}\mathbf{y}^{(0)} = \begin{bmatrix} 0.1 & 0.9 \\ 0.7 & 0.3 \end{bmatrix} \begin{bmatrix} a \\ 1-a \end{bmatrix} = \begin{bmatrix} 0.9 - 0.8a \\ 0.3 + 0.4a \end{bmatrix}$$

$$\mathbf{y}^{(8)} = \mathbf{P}^8\mathbf{y}^{(0)} = \begin{bmatrix} 0.4463 & 0.5537 \\ 0.4294 & 0.5706 \end{bmatrix} \begin{bmatrix} a \\ 1-a \end{bmatrix} = \begin{bmatrix} 0.5537 - 0.1074a \\ 0.5706 - 0.1412a \end{bmatrix}$$

$$\mathbf{y}^{(16)} = \mathbf{P}^{16}\mathbf{y}^{(0)} = \begin{bmatrix} 0.437 & 0.563 \\ 0.437 & 0.563 \end{bmatrix} \begin{bmatrix} a \\ 1-a \end{bmatrix} = \begin{bmatrix} 0.563 - 0.126a \\ 0.563 - 0.126a \end{bmatrix}$$

Thus two points are to be made. The vectors $\mathbf{y}^{(1)}$, $\mathbf{y}^{(8)}$, $\mathbf{y}^{(16)}$ approach a vector $\mathbf{y}$ which has equal components, and the matrices $\mathbf{P}$, $\mathbf{P}^8$, and $\mathbf{P}^{16}$ approach a matrix with equal rows.

The proof of the last theorem is much more complicated than the previous two proofs. We have not used the condition that the Markov chain is regular to this point, but it is used in an essential way in the proof of Theorem 2.5.

PROOF OF THEOREM 2.5   We must show that as n increases the vectors $\mathbf{y}^{(n)}$ approach a vector $\mathbf{y}$ and that all the components of $\mathbf{y}$ are equal. Our proof accomplishes this goal by first showing that the largest components of $\mathbf{y}^{(0)}$, $\mathbf{y}^{(1)}$, . . . , are never increasing as n increases and that the smallest components of $\mathbf{y}^{(0)}$, $\mathbf{y}^{(1)}$, . . . , are never decreasing as n increases. The conclusion to the proof is provided by showing that the difference between the largest and smallest components approaches 0 as n increases. We conclude that all the components of $\mathbf{y}^{(n)}$ are nearly equal as they are squeezed between the largest and smallest components of $\mathbf{y}^{(n)}$.

We begin by introducing the notation $s^{(n)}$ and $L^{(n)}$, respectively, for the smallest and largest components of $\mathbf{y}^{(n)}$ Note that since $\mathbf{y}^{(n+1)} = \mathbf{P}^{n+1}\mathbf{y}^{(0)}$ $= \mathbf{P}\mathbf{y}^{(n)}$, we have for the jth component of $\mathbf{y}^{(n+1)}$

$$y_j^{(n+1)} = p_{j,1}y_1^{(n)} + p_{j,2}y_2^{(n)} + \cdots + p_{j,k}y_k^{(n)}$$
$$\leq p_{j,1}L^{(n)} + p_{j,2}L^{(n)} + \cdots + p_{j,k}L^{(n)}$$

since each component of $\mathbf{y}^{(n)}$ is no larger than $L^{(n)}$. But the sum $p_{j,1} + p_{j,2} + \cdots + p_{j,k} = 1$, and so the inequality states that $y_j^{(n+1)} \leq L^{(n)}$. Now j can be any integer $1, 2, \ldots, k$ so we conclude that $L^{(n+1)} \leq L^{(n)}$. Similarly it can be shown that $s^{(n)} \leq s^{(n+1)}$.

Next we must show that $L^{(m)} - s^{(m)}$ approaches 0 as m increases. Because $s^{(0)} \leq s^{(m)} \leq s^{(m+n)} \leq \cdots \leq L^{(m+n)} \leq L^{(m)} \leq L$, we need only show that this difference approaches 0 for certain larger and larger values of m and need not consider all values of m.

The regularity of the Markov chain enters at this point. We know that, for some positive integer m, $\mathbf{P}^m$ has only positive elements. Let such an m be given and let a be the smallest element of $\mathbf{P}^m (0 < a \leq 1/k$ since there are k elements in each row of P^m and they must sum to 1).

Two inequalities which tell us how fast the difference $L^{(m)} - s^{(m)}$ must approach 0 are now stated. These inequalities are:

$$s^{(n+m)} \geq (1-a)s^{(n)} + aL^{(n)} = s^{(n)} + a(L^{(n)} - S^{(n)})$$

$$L^{(n+m)} \leq (1-a)L^{(n)} + as^{(n)} = L^{(n)} - a(L^{(n)} - s^{(n)})$$

(2.1)

Finding the difference $L^{(n+m)} - s^{(n+m)}$ in terms of $L^{(n)} - s^{(n)}$ from these inequalities, we subtract the inequality for $s^{(n+m)}$ from the inequality for $L^{(n+m)}$ to obtain:

$$L^{(n+m)} - s^{(n+m)} \leq (1-a)L^{(n)} + as^{(n)} - (1-a)s^{(n)} - aL^{(n)}$$
$$= (1-2a)(L^{(n)} - s^{(n)})$$

We can now consider other differences such as

$$L^{(n+2m)} - s^{(n+2m)} \leq (1-2a)(L^{(n+m)} - s^{(n+m)}) \leq (1-2a)^2(L^{(n)} - s^{(n)})$$

and in general for $h = 1, 2, \ldots$

$$L^{(n+hm)} - s^{(n+hm)} \leq (1-2a)^h(L^{(n)} - s^{(n)})$$

Because $0 \leq 1 - 2a < 1$, the terms $(1-2a)^h$ approach 0 as h increases, and thus $L^{(n+hm)} - s^{(n+hm)}$ approaches 0 as well, requiring the components of $\mathbf{y}^{(n+hm)}$ to become equal as h increases.

It remains only to establish the inequalities (2.1) for $s^{(n+m)}$ and $L^{(n+m)}$. Since both demonstrations use the same technique, we will give a demonstration for $s^{(n+m)}$ only.

Again consider any component of $\mathbf{y}^{(n+m)}$, say $y_j^{(n+m)}$. Since $\mathbf{y}^{(n+m)} = \mathbf{P}^{(n+m)}\mathbf{y}^{(0)} = \mathbf{P}^m\mathbf{P}^n\mathbf{y}^{(0)} = \mathbf{P}^m\mathbf{y}^{(n)}$, we can write

$$y_j^{(n+m)} = p_{j,1}^{(m)}y_1^{(n)} + p_{j,2}^{(m)}y_2^{(n)} + \cdots + p_{j,k}^{(m)}y_k^{(n)}$$

Each of the numbers $p_{j,1}{}^{(m)}$, $p_{j,2}{}^{(m)}$, ..., $p_{j,k}{}^{(m)}$ is at least as large as a (a was the value of the smallest element of $\mathbf{P}^m$), and each of the numbers $y_1{}^{(n)}$, ..., $y_k{}^{(n)}$ is at least as large as $s^{(n)}$, while one of them must equal $L^{(n)}$. Two crucial facts can be stated from this:

1. The component equal to $L^{(n)}$, say $y_i{}^{(n)}$, is multiplied by $p_{j,i}{}^{(m)} \geq a$, so that

$$p_{j,i}{}^{(m)} y_i{}^{(n)} = aL^{(n)} + (p_{j,i}{}^{(m)} - a)L^{(n)}$$

2. For all other terms in the sum above we have

$$p_{j,h}{}^{(m)} y_h{}^{(n)} \geq p_{j,h}{}^{(m)} s^{(n)}$$

Putting items 1 and 2 together, we have
$$y_j{}^{(n+m)} \geq p_{j,1}{}^{(m)} s^{(n)} + \cdots + p_{j,i-1}{}^{(m)} s^{(n)} + aL^{(n)} + (p_{j,i}{}^{(m)} - a)L^{(n)}$$
$$+ \cdots + p_{j,k}{}^{(m)} s^{(n)}$$

Next we will replace the term $(p_{j,i}{}^{(m)} - a)L^{(n)}$ by $(p_{j,i}{}^{(m)} - a)s^{(n)}$, and since this does not increase the right side of the last inequality above, we can write

$$y_j{}^{(n+m)} \geq p_{j,1}{}^{(m)} s^{(n)} + \cdots + p_{j,i-1}{}^{(m)} s^{(n)} + aL^{(n)} + (p_{j,i}{}^{(m)} - a)s^{(n)}$$
$$+ \cdots + p_{j,k}{}^{(m)} s^{(n)}$$

Next we combine all the coefficients of $s^{(n)}$ into one, producing the expression

$$p_{j,1}{}^{(m)} + \cdots + p_{j,i-1}{}^{(m)} + p_{j,i}{}^{(m)} - a + \cdots + p_{j,k}{}^{(m)}$$

which equals $1 - a$, since the sum of the elements of the jth row of $\mathbf{P}^m$ equals 1. The last inequality for $y_j{}^{(n+m)}$ can now be expressed simply

$$y_j{}^{(n+m)} \geq (1 - a)s^{(n)} + aL^{(n)}$$

The inequality for $y_j{}^{(n+m)}$ holds for every component of $\mathbf{y}^{(n+m)}$, and so it must hold for the smallest one as well; thus

$$s^{(n+m)} \geq (1 - a)s^{(n)} + aL^{(n)} = s^{(n)} + a(L^{(n)} - s^{(n)})$$

as desired.

This concludes the proof of Theorem 2.5.

Example 2.31   The matrix

$$\mathbf{P} = \begin{bmatrix} 1/2 & 1/3 & 1/6 \\ 1/3 & 1/6 & 1/2 \\ 1/6 & 1/2 & 1/3 \end{bmatrix}$$

has all positive elements, the smallest element $a = 1/6$. The vector

$$\mathbf{y}^{(0)} = \begin{bmatrix} 0 \\ 1 \\ 2 \end{bmatrix}$$

has smallest component 0 and largest component 2. The difference between the largest and smallest components is 2.

The inequalities (2.1) given in the proof of Theorem 2.5 assert that

$$s^{(1)} \geq s^{(0)} + a(L^{(0)} - s^{(0)}) = 0 + \tfrac{1}{6}(2) = \tfrac{1}{3}$$
$$L^{(1)} \leq L^{(0)} - a(L^{(0)} - s^{(0)}) = 2 - \tfrac{1}{6}(2) = \tfrac{5}{3}$$

and

$$L^{(1)} - s^{(1)} \leq (1 - 2a)(L^{(0)} - s^{(0)}) = \tfrac{2}{3} \cdot 2 = \tfrac{4}{3}$$

The vector $\mathbf{y}^{(1)} = \mathbf{P}\mathbf{y}^{(0)}$ is

$$\mathbf{y}^{(1)} = \begin{bmatrix} \tfrac{2}{3} \\ \tfrac{7}{6} \\ \tfrac{7}{6} \end{bmatrix}$$

and so $s^{(1)} = \tfrac{2}{3}$, $L^{(1)} = \tfrac{7}{6}$, $L^{(1)} - s^{(1)} = \tfrac{7}{6} - \tfrac{2}{3} = \tfrac{1}{2}$. Thus the smallest component is larger than $\tfrac{1}{3}$, the largest component is smaller than $\tfrac{5}{3}$, and their difference is smaller than $\tfrac{4}{3}$.

The inequalities are verified for this example.

Exercise 2.10   1.   Note the truth of Theorem 2.5 and 2.6 for the case when

$$\mathbf{P} = \begin{bmatrix} \tfrac{1}{4} & \tfrac{1}{8} & \tfrac{1}{8} & \tfrac{1}{3} & \tfrac{1}{6} \\ \tfrac{1}{4} & \tfrac{1}{8} & \tfrac{1}{8} & \tfrac{1}{3} & \tfrac{1}{6} \\ \tfrac{1}{4} & \tfrac{1}{8} & \tfrac{1}{8} & \tfrac{1}{3} & \tfrac{1}{6} \\ \tfrac{1}{4} & \tfrac{1}{8} & \tfrac{1}{8} & \tfrac{1}{3} & \tfrac{1}{6} \\ \tfrac{1}{4} & \tfrac{1}{8} & \tfrac{1}{8} & \tfrac{1}{3} & \tfrac{1}{6} \end{bmatrix} \quad \text{and} \quad \mathbf{y}^{(0)} = [6, 0, 2, 4, 1]$$

(What is special about the matrix $\mathbf{P}$?)

2.   For the transition matrix of Exercise 2.10.1 find the product $\mathbf{p}^{(0)}\mathbf{P}$ for

a. $\mathbf{p}^{(0)} = [1, 0, 0, 0, 0]$
b. $\mathbf{p}^{(0)} = [\tfrac{1}{2}, \tfrac{1}{4}, \tfrac{1}{8}, \tfrac{1}{16}, \tfrac{1}{16}]$

3.   Let

$$\mathbf{P} = \begin{bmatrix} \tfrac{1}{2} & \tfrac{1}{3} & \tfrac{1}{6} \\ \tfrac{1}{4} & 0 & \tfrac{3}{4} \\ \tfrac{1}{8} & \tfrac{1}{2} & \tfrac{3}{8} \end{bmatrix}$$

and $\mathbf{y} = [1, 3, 7]$.

a.   Find $s^{(1)}$ and $L^{(1)}$ as defined in the proof of Theorem 2.5.
b.   Find an m and the value of a which are used in the proof of Theorem 2.5.
c.   Find $s^{(1+m)}$ and $L^{(1+m)}$ (see item b).
d.   Compare your answers to items a and b with the inequalities (2.1) stated in the proof of Theorem 2.5.

4.   For the matrix $\mathbf{P}$ and vector $\mathbf{y}^{(0)}$ of Example 2.31 carry out the steps

used to obtain the inequality (2.1) for $s^{(1)}$ as given in the proof of Theorem 2.5.

Explain why the value for $s^{(1)}$ as found in Example 2.31 is actually larger than the value guaranteed by the inequality.

2.11 ABSORBING MARKOV CHAINS: 1

The theory presented in Secs. 2.8 and 2.10 is not sufficient to take care of many interesting and useful Markov chain models. For example, *absorbing* Markov chains have many applications both practical and theoretical, and as we shall explain later, such Markov chains are not regular.

An *absorbing Markov chain* has one or more states, called *absorbing states*, for which transitions from such a state to any other state cannot occur. If state i is an absorbing state, then $p_{i,i} = 1$, and consequently, $p_{i,j} = 0$ for all $j \neq i$. When state i occurs, so that $X_n = i$ for some n, we say the Markov chain has been *absorbed* at state i, for then $X_{n+1} = i$, $X_{n+2} = i$, ... the state can never be different from i at any later time.

Example 2.32   In the game of matching pennies (see Exercise 2.5.5) in which one player has two pennies and a second player has one penny, a penny passes from the loser to the winner on each play of the game. The winner is determined by prearrangement according to whether or not two pennies—one for each player—match when flipped (matching means both coins show heads or both show tails). The game ends when one player loses all his pennies.

If X_0 denotes the number of pennies owned by an individual at the beginning of the game, and X_n the number owned after the nth play of the game, then this Markov chain has states 0, 1, 2, 3. The Markov chain model does not stop when the game ends but becomes absorbed at 0 if the individual loses all his pennies and absorbed at 3 if he wins all the pennies.

The transition matrix for this Markov chain (assuming the pennies all have a 50:50 chance of coming up heads when flipped) is

$$\mathbf{P} = \begin{bmatrix} 1 & 0 & 0 & 0 \\ 1/2 & 0 & 1/2 & 0 \\ 0 & 1/2 & 0 & 1/2 \\ 0 & 0 & 0 & 1 \end{bmatrix}$$

The states 0 and 3 are absorbing states as indicated by the transition probabilities $p_{0,0} = 1$ and $p_{3,3} = 1$.

The state distributions for an absorbing Markov chain can be computed using the general methods of vector and matrix products as given in Secs. 2.5 and 2.6. The results for finding limiting probabilities in Secs. 2.8 and 2.10 do not apply, because no matter how large the value n may be, if i is an absorbing state, $p_{i,j}^{(n)} = 0$ for $j \neq i$ because transitions from state i to other states never occur; thus *an absorbing Markov chain is never regular*. It is perfectly natural to ask, however, for the probability of ever being absorbed at state i—in Example 2.32 we want to know the probability of losing all the pennies—and this is a question about limiting probabilities. The limiting state distributions do exist in general, and for any absorbing state i, the transition probabilities $p_{j,i}^{(n)}$ do approach a value, say $b_{j,i}$, as n increases. This is a fact which we will not prove here.

We will now show that the probabilities $b_{j,i}$ must satisfy a certain system of equations. For each state j which is not absorbing, and for each absorbing state i, $b_{j,i}$ is the probability of ever being absorbed at state i starting from the state j. More explicitly, $p_{j,i}^{(n)} = P(X_n = i | X_0 = j)$, and as n increases, we write

$$b_{j,i} = P(X_n = i \text{ for some } n | X_0 = j)$$

for the limiting probability.

We now consider the various possible values for X_1 and write another equation for $b_{j,i}$

$$b_{j,i} = p_{j,i} + \sum_{\substack{k, \\ k \text{ not absorbing}}} P(X_n = i \text{ for some } n, X_1 = k | X_0 = j)$$

notice that the sum is over only states which are not absorbing. In words, this equation states that the probability of ever being absorbed at state i starting at state j equals the probability $p_{j,i}$ of being absorbed at time 1, $X_1 = i$, plus the probabilities for making a transition to some state k which is not absorbing and (later) being absorbed at state i.

Each term of the sum can be simplified by writing

$$\begin{aligned} P(X_n = i \text{ for some } n, X_1 = k | X_0 = j) \\ = P(X_1 = k | X_0 = j)P(X_n = i \text{ for some } n | X_1 = k, X_0 = j) \\ = p_{j,k}P(X_n = i \text{ for some } n | X_1 = k, X_0 = j) \end{aligned}$$

We can drop the condition $X_0 = j$ in the last line above because of the Markov property. The probability which results is

$$p_{j,k}P(X_n = i \text{ for some } n | X_1 = k)$$

Now k is not absorbing, and to ask for the probability of being absorbed at i given the current state is k is the same as

$$P(X_n = i \text{ for some } n | X_0 = k) = b_{k,i}$$

We combine these facts from above to write

$$b_{j,i} = p_{j,i} + \sum_{\substack{k, \\ k \text{ not absorbing}}} p_{j,k}b_{k,i}$$

Such an equation is determined for each pair of states j and i, where j is not absorbing and i is absorbing.

EXAMPLE 2.32 Continued

The absorbing states are 0 and 3, while states 1 and 2 are not absorbing. The equations for the probability of ever being absorbed at state 0 are

$$b_{1,0} = \frac{1}{2} + \frac{1}{2}b_{2,0}$$
$$b_{2,0} = \frac{1}{2}b_{1,0}$$

so that

$$b_{1,0} = \frac{1}{2} + \frac{1}{2}(\frac{1}{2}b_{1,0}) \quad \text{or} \quad b_{1,0} = \frac{2}{3} \quad \text{and} \quad b_{2,0} = \frac{1}{3}$$

The similar equations for ever being absorbed at state 3 are:

$$b_{1,3} = \frac{1}{2}b_{2,3}$$
$$b_{2,3} = \frac{1}{2} + \frac{1}{2}b_{1,3}$$

and thus

$$b_{2,3} = \frac{1}{2} + \frac{1}{4}b_{2,3} \quad \text{so} \quad b_{2,3} = \frac{2}{3} \quad \text{while} \quad b_{1,3} = \frac{1}{3}$$

Notice that $b_{1,0} + b_{1,3} = 1$ and that $b_{2,0} + b_{2,3} = 1$; that is, the Markov chain is sure to be absorbed, or the game is sure to end.

In Example 2.32, absorption is sure to occur. This is a fact which is generally true but one which we will not prove here. Specifically, if from every state which is not absorbing there is a positive probability of being absorbed eventually, then the Markov chain is sure to be absorbed eventually, or in other terms

$$\sum_{\substack{i,\\ i \text{ absorbing}}} b_{j,i} = 1 \quad \text{for all states } j \text{ which are not absorbing}$$

Summary

The probabilities of eventually being absorbed at state i starting at state j, $b_{j,i}$, for all states j which are not absorbing satisfy the system of equations

$$b_{j,i} = p_{j,i} + \sum_{\substack{k,\\ k \text{ not absorbing}}} p_{j,k}b_{k,i}$$

Exercise 2.11

1.   In the game of matching pennies, if there are four pennies between the two players the states are 0, 1, 2, 3, 4, and the transition matrix for the model is

$$\mathbf{P} = \begin{bmatrix} 1 & 0 & 0 & 0 & 0 \\ \frac{1}{2} & 0 & \frac{1}{2} & 0 & 0 \\ 0 & \frac{1}{2} & 0 & \frac{1}{2} & 0 \\ 0 & 0 & \frac{1}{2} & 0 & \frac{1}{2} \\ 0 & 0 & 0 & 0 & 1 \end{bmatrix}$$

Find the probabilities $b_{1,0}$, $b_{2,0}$, and $b_{3,0}$ for this Markov chain.

2.   In a certain corporation the management trainees are evaluated at regular times each year. At each evaluation a trainee at the lowest level (state 1) has probability $\frac{1}{3}$ of remaining at that level, probability $\frac{1}{2}$ of being placed as a second-level trainee (state 2), and probability $\frac{1}{6}$ that he will be transferred to a permanent assignment in sales (state 3). A trainee at the second level has a probability of $\frac{1}{2}$ of remaining at the second level, a probability of $\frac{1}{10}$ of being given a permanent assignment in sales, and a probability of $\frac{4}{10}$ of being given a permanent assignment in management (state 4).

a.   Find the transition matrix and the absorbing states for this Markov chain.
b.   What is the probability that a lowest-level trainee is eventually assigned to sales? To management?
c.   What is the probability that a second-level trainee is eventually assigned to sales? To management?

3.   In any Markov chain it is possible to introduce absorbing states for the purpose of finding certain probabilities of interest. In Application 2.1, it might be of interest to know the probability that in a line of descendants of a middle-class (state 3) individual the highest social class (state 5) is attained with no intervening generation being in the lowest social class (state 1). This can be computed if we consider a Markov chain with transitions between states exactly as in the original Markov chain but for states 1 and 5 which are made absorbing. Then this absorbing Markov chain will be absorbed at the first occurrence of states 1 or 5. $b_{3,1}$ will be the probability that the lowest state occurs before the highest, and $b_{3,5}$ gives the desired probability. Find the probability $b_{3,5}$ for the model of social mobility. The original transition matrix is

$$\mathbf{P} = \begin{bmatrix} 0.46 & 0.38 & 0.13 & 0.02 & 0.01 \\ 0.25 & 0.45 & 0.23 & 0.06 & 0.01 \\ 0.13 & 0.31 & 0.41 & 0.12 & 0.03 \\ 0.07 & 0.18 & 0.34 & 0.33 & 0.08 \\ 0.03 & 0.07 & 0.30 & 0.30 & 0.30 \end{bmatrix}$$

2.12   ABSORBING MARKOV CHAINS: 2

In Sec. 2.11 we considered questions about limiting probabilities for absorbing Markov chains. Because absorbing Markov chains are used as models for such things as the game of matching pennies, as well as for more practical situations, it is often desirable to know how many transitions occur before the Markov chain is absorbed. The number of transitions is, of course, a random variable so what we really mean is that the expected number is of interest. Actually we will see that we can find more detailed information about the transitions which occur prior to the absorption of the Markov chain. For example we can find the expected value of the number of times that each nonabsorbing state occurs (a state j is nonabsorbing if $p_{j,j} < 1$).

Let N_j be the random variable equal to the number of times that the nonabsorbing state j occurs. The probability distribution for N_j will depend on the state distribution at 0, for if $X_0 = j$, then N_j must be at least 1 or more, while if $X_0 = i \neq j$, there is no guarantee that the state j will ever occur—absorption may occur before j occurs. We will consider the various expected values of N_j when $P(X_0 = i) = 1$, and i is equal to various nonabsorbing states.

When the initial state is i, $P(X_0 = i) = 1$, we will write $n_{i,j}$ for the expected value of N_j. We wish to show that the numbers $n_{i,j}$ satisfy a system of equations similar to the equations we found in Sec. 2.11.

We are going to write

$$N_j = Z_0 + Z_j$$

where

$$Z_0 = \begin{cases} 1 & \text{if } X_0 = j \\ 0 & \text{if } X_0 \neq j \end{cases}$$

and Z_j equals the number of occurrences of j after time 0. To find $n_{i,j}$ we are going to determine the expected value of Z_0 and Z_j separately, then add them to find the expected value of N_j.

The expected value of Z_0 when $P(X_0 = i) = 1$ is very easy to determine, since if $i = j$ it is 1 and if $i \neq j$ it is 0. The expected value of Z_j is more complicated. We want to convince you that the expected value of Z_j equals

$$\sum_{\substack{k, \\ k \text{ nonabsorbing}}} p_{i,k} n_{k,j}$$

First we claim that when $X_1 = k$ the expected value of Z_j, the number of occurrences of j, equals $n_{k,j}$. We reason that from state k at time 1 the future occurrences of state j are just the same as if we were starting at state k at time 0. Remember that the transition probabilities do not change with time so the only determining factor is the starting state when you begin counting occurrences of state j.

A more difficult result to "prove" but one which should be plausible is that if (say) $p_{i,k} = \frac{1}{2}$, then with probability $\frac{1}{2}$ the number of occurrences of state j will, on the average, be $n_{k,j}$. Thus when $X_0 = i$, $\frac{1}{2}n_{k,j}$ will be the part of the expected value of Z_j which results when we also have $X_1 = k$. More generally, $p_{i,k} n_{k,j}$ is the portion of the expected value of Z_j resulting from the event $X_1 = k$.

It is from this line of reasoning that a real proof can be given. The details of that proof contribute nothing to our understanding of why the result is true, and we shall not show the details.

Applying Theorem 1.10 to the relation $N_j = Z_0 + Z_j$, and using the expected values given above for Z_0 and Z_j, we write

$$n_{i,j} = \begin{cases} 1 & \text{if } i = j \\ 0 & \text{if } i \neq j \end{cases} + \sum_{\substack{k, \\ k \text{ nonabsorbing}}} p_{i,k} n_{k,j}$$

Example 2.33   Consider the game of matching pennies once more. The transition matrix was

$$\mathbf{P} = \begin{bmatrix} 1 & 0 & 0 & 0 \\ \frac{1}{2} & 0 & \frac{1}{2} & 0 \\ 0 & \frac{1}{2} & 0 & \frac{1}{2} \\ 0 & 0 & 0 & 1 \end{bmatrix}$$

Suppose you begin this game with one penny; what is the expected number of times during the game that you will have one penny? Two pennies? The first is $n_{1,1}$ (including the fact that you begin with one penny so $N_1 \geq 1$ for sure), and the second is $n_{1,2}$. To find these expected values we solve the equations below.

$$n_{1,1} = 1 + p_{1,1} n_{1,1} + p_{1,2} n_{2,1} = 1 + \frac{1}{2} n_{2,1}$$
$$n_{1,2} = p_{1,1} n_{1,2} + p_{1,2} n_{2,2} = \frac{1}{2} n_{2,2}$$
$$n_{2,1} = p_{2,1} n_{1,1} + p_{2,2} n_{2,1} = \frac{1}{2} n_{1,1}$$
$$n_{2,2} = 1 + p_{2,1} n_{1,2} + p_{2,2} n_{2,2} = 1 + \frac{1}{2} n_{1,2}$$

The first and third of these equations assert that

$$n_{1,1} = 1 + \frac{1}{2} n_{2,1} = 1 + \frac{1}{2}(\frac{1}{2} n_{1,1}) = 1 + \frac{1}{4} n_{1,1}$$

or

$$n_{1,1} = \frac{4}{3}$$

We use the second and fourth equations to find that

$$n_{1,2} = \frac{2}{3}$$

If we start the game with one penny, we will have one penny $\frac{4}{3}$ times on the average and two pennies $\frac{2}{3}$ times on the average during the game. This must mean that the average game lasts only two plays, i.e., $\frac{4}{3} + \frac{2}{3} = 2$.

The expected number of transitions until the Markov chain is absorbed must be the expected value of the sum of the number of times each nonabsorbing state occurs. The $\sum_j N_j$ where j is not an absorbing state is the number of transitions until the Markov chain is absorbed. Theorem 1.10 again is used to give the expected value of this sum as $\sum_j n_{i,j}$ when the initial state is i.

The expected number of times the state j occurs in an absorbing Markov chain when the initial state is i, $n_{i,j}$, determines a set of numbers (each pair of nonabsorbing states i and j define such a number) which satisfy the equations

$$n_{i,j} = \sum_k p_{i,k} n_{k,j} + \begin{cases} 1 & \text{if } i = j \\ 0 & \text{if } i \neq j \end{cases}$$

where the sum is over all nonabsorbing states.

The expected number of transitions until the Markov chain is absorbed if the initial state is i is the sum

$$\sum_k n_{i,k}$$

Exercise 2.12 1. For the absorbing Markov chain in Exercise 2.11.1 find $n_{1,1}$, $n_{1,2}$, and $n_{1,3}$.

2. For the absorbing Markov chain in Exercise 2.11.2 find the expected number of times a new trainee (lowest-level) will be evaluated before he is given a permanent assignment.

3. For the absorbing Markov chain in Exercise 2.11.3 consider a line of descendants of a middle-class (state 3) individual. Find the expected number of individuals in this line of descendants who will also be middle class until for the first time a member of the line of descendants is either class 1 or class 5.

4. Using the results of Exercise 2.12.1 for comparison perform an experiment to verify those values. Using your roommate as an opponent play the game of matching pennies (using a total of four pennies between you 25 times). It is suggested that since the player who starts with one penny wins only $\frac{1}{4}$ of the time (see Exercise 2.11.1) you should start with three pennies and have your roommate start with one penny. That way you can beat your roommate more often.

2.13 APPLICATIONS

Both applications of absorbing Markov chains presented here use only the results from Sec. 2.12 for the expected number of occurrences of nonabsorbing states. The first application is a model for "avoidance conditioning" in rats, an experiment that was analyzed in Sec. 1.12. The second application concerns quality control for a production process.

Application 2.5   A Learning Model Returns

Application 1.4 concerned a model for "avoidance conditioning" in rats. We will see that an absorbing Markov chain provides a model for the experiment described there and that we are now capable of a deeper analysis of that model.

As a review of the basic facts from that model, we mention here that there were three states of learning for the rat; there was an initial state 0 in which the rat was completely unaware, an intermediate state 1 reached from state 0 by learning to fear the bell, and a final state 2 in which the rat was able to always avoid shock. This is a Markov chain model in which state 2 is absorbing. The transition probabilities derived in Sec. 1.12 are given in the matrix

$$\mathbf{P} = \begin{bmatrix} 1 - \alpha & \alpha & 0 \\ 0 & 1 - \alpha p & \alpha p \\ 0 & 0 & 1 \end{bmatrix}$$

Recall that α is the probability of learning if a shock is received and that p is the probability of receiving shock on a given trial when in state 1.

Since there is only one absorbing state, state 2, both of the probabilities $b_{0,2}$ and $b_{1,2}$ are equal to 1. We will thus focus on finding the expected number of occurrences of states 0 and 1 using the equations derived in Sec. 2.12 for that purpose.

These equations are

$$n_{0,0} = p_{0,0}n_{0,0} + p_{0,1}n_{1,0} + 1$$
$$n_{0,1} = p_{0,0}n_{0,1} + p_{0,1}n_{1,1}$$
$$n_{1,0} = p_{1,0}n_{0,0} + p_{1,1}n_{1,0}$$
$$n_{1,1} = p_{1,0}n_{0,1} + p_{1,1}n_{1,1} + 1$$

which are simplified when the actual values from $\mathbf{P}$ are used. We then have

$$n_{0,0} = (1 - \alpha)n_{0,0} + \alpha n_{1,0} + 1$$
$$n_{0,1} = (1 - \alpha)n_{0,1} + \alpha n_{1,1}$$
$$n_{1,0} = (1 - \alpha p)n_{1,0}$$
$$n_{1,1} = (1 - \alpha p)n_{1,1} + 1$$

The third equation says $n_{1,0} = 0$ since $1 - \alpha p < 1$ (because $0 < p < 1$ and $0 < \alpha \leq 1$). Then the first equation gives us $n_{0,0} = 1/\alpha$. The fourth equation states that $n_{1,1} = 1/(\alpha p)$, and the second equation gives the same value for $n_{0,1}$. Thus we have the expected values

$$n_{0,0} = \frac{1}{\alpha} \qquad n_{0,1} = \frac{1}{\alpha p}$$

Beginning in state 0, the expected number of trials required before the rat learns to avoid shock is then

$$n_{0,0} + n_{0,1} = \frac{1}{\alpha} + \frac{1}{\alpha p} = \frac{1 + p}{\alpha p}$$

We can also determine the expected number of shocks administered to the rat during the course of the experiment. To do this we use a different Markov chain model. The new model uses four states rather than three and differentiates the outcomes while in state 1 as to whether a shock is or is not received by the rat. We might write the state of this model as 0, $1s$, $1a$, 2. The state $1s$ is the outcome when a shock is received and the rat is in learning state 1, while $1a$ indicates the outcome when a shock is avoided and the rat is in learning state 1.

With this state space we are able to record more information at each trial. We can readily verify that

$$P(X_n = 1a | X_{n-1} = 1a) = P(X_n = 1a | X_{n-1} = 1s) = 1 - p$$
$$P(X_n = 2 | X_{n-1} = 1a) = P(X_n = 2 | X_{n-1} = 1s) = \alpha p$$

What about $P(X_n = 1a | X_{n-1} = 0)$ and $P(X_n = 1s | X_{n-1} = 0)$? Since learning has taken place, we conclude a shock has been received, thus

$$P(X_n = 1a | X_{n-1} = 0) = 0$$
$$P(X_n = 1s | X_{n-1} = 0) = \alpha$$

The transition matrix for this model is therefore

$$\mathbf{P} = \begin{bmatrix} 1 - \alpha & 0 & \alpha & 0 \\ 0 & 1 - p & (1 - \alpha)p & \alpha p \\ 0 & 1 - p & (1 - \alpha)p & \alpha p \\ 0 & 0 & 0 & 1 \end{bmatrix}$$

and the relevant equations are now

$$n_{0,0} = (1 - \alpha)n_{0,0} + \alpha n_{1s,0} + 1$$
$$n_{0,1a} = (1 - \alpha)n_{0,1a} + \alpha n_{1s,1a}$$
$$n_{0,1s} = (1 - \alpha)n_{0,1s} + \alpha n_{1s,1s}$$
$$n_{1a,0} = (1 - p)n_{1a,0} + (1 - \alpha)p n_{1s,0}$$
$$n_{1a,1a} = (1 - p)n_{1a,1a} + (1 - \alpha)p n_{1s,1a} + 1$$
$$n_{1a,1s} = (1 - p)n_{1a,1s} + (1 - \alpha)p n_{1s,1s}$$
$$n_{1s,0} = (1 - p)n_{1a,0} + (1 - \alpha)p n_{1s,0}$$
$$n_{1s,1a} = (1 - p)n_{1a,1a} + (1 - \alpha)p n_{1s,1a}$$
$$n_{1s,1s} = (1 - p)n_{1a,1s} + (1 - \alpha)p n_{1s,1s} + 1$$

From the fourth and seventh equations we conclude

$$n_{1a,0} = (1 - \alpha)n_{1s,0} \qquad \text{which leads to} \qquad n_{1s,0} = (1 - \alpha)n_{1s,0}$$

or

$$n_{1a,0} = n_{1s,0} = 0 \qquad \text{if } 0 < \alpha$$

Then the first equation produces as before $n_{0,0} = 1/\alpha$, while the second and third equations become $n_{0,1a} = n_{1s,1a}$ and $n_{0,1s} = n_{1s,1s}$, respectively. Finally from the fifth, sixth, eighth, and ninth equations we determine

$$n_{1s,1a} = \frac{1-p}{\alpha p} \quad \text{and} \quad n_{1s,1s} = \frac{1}{\alpha}$$

Thus

$$n_{0,1a} = \frac{1-p}{\alpha p} \quad \text{and} \quad n_{0,1s} = \frac{1}{\alpha}$$

so that the expected number of shocks received by the rat is

$$n_{0,0} + n_{0,1s} = \frac{1}{\alpha} + \frac{1}{\alpha} = \frac{2}{\alpha}$$

The expected number of avoidances while in state 1 is

$$n_{0,1a} = \frac{1-p}{\alpha p}$$

The total number of shocks which the rat receives is an observable event during the experiment and provides useful information to make comparisons of the experimental data and the prediction given by the model. The decision as to whether the data and the prediction are in sufficient agreement is a problem which falls within the realm of statistics and will not be considered here.

Exercise 2.13   1.   In the model for avoidance conditioning suppose that p, the probability of receiving a shock in state 1, is $3/4$ and that α, the probability of learning, is $1/10$. Verify the expected values $n_{0,0}$, $n_{0,1a}$, and $n_{0,1s}$ by solving the equations directly.

a. What is the expected number of shocks the rat receives in this case?
b. What is the expected number of trials until the rat learns?

2.   Verify that the probability of the rat eventually learning is 1.
3.   Find the probability that the rat never accidentally avoids shock (is never in state 1a). (*Hint*: By making state 1a absorbing this probability can be computed by using the results of Sec. 2.11.)

Application 2.6   Quality Control

Testing the quality of items being manufactured is a necessary step in any mass-production process. When the outcoming quality does not meet specifications, contractual agreements are not being fulfilled and serious business problems are sure to come. It is thus the purpose of quality testing to indicate when the production process is no longer in "control," meaning that the proportion of defectives appears to be too great.

Thus the manufacturer is willing to ship goods with a certain proportion of defectives, because his contracts allow it, and it is too expensive or even impossible to check the quality of every item produced. When the testing scheme indicates that the proportion of defectives is too high, it becomes necessary to locate the step or steps in the production process which are responsible for defective manufacture and to correct the problem.

Schemes for deciding which items should be tested depend greatly on the type of manufacturing process. Our scheme will be a plausible one for testing items as they come off the production line. The rules for this scheme are given below:

1. Test every item until there have been four successive good ones.
2. After testing four successive good items, test only every tenth one as long as the item last tested is good.
3. Whenever a bad item is detected, begin testing every item until four successive good items are found.
4. After a bad item is found, if two additional bad items are found before four successive good items, then the process is suspected of being out of control and is stopped.

We will develop a model for this testing scheme when the acceptable proportion of bad items is actually being produced. Suppose that proportion is b. Each item that is tested then has probability b of being bad.

Remember that a Markov chain has transitions which only depend on the current state and not on any past states. We will have to use a clever method for labeling states so as to capture rules 1 to 4 in our model and still have the transitions depend only on the current state.

Our model will have the state space $\{s,0',1',2',3',0,1,2,3,4\}$. The possible transitions between these states are indicated in Fig. 2.8 below with a word description of the states following.

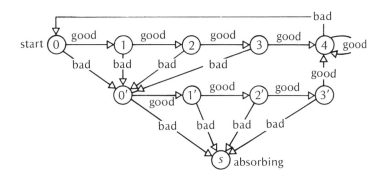

Figure 2.8   The states and possible transitions in the model

To implement rules 1 to 4, we must be able to determine the number of successive good items last tested up to a maximum of four. After four successive good items, the testing scheme does not change until a bad item is found. Upon detecting a bad item while the process is in state 4, there is a transition to state 0 indicating that the last item tested was bad. Next the states 1,2,3,4 occur in that order if four good items are tested, but if a bad item is found before reaching state 4, a transition to state $0'$ occurs. The use of the prime $'$ with the integer allows us to record successive good items and at the same time record the proviso that a bad item was tested before state 4 was reached. The states $1',2',3',4$ occur in that order if four good items follow, but if another bad item is found, rule 4 requires that we stop the process. This action we consider as an absorbing state and have called it s.

This is a complicated description of states; so we will give the transition matrix and label its rows and columns and then discuss the transition probabilities after we have seen the matrix.

$$\mathbf{P} = \begin{array}{c|cccccccccc} & s & 0' & 1' & 2' & 3' & 0 & 1 & 2 & 3 & 4 \\ \hline s & 1 & 0 & 0 & 0 & 0 & 0 & 0 & 0 & 0 & 0 \\ 0' & b & 0 & 1-b & 0 & 0 & 0 & 0 & 0 & 0 & 0 \\ 1' & b & 0 & 0 & 1-b & 0 & 0 & 0 & 0 & 0 & 0 \\ 2' & b & 0 & 0 & 0 & 1-b & 0 & 0 & 0 & 0 & 0 \\ 3' & b & 0 & 0 & 0 & 0 & 0 & 0 & 0 & 0 & 1-b \\ 0 & 0 & b & 0 & 0 & 0 & 0 & 1-b & 0 & 0 & 0 \\ 1 & 0 & b & 0 & 0 & 0 & 0 & 0 & 1-b & 0 & 0 \\ 2 & 0 & b & 0 & 0 & 0 & 0 & 0 & 0 & 1-b & 0 \\ 3 & 0 & b & 0 & 0 & 0 & 0 & 0 & 0 & 0 & 1-b \\ 4 & 0 & 0 & 0 & 0 & 0 & b & 0 & 0 & 0 & 1-b \end{array}$$

Notice that from any state there are at most two possibilities corresponding to testing a good or bad item, and the probabilities in each such case are b and $1 - b$. Only the state s differs since it is absorbing. The state 0 means either we are just starting or while in state 4 a bad item has been found.

Two immediate questions which come up are these:

1. How often does this scheme cause us to stop when in fact the acceptable proportion b of bad items is being produced?
2. What fraction of the items do we test when the acceptable proportion b of bad items is being produced?

The answer to question 1 is given by finding the expected number of transitions before stopping and then computing the average number of items which have been produced during that time. Since one item is produced for each time the process is in a state other than s and 4, we can easily determine the average number of items produced before stopping.

Solving the system of equations for the expected number of occurrences of states $0',1',2',3',0,1,2,3,4$ is not an easy task since 81 equations and 81 unknowns are involved. The specific expected values which we are interested in correspond to the initial state 0 assuming that the scheme begins testing each item until four successive good ones are found at the initiation of testing.

The expressions in terms of b for these expected values are given in Table 2.9 along with the numerical values for several values of b. These numbers also give us a better feeling for the way this scheme works since we see how often the various states occur.

Term	Formula	$b = 0.01$	$b = 0.05$	$b = 0.10$
$n_{0,0'}$	$\dfrac{1}{1 - (1 - b)^4}$	25.38	5.39	2.91
$n_{0,1'}$	$\dfrac{1 - b}{1 - (1 - b)^4}$	25.12	5.12	2.62
$n_{0,2'}$	$\dfrac{1 - b^2}{1 - (1 - b)^4}$	24.87	4.87	2.36
$n_{0,3'}$	$\dfrac{1 - b^3}{1 - (1 - b)^4}$	24.62	4.62	2.12
$n_{0,0}$	$\dfrac{1}{[1 - (1 - b)^4]^2}$	644.05	29.06	8.46
$n_{0,1}$	$\dfrac{1 - b}{[1 - (1 - b)^4]^2}$	637.61	27.61	7.61
$n_{0,2}$	$\dfrac{(1 - b)^2}{[1 - (1 - b)^4]^2}$	631.23	26.23	6.85
$n_{0,3}$	$\dfrac{(1 - b)^3}{[1 - (1 - b)^4]^2}$	624.92	24.92	6.16
$n_{0,4}$	$\dfrac{1}{b[1 - (1 - b)^4]^2} - \dfrac{1}{b}$	64,311.52	561.26	74.55

Table 2.9   The expected number of occurrences of each state before stopping

Since one item is produced corresponding to each occurrence of the states $0',1',2',3',0,1,2,3$, and 10 items are produced for each occurrence of state 4, we need only add the expected values in the first eight rows of Table 2.9 to 10 times the expected value in the last row to find the expected number of items produced before stopping. Thus

$$n_{0,0'} + n_{0,1'} + n_{0,2'} + n_{0,3'} + n_{0,0} + n_{0,1} + n_{0,2} + n_{0,3} + 10n_{0,4}$$

is this expected value.

In the three cases given in Table 2.9

if $b = 0.01$, the expected number produced is 645,752

if $b = 0.05$, the expected number produced is 5,740

if $b = 0.10$, the expected number produced is 785

Finally we consider question 2. How many items are tested as compared to the number of items produced? Each occurrence of a state in the Markov chain corresponds to an item tested. The expected number of items tested is just the sum of the appropriate entries of Table 2.9,

if $b = 0.01$, it is 66,948
if $b = 0.05$, it is 689
if $b = 0.10$, it is 114

Comparison of the expected number of items tested to the expected number of items produced gives an answer to question 2. The fraction of items tested

if $b = 0.01$, is 0.104
if $b = 0.05$, is 0.120
if $b = 0.10$, is 0.145

Thus we see that when the actual production is 1 percent defectives, $b = 0.01$, this scheme will stop after an average of 66,948 tests during which time an average of 645,752 items have been produced. If however 5 percent defectives, $b = 0.05$, are being produced, then only 698 tests are made on the average and an average of 5740 items are produced before the scheme will stop the manufacturing process.

Other questions can be asked also. For example, how many bad items do we find by our testing scheme? See Exercises 2.13.4 to 2.13.6 for an interesting question related to the bad items found in the scheme.

Exercise 2.13   *All exercises should be answered for $b = 0.01$, $b = 0.05$, and $b = 0.10$.*

4.   From Table 2.9 determine the expected number of bad items detected before stopping.
5.   If the bad items detected by the testing are removed what is the outgoing proportion of bad items?
6.   Suppose the testing process destroys the item tested. What will be the outgoing proportion of bad items?
7.   Suppose the testing process destroys the item tested. What will be the average number of items available for sale before stopping?

2.14   FOOD FOR THOUGHT—PERIODIC MARKOV CHAINS

We have seen that absorbing Markov chains are not regular; yet there are interesting questions about limiting probabilities which we can answer. In this section we will present periodic Markov chains, another type of Markov chain which is not regular. Questions about limiting probabilities can be answered for these Markov chains also.

Example 2.34   Consider a rat placed in the maze shown in Fig. 2.9. Notice that this maze is really very simple. A rat placed in the center compartment, labeled 1, is able to exit to either compartment 4 or compartment 5. Compartment 3 is entered only from compartment 5, and compartment 2 is entered only from compartment 4. The maze is really like two deadend corridors joined at compartment 1.

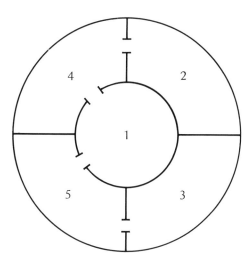

Figure 2.9   A maze for a rat

Suppose that a model for the movement of the rat between compartments of this maze is based on an assumption that the rat is equally likely to use each of the possible exits when leaving any compartment of the maze. The conditional probabilities for movement from one compartment to another are transition probabilities. The states are the labels (numbers) used to denote each compartment. The model is a Markov chain, and the transition matrix is

$$\mathbf{P} = \begin{bmatrix} 0 & 0 & 0 & 1/2 & 1/2 \\ 0 & 0 & 0 & 1 & 0 \\ 0 & 0 & 0 & 0 & 1 \\ 1/2 & 1/2 & 0 & 0 & 0 \\ 1/2 & 0 & 1/2 & 0 & 0 \end{bmatrix}$$

Now you can see that from compartments 1, 2, and 3 the next compartment must be either 4 or 5, while from compartments 4 and 5 the next compartment must be either 1, 2, or 3. If the rat initially is in state 1, $X_0 = 1$, then only on even-numbered transitions will the state be either 1, 2, or 3 and only on odd-numbered transitions will the state be 4 or 5. The Markov chain considered here is *not regular*. It is never possible to make a transition from state 1 to any of the other states in exactly n steps.

The matrix $\mathbf{P}^2$ reveals the fact that transitions between states 1, 2, and 3 require two steps as do transitions between states 4 and 5.

$$\mathbf{P}^2 = \begin{bmatrix} \frac{1}{2} & \frac{1}{4} & \frac{1}{4} & 0 & 0 \\ \frac{1}{2} & \frac{1}{2} & 0 & 0 & 0 \\ \frac{1}{2} & 0 & \frac{1}{2} & 0 & 0 \\ 0 & 0 & 0 & \frac{3}{4} & \frac{1}{4} \\ 0 & 0 & 0 & \frac{1}{4} & \frac{3}{4} \end{bmatrix}$$

Now if an even number of transitions are to be made, say $2m$ transitions, we use the matrix $\mathbf{P}^2$ to find $\mathbf{P}^{2m}$ since

$$\mathbf{P}^{2m} = (\mathbf{P}^2)^m$$

Then, since $\mathbf{P}^2$ only allows transitions between states 1, 2, 3 and between states 4, 5, the matrix $\mathbf{P}^{2m}$ will only allow those transitions. When an odd number of transitions are made, say $2m + 1$, then

$$\mathbf{P}^{2m+1} = \mathbf{P}\mathbf{P}^{2m} = \mathbf{P}(\mathbf{P}^2)^m$$

so that after $2m$ transitions having started in states 1, 2, or 3, the process will be in one of the states 1, 2, 3, but then, after one more transition, the state must be either 4 or 5. Similar arguments show that starting in either state 4 or 5 the process must be in one of the states 1, 2, 3 after $2m + 1$ transitions.

It is possible to show that n-step transitions between the states 1, 2, and 3 can be determined by the matrix

$$\mathbf{P}^* = \begin{bmatrix} \frac{1}{2} & \frac{1}{4} & \frac{1}{4} \\ \frac{1}{2} & \frac{1}{2} & 0 \\ \frac{1}{2} & 0 & \frac{1}{2} \end{bmatrix}$$

formed by the first three rows and three columns of $\mathbf{P}^2$, this is found in the upper-left corner of $\mathbf{P}^2$ (see Exercise 2.14.1).

Example 2.34 involves a Markov chain which is periodic. One way to define periodic Markov chains is now stated.

DEFINITION 2.13   A Markov chain is called *periodic* if there is an integer $d > 1$ so that for every state i the only n-step transition probabilities from i to i, $p_{i,i}^{(n)}$, which are positive are those for which $n = kd$ for $k = 1, 2, \ldots$. The largest integer d for which this is true is called the *period* of the Markov chain.

According to this definition a Markov chain with period d can return to its initial state only after some multiple of d transitions. In Example 2.34 we saw that it was possible to return to the original state only after an even number of transitions, here $d = 2$. The period could not be greater than 2 since $p_{1,1}^{(2)} > 0$.

EXAMPLE 2.34 Continued

When we compute the matrices $\mathbf{P}^4$, $\mathbf{P}^6$, and so on, the $2k$-step transition probabilities are determined by the $2k$th power of the matrix (see exercise 2.14.1)

$$\mathbf{P}^* = \begin{bmatrix} 1/2 & 1/4 & 1/4 \\ 1/2 & 1/2 & 0 \\ 1/2 & 0 & 1/2 \end{bmatrix}$$

$\mathbf{P}^*$ is formed from the elements in the upper-left corner of the matrix $\mathbf{P}^2$ and is a transition matrix. A Markov chain with the transition matrix $\mathbf{P}^*$ is regular; thus the matrices $\mathbf{P}^{*n}$ approach a limiting matrix as n increases (Theorems 2.4, 2.5, and 2.6).

The same arguments apply to transitions between the states 4 and 5 which are determined from the matrix $\mathbf{P}^{**}$ formed by the elements in the lower-right corner of the matrix $\mathbf{P}^2$,

$$\mathbf{P}^{**} = \begin{bmatrix} 3/4 & 1/4 \\ 1/4 & 3/4 \end{bmatrix}$$

We thus find the limiting matrices $\mathbf{B}^*$ and $\mathbf{B}^{**}$

$$\mathbf{B}^* = \begin{bmatrix} 1/2 & 1/4 & 1/4 \\ 1/2 & 1/4 & 1/4 \\ 1/2 & 1/4 & 1/4 \end{bmatrix} \quad \text{and} \quad \mathbf{B}^{**} = \begin{bmatrix} 1/2 & 1/2 \\ 1/2 & 1/2 \end{bmatrix}$$

From these two limiting matrices, and from the fact that $\mathbf{P}^{2k}$ can be found from $(\mathbf{P}^*)^k$ and $(\mathbf{P}^{**})^k$, we conclude that $\mathbf{P}^{2k}$ approaches a limiting matrix as k increases. This limiting matrix is

$$\mathbf{B} = \begin{bmatrix} 1/2 & 1/4 & 1/4 & 0 & 0 \\ 1/2 & 1/4 & 1/4 & 0 & 0 \\ 1/2 & 1/4 & 1/4 & 0 & 0 \\ 0 & 0 & 0 & 1/2 & 1/2 \\ 0 & 0 & 0 & 1/2 & 1/2 \end{bmatrix}$$

Finally, if $\mathbf{P}^{2k}$ approaches a limiting matrix, then

$$\mathbf{P}^{2k+1} = \mathbf{P}\mathbf{P}^{2k}$$

will also approach a limiting matrix since $\mathbf{P}\mathbf{P}^{2k}$ must approach $\mathbf{P}\mathbf{B}$ which is

$$\mathbf{P}\mathbf{B} = \begin{bmatrix} 0 & 0 & 0 & 1/2 & 1/2 \\ 0 & 0 & 0 & 1/2 & 1/2 \\ 0 & 0 & 0 & 1/2 & 1/2 \\ 1/2 & 1/4 & 1/4 & 0 & 0 \\ 1/2 & 1/4 & 1/4 & 0 & 0 \end{bmatrix}$$

The matrices **B** and **PB** give a complete description of the transition probabilities as n increases. Even though the Markov chain is not regular, we are able to use the results for regular Markov chains to describe the transition probabilities as n increases.

THEOREM 2.8   In a periodic Markov chain with period d, the matrices $\mathbf{P}^{dk}$, as k increases, approach a limiting matrix **B**, and for each integer m, $0 < m < d$, the matrices $\mathbf{P}^{kd+m}$ approach the limiting matrix $\mathbf{P}^m\mathbf{B}$, as k increases.

Theorem 2.8 states that in a periodic Markov chain the probabilities still have a "limit" in a certain sense.

Exercise 2.14   1.   (a) Verify that in finding $\mathbf{P}^4$ from $\mathbf{P}^2$ in Example 2.34 the results are determined from $(\mathbf{P}^*)^2$ and $(\mathbf{P}^{**})^2$. (b) Note that in the matrix $\mathbf{P}^2$ of Example 2.34 the only positive transition probabilities from states 1, 2, 3 are to states 1, 2, 3. Give an argument to show that in the matrix $\mathbf{P}^{2k}$ the transition probabilities from the states 1, 2, 3 can be determined from the matrix $\mathbf{P}^*$. (*Hint*: $\mathbf{P}^{2k} = \mathbf{P}^2 \cdot \mathbf{P}^2 \ldots \mathbf{P}^2$ the product of $\mathbf{P}^2$ taken k times.)

2.   For the maze of Example 2.34 answer the following questions.

a.   If the rat is initially in compartment 1, what is the probability that on step 100 he is back in compartment 1?

b.   After 101 steps what is the probability of being in compartment 1 if $\mathbf{P}^{(0)} = [1, 0, 0, 0, 0]$?

c.   After 101 steps what is the probability the rat is in compartment 1 if $\mathbf{P}^{(0)} = [0, 0, 0, 0, 1]$?

3.   Let

$$\mathbf{P} = \begin{bmatrix} 0 & 0 & 0 & 1/3 & 1/3 & 1/3 \\ 0 & 0 & 0 & 1/2 & 0 & 1/2 \\ 0 & 0 & 0 & 2/3 & 1/3 & 0 \\ 2/3 & 1/3 & 0 & 0 & 0 & 0 \\ 1/3 & 1/3 & 1/3 & 0 & 0 & 0 \\ 0 & 1/3 & 2/3 & 0 & 0 & 0 \end{bmatrix}$$

a.   Find $\mathbf{P}^2$.

b.   Find the fixed vector for the matrix of the first three rows and columns of $\mathbf{P}^2$ and the fixed vector for the matrix of the last three rows and columns of $\mathbf{P}^2$.

c.   Find the limiting matrix for $\mathbf{P}^{2n}$ as n increases.

d.   Find the limiting matrix for $\mathbf{P}^{2n+1}$ as in increases.

4.   Write out and analyze the transition matrix for the maze in the

diagram assuming the rat is equally likely to exit from a compartment via any door.

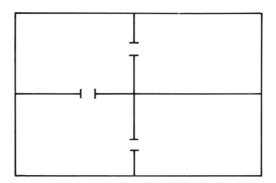

COMPUTER PROJECTS
Project 1:   A Markov Chain Computing Package

Computation of n-step transition probabilities and fixed vectors is a routine task which a computer can do—more accurately than a human. The application of Markov chain models and the interpretation of results obtained from those models require human intelligence. You are to write a computer program which will do the routine computations for a Markov chain with:

1.   a specified initial state distribution $\mathbf{p}^{(0)}$
2.   a specified transition matrix $\mathbf{P}$

The program should be able to handle a transition matrix and initial state distribution for a Markov chain with any number of states up to some specified largest possible number. This largest possible number depends on the computer memory available; consult your instructor for advice about this number.

   You should be able to determine the following information from the computer output:

1.   whether or not the Markov chain is regular
2.   the state distributions at n, $\mathbf{p}^{(n)}$ for as many values of n as desired
3.   the fixed vector for the transition matrix $\mathbf{P}$ (if the Markov chain is regular)

   It can be proved that if a Markov chain with N states is regular, then the matrix $\mathbf{P}^N$ must contain only positive elements, thus the answer to question 1 can be determined by checking the elements of $\mathbf{P}^N$. An efficient method for finding $\mathbf{P}^N$ and determining if all the elements are positive may be superior to determining each matrix $\mathbf{P}^2$, $\mathbf{P}^3$, . . . , $\mathbf{P}^N$ until one with only positive elements is found.

The printed output should only give those state distributions which are desired, such as the even numbered ones, every fifth one, or perhaps only one specific state distribution.

Project 2:   Simulating a Markov Chain

The exposition and mathematical development in this chapter do not give you intuition about the actual sequence of outcomes that are observed for a Markov chain. For this project you are to simulate a Markov chain by implementing the simulation procedure described in Project 2 of Chapter 1 to produce the "random" outcomes.

You may choose the Markov chain you simulate. (*Suggestion*: Simulate the Markov chain discussed in Application 2.6. By using two different proportions of defectives, say 0.05 and 0.10, you will be able to see from the simulation the mechanics of that quality control scheme.)

Project 3:   Some Theory Explored

In Sec. 2.10 a discussion of limiting matrices was given. You may have wondered how large the number n must be for the limiting matrix and $\mathbf{P}^n$ to differ by only a very little (see the last paragraph below). For this project you will do some experimental work related to that question.

First you must be sure that the Markov chain is regular to be assured that a limiting matrix exists—see the remarks about determining if the Markov chain is regular which are given in Project 1. Next, you need not determine the n-step transition matrix for each successive value of n to observe the rapidity of the convergence. The matrices $\mathbf{P}^5$, $\mathbf{P}^{10,}$ $\mathbf{P}^{15}$, . . . , for example, also give you a good idea of how quickly the entire sequence $\mathbf{P}$, $\mathbf{P}^2$, $\mathbf{P}^3$, . . . approaches the limit. Much less computer time is required by finding only multiples of the fifth power of $\mathbf{P}$. Of course you could also use only $\mathbf{P}^{10}$, $\mathbf{P}^{20}$, . . . if you choose.

You should repeat these computations for Markov chains having 3, 5, and 10 states and for several choices of matrices $\mathbf{P}$ for each number of states. This will illustrate the difference in rapidity of convergence with differing numbers of states and with differing transition matrices for the same number of states.

To determine when $\mathbf{P}^n$ is sufficiently near the limit matrix $\mathbf{B}$, use the criterion suggested by Theorem 2.7 that the elements in each column of $\mathbf{P}^n$ will be nearly equal. Terminate the computation when the largest difference between two elements of the same column is no more than 0.001.

3   LINEAR PROGRAMING

3.1   INTRODUCTION

It is not necessary to know probability theory for the study of this chapter. Thus it is possible to read this chapter before Chapters 1 and 2 provided you are familiar with the concepts of vectors and matrices as they are discussed in Secs. 2.3 and 2.4.

The mathematics discussed in this chapter concerns methods for solving a type of problem which commonly arises in business and industry. One example of such a problem is motivated by the desire to reduce production costs to a minimum. Typically a production process can be carried out using any one of many different combinations of values for the variables which determine that process. It is also typical that certain combinations of values for the variables will entail higher production costs than other combinations. We call it a *programing problem* when we seek to determine a combination of values which results in the minimum (smallest) cost. Whenever the entire problem can be specified using only linear equations and linear inequalities in the variables, we say it is a *linear programing problem.*

This description is vague and abstract so an illustrative example of a linear programing problem and its solution will occupy the remainder of this section. We should realize that no one example can simultaneously embody all the interesting ideas or illustrate the full range of applicability for a theory. The example we present here is chosen for its simplicity in order that our approach to the problem might be clear. It is possible that you might find the solution to this problem by a simpler thought process than the one we follow; however, we ask for your patience as we illustrate a general method for solving linear programing problems.

THE PROBLEM   A manufacturer of prepared foods markets a fortified dry cereal called Brunch. Brunch is called fortified because various vitamins are chemically added to the natural grains. The claim for Brunch is that 1 cup provides 100 percent of the minimum daily adult requirement (MDAR) of vitamins A, B_1, and B_{12}. (This is not completely realistic as the list of vitamins is usually much longer.) The manufacturer has two sources of multiple vitamins which can be used in the production process. We will refer to these as brand X and brand Y.

Table 3.1 shows the percentage of MDAR for each of the three vitamins which is provided by each of the two multiple-vitamin sources.

Vitamin	Brand X percentage	Brand Y percentage
A	2000	2000
B_1	1500	3000
B_{12}	2500	1500

Table 3.1   Percentage of MDAR per gram of bulk multiple vitamins

We assume that the manufacturing process allows for the addition of vitamins in any quantity and that it is possible to obtain a blend of the two brands in any desired proportion. Our computations will also ignore any possible nutrient value of the cereal itself (this assumption is probably not too unrealistic).

To visually illustrate our analysis of the problem, a graph will be used with the variable x denoting the number of grams of brand X vitamins and y the number of grams of brand Y vitamins which are added to each cup of Brunch. From Table 3.1 we see that the proportion of MDAR for vitamin A provided in x g (grams) of brand X is $20x$ and in y g of brand Y is $20y$. To obtain exactly 1 MDAR per cup of Brunch, x and y must satisfy the equation

$$20x + 20y = 1$$

We will consider all combinations of x and y for which *at least* 1 MDAR for vitamin A is supplied, since the claims for a fortified cereal remain true and we want to allow for any possible solution. To express this set of possible solutions, we write an inequality which x and y must satisfy, namely

$$20x + 20y \geq 1$$

Notice that a negative value for x or y is meaningless for this manufacturing process. We must require that any proposed solution involve only nonnegative values of x and y. To express that condition, two more inequalities for x and y are written; they are

$$x \geq 0 \quad \text{and} \quad y \geq 0$$

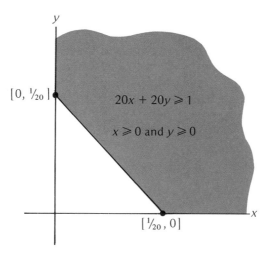

Figure 3-1   Combinations of brand X and brand Y which provide at least 1 MDAR of vitamin A

Figure 3.1 provides a graphic representation for the values of x and y which will provide the vitamin A requirement.

The requirement for vitamin B_1 leads to a similar consideration of possible solutions. Table 3.1 supplies the information that x g of brand X provides $15x$ MDAR of B_1 and y g of brand Y provides $30y$ MDAR. Thus the values for x and y which provide *at least* 1 MDAR of B_1 must satisfy the inequality

$$15x + 30y \geq 1$$

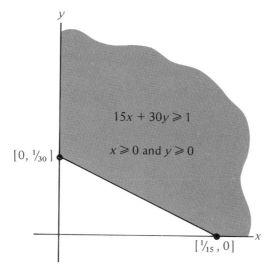

Figure 3-2   Combinations of brand X and brand Y which provide at least 1 MDAR of vitamin B_1

From this inequality and the requirement $x \geq 0$ and $y \geq 0$, we obtain another set of possible solutions shown in Fig. 3.2. When it is required that both vitamin A and B_1 requirements be satisfied, we must use a solution which is in the intersection of the blue regions in Figs. 3.1 and 3.2. Figure 3.3 shows this set of possible solutions. The point labeled M indicates the only value of x and y for which both MDAR requirements are exactly satisfied. M is the point of intersection of the two lines, and it can be found by solving the system of equations

$$20x + 20y = 1$$
$$15x + 30y = 1$$

(See Exercise 3.1.1.)

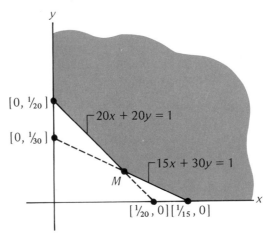

Figure 3.3   The combinations of brand X and brand Y which provide at least 1 MDAR of vitamins A and B_1

Of course it is necessary to express the vitamin B_{12} requirement in a similar manner. To do so we determine the inequality for those possible solutions and find it to be

$$25x + 15y \geq 1$$

Intersecting this region of possible solutions with those already shown in Fig. 3.3, we obtain the result displayed in Fig. 3.4.

Notice that the point M lies above the line $25x + 15y = 1$ so that the only exact solution for the MDAR of vitamins A and B_1, corresponding to the point M, provides more vitamin B_{12} than the MDAR. This means there is no combination of brand X and brand Y which simultaneously provides the exact MDAR of vitamins A, B_1, and B_{12}.

Figure 3.4 illustrates the possible values of x and y which will produce a fortified cereal. The set of values for which all the requirements are satisfied

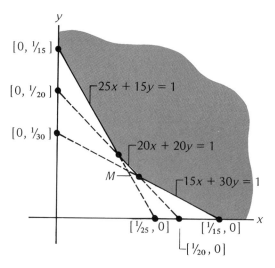

Figure 3.4   Combinations of brand X and brand Y which provide at least the MDAR of vitamins A, B_1, and B_{12}

is called the set of *feasible solutions* to the problem, and any solution in this set is called a feasible solution.

THE MINIMUM PROBLEM   The entire problem has not yet been stated. To complete the problem, we must consider the cost of each feasible solution. For that purpose suppose that 1 kg of brand X costs \$1 so that x g costs $x(1/1000)$ dollars and that 1 kg of brand Y costs \$1.50, so that y g costs $y(1.5/1000)$ dollars. Then the cost of a feasible solution can be expressed as a function of x and y, namely

$$c(x, y) = x\frac{1}{1000} + y\frac{1.5}{1000}$$

The *minimum problem* is to determine which of the feasible solutions results in the minimum cost. To investigate this question, we consider those values of x and y for which the cost is the same. If the cost is fixed at any particular value, $c(x, y) = c_0$, then an equation in x and y is determined. If for example $c_0 = 1$, then this equation is

$$x\frac{1}{1000} + y\frac{1.5}{1000} = 1$$

and the values of x and y which satisfy this equation lie on a line. Figure 3.5 compares these lines for three different values of c_0.

   We observe two things in Fig. 3.5. First note that the several lines are parallel, and second note that as the cost decreases the corresponding line is

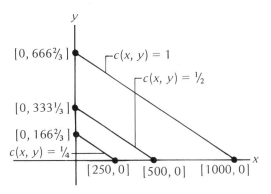

Figure 3.5   The lines of equal cost when $c_0 = 1$, $\frac{1}{2}$, and $\frac{1}{4}$

closer to the origin of the coordinate system. The importance of these two facts is illustrated by Fig. 3.6. When the line corresponding to a certain cost is superimposed upon the set of feasible solutions, we see that the minimum cost corresponds to the line which is as close to the origin as possible and which still intersects the set of feasible solutions.

The cost c_1 results in many feasible solutions, but clearly a smaller cost corresponding to a line closer to the origin can be found. The cost c_3 is too small since no feasible solutions lie on that line. The cost c_2 corresponds to the minimum cost since no line parallel to $c(x, y) = c_2$ and closer to the origin will intersect the set of feasible solutions. From our graphical pre-

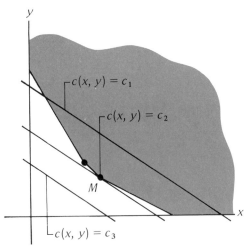

Figure 3.6   Lines of equal cost superimposed on the set of feasible solutions

sentation in Fig. 3.6 we see that the point M gives the solution to the minimum problem.

To find the point M we recall that in Fig. 3.3 it was noted to be the point determined by the solution to the system of equations

$$20x + 20y = 1$$
$$15x + 30y = 1$$

The solution is $x = \frac{1}{30}$, $y = \frac{1}{60}$, so that to each cupful of Brunch we should add $\frac{1}{30}$ g of brand X and $\frac{1}{60}$ g of brand Y to obtain a fortified cereal at the minimum cost.

Exercise 3.1

1.   Verify the values of x and y given above as the coordinates of the point M.
2.   Find the cost of vitamin supplement for 1 cup of Brunch when the x and y determined by the point M are used for the blend.
3.   Find the exact MDAR of each vitamin which is supplied by the solution given by the point M.
4.   Find the cost of vitamin supplement for 1 cup of Brunch when the x and y determined by the point of intersection of the lines $25x + 15y = 1$ and $20x + 20y = 1$ is used instead of the point M.
5.   Suppose only brand X is available as a supplement. How much brand X must be used, and what is the cost per cup of Brunch?
6.   Suppose brand X costs $1.50 per kilogram and brand Y costs $1 per kilogram. Find the feasible solution for which the cost is a minimum.
7.   If brand X costs $2 and brand Y $3 per kilogram, find the feasible solution for which the cost is a minimum.
8.   Find prices for brand X and brand Y so that the feasible solution of minimum cost uses only brand X.
9.   Determine the set of feasible solutions when the vitamin content for each gram of brand X and brand Y is that given below.

Vitamin	Brand X, percentage	Brand Y, percentage
A	1000	1500
B_1	800	2000
B_{12}	2000	800

3.2   VECTOR GEOMETRY

In Sec. 3.1 we considered a problem which involved some geometric concepts in the plane. Since many problems of interest cannot be stated using only two variables (and so we cannot always rely on pictures), we will formally develop the concepts used in Sec. 3.1, as well as the generalizations of those concepts to higher dimensional spaces. Vectors and matrices play an important role in expressing the main ideas, so a quick review of Secs. 2.3 and 2.4 would be helpful if your knowledge of vectors and matrices needs to be refreshed.

DIRECTION   Whenever a point other than the origin in a coordinate system is specified, a *direction* in the coordinate system is determined. This is the direction from the origin to the specified point. When a nonzero vector **v** is used to identify a point, the direction from the origin to the point **v** is associated with the vector **v**.

Example 3.1   The vector $\mathbf{v} = [4, 1]$ determines a point in the plane as shown in Fig. 3.7. The arrow drawn from the origin to the point $[4, 1]$ specifies a direction in the plane; that direction is associated with the vector $[4, 1]$.

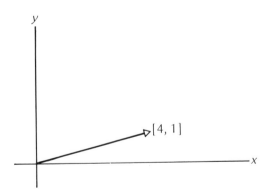

Figure 3.7   The point $[4, 1]$ and the direction determined

Example 3.2   In three dimensions the vector $[2, 3, 3]$ also determines a direction as well as a point. Figure 3.8 illustrates the point $[2, 3, 3]$ and the direction determined from the origin to the point $[2, 3, 3]$.

To refer to the direction from the origin to the point denoted by **v**, we will say *the direction of the vector* **v**. While **v** determines a direction, it must be mentioned that there are many vectors which determine the same direction.
    In fact, if two vectors **u** and **v** are proportional; that is, if $\mathbf{u} = c\mathbf{v}$, and $c > 0$, then **u** and **v** determine the same direction as our next example illustrates.

Example 3.3   The vectors $c[4, 1]$ for all values of $c > 0$ each determine the same direction in the plane. See Fig. 3.9.

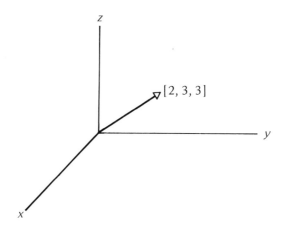

Figure 3.8   The point [2, 3, 3] and the direction determined

It is also useful to denote the direction from one point to another point by specifying a vector. Thus the direction from a point **v** to a point **u** is the direction of the vector **u** − **v**.

Example 3.4   Consider the direction from the point [1, 2] to the point [3, 3]. The arrow from [1, 2] to [3, 3] in Fig. 3.10 indicates this direction, which is the same as the direction of the vector [3, 3] − [1, 2] = [2, 1].

PERPENDICULARITY   The idea of perpendicular vectors is the next concept we wish to consider. We will define perpendicularity of two vectors in terms of the vector dot product, but before presenting that definition a short discussion of vector dot products is necessary.

In Chapter 2 when forming the dot product a distinction was made between the vectors by writing a row vector on the left and a column vector

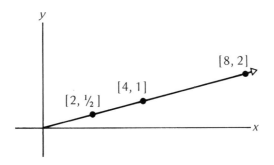

Figure 3.9   The direction of the vectors $c[4, 1]$ when $c > 0$

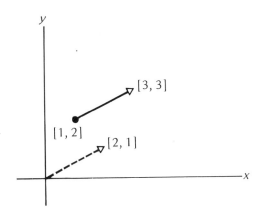

Figure 3.10   The direction from [1, 2] to [3, 3]

on the right. That distinction was useful in Chapter 2 but is inconvenient here since we will form the dot product of two vectors each of which determines a point in the same coordinate system (and for that reason it is only confusing to write one of them as a row vector and the other as a column vector).

We now extend Definition 2.5 by defining the dot product for any two n-dimensional vectors as follows:

The dot product of two n-dimensional vectors $\mathbf{u} = [u_1, u_2, \ldots, u_n]$ *and* $\mathbf{v} = [v_1, v_2, \ldots, v_n]$. *is the number*

$$\mathbf{u} \cdot \mathbf{v} = u_1 v_1 + u_2 v_2 + \cdots + u_n v_n = \sum_{i=1}^{n} u_i v_i$$

DEFINITION 3.1   The vectors $\mathbf{u}$ and $\mathbf{v}$ are called *perpendicular* if their dot product is zero

$$\mathbf{u} \cdot \mathbf{v} = 0$$

Example 3.5   In the plane any point on the x axis is represented by a vector of the form $[x, 0]$, and similarly any point on the y axis is given by a vector of the form $[0, y]$. Notice that the dot product of $[x, 0]$ and $[0, y]$ is

$$[x, 0] \cdot [0, y] = x \cdot 0 + 0 \cdot y = 0$$

so that any two such vectors are perpendicular by Definition 3.1.

Example 3.6   The vectors $[3, 2]$ and $[-2, 3]$ are perpendicular, as you can easily verify from Definition 3.1. These vectors are shown in Fig. 3.11 to illustrate their perpendicularity.

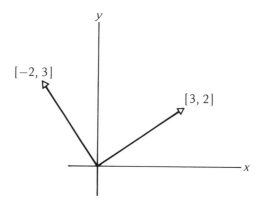

Figure 3.11   Perpendicular vectors

LINEAR EQUATIONS   In Sec. 3.1 it was stated that linear programing dealt with problems described by linear equations and linear inequalities. We next wish to investigate some useful facts about linear equations.

DEFINITION 3.2   Let **a** and **v** be two n-dimensional vectors, and b be any real number. Then if **a** is not the zero vector, the equation

$$\mathbf{a} \cdot \mathbf{v} = a_1 v_1 + a_2 v_2 + \cdots + a_n v_n = b$$

is called a *linear equation*, and the set of all vectors **v** which satisfy this equation for a fixed vector **a** and a fixed number b is called a *hyperplane*.

When **a** and **v** are two-dimensional vectors, the linear equation is $a_1 v_1 + a_2 v_2 = b$, which is the equation of a line (hence the term *linear equation*). If the vectors **a** and **v** are three-dimensional vectors, the linear equation $a_1 v_1 + a_2 v_2 + a_3 v_3 = b$ is the equation of a plane. The term *hyperplane* is a generalization of the term *plane* and is used primarily in four or more dimensions to refer to the set of solutions of a linear equation.

Example 3.7   The dot product

$$[3, -2] \cdot [x, y] = 4$$

determines the linear equation $3x - 2y = 4$. This is the equation of a line in the plane. The line is also called a hyperplane according to Definition 3.2.

Example 3.8   The equation of any line in the plane can be written as a dot product. The general equation of a line, $ax + by = c$, is expressed by the dot product

$$[a, b] \cdot [x, y] = c$$

Thus any line in the plane is a hyperplane according to Definition 3.2.

Example 3.9   The dot product

$$[1, -1, 2] \cdot [x, y, z] = 2$$

determines the linear equation $x - y + 2z = 2$. This is the equation of a plane in three-dimensional space. This plane is also called a hyperplane.

For any two vectors, say $\mathbf{v}'$ and $\mathbf{v}''$, which are in the hyperplane given by $\mathbf{a} \cdot \mathbf{v} = b$, the vector $\mathbf{a}$ is perpendicular to the difference of these vectors, $\mathbf{v}' - \mathbf{v}''$. Several examples will demonstrate this perpendicularity which is important to understanding the geometry of hyperplanes.

Example 3.10   The line determined by the linear equation $[3, 2] \cdot [x, y] = 0$ and the vector $[3, 2]$ are shown in Fig. 3.12. Notice that $[3, 2]$ is perpendicular to all vectors which lie on the line through the origin determined by the equation $[3, 2] \cdot [x, y] = 0$.

Next consider the graph of the equation $3x + 2y = 6$ which is also shown in Fig. 3.12. The two equations $3x + 2y = 0$ and $3x + 2y = 6$ determine parallel lines, and the vector $[3, 2]$ is perpendicular to any vector which determines a direction between two points on the line $3x + 2y = 6$. This fact is illustrated by the two points $\mathbf{v}'$ and $\mathbf{v}''$ in Fig. 3.12. Intuitively the vector $[3, 2]$ is perpendicular to the line $3x + 2y = 6$.

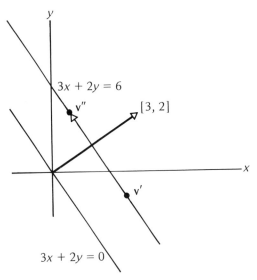

Figure 3.12   The graphs of $3x + 2y = 0$, $3x + 2y = 6$, and the vector $[3, 2]$

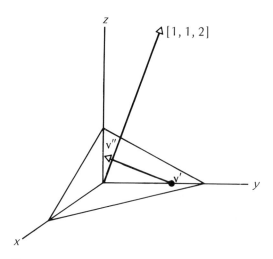

Figure 3.13   The plane $x + y + 2z = 1$, the vector $[1, 1, 2]$, and the direction from $\mathbf{v}'$ to $\mathbf{v}''$

Example 3.11   The graph of the linear equation $[1, 1, 2] \cdot [x, y, z] = 1$ and the vector $[1, 1, 2]$ are shown in three dimensions in Figure 3.13. Notice here the perpendicularity of $[1, 1, 2]$ to the vector giving the direction between the two points $\mathbf{v}'$, $\mathbf{v}''$ in the plane determined by $x + y + 2z = 1$. The intuitive idea is that the vector $[1, 1, 2]$ is perpendicular to any line which lies in the plane $x + y + 2z = 1$.

The general fact of importance here is that for the linear equation $\mathbf{a} \cdot \mathbf{v} = 0$ the hyperplane contains the origin and that all vectors which lie in this hyperplane are perpendicular to $\mathbf{a}$. The hyperplane determined by the equation $\mathbf{a} \cdot \mathbf{v} = b$ is parallel to the hyperplane determined by $\mathbf{a} \cdot \mathbf{v} = 0$, and if $\mathbf{v}'$, $\mathbf{v}''$ are any two points in the hyperplane $\mathbf{a} \cdot \mathbf{v} = b$, then $\mathbf{a}$ *is perpendicular to the vector giving the direction from* $\mathbf{v}'$ *to* $\mathbf{v}''$. This vector is $\mathbf{v}'' - \mathbf{v}'$, so we have

$$\mathbf{a} \cdot (\mathbf{v}'' - \mathbf{v}') = 0$$

Exercise 3.2   (Exercise 3.2.7 asks for a proof of this perpendicularity.)

1.   Give a vector which denotes the direction

a.   from $[3, 6]$ to $[6, -3]$ d.   from $[3, 4]$ to $[0, 0]$
b.   from $[4, 1]$ to $[0, 2]$ e.   from $[6, -1, 2]$ to $[7, 4, 4]$
c.   from $[5, -1]$ to $[3, 0]$ f.   from $[2, 2, -3]$ to $[5, 4, -3]$

2.   Which of the following vectors are perpendicular?

a. $[3, 6]$ and $[-3, -6]$ d. $[-1, 2, 7]$ and $[0, 7, -2]$
b. $[3, 6]$ and $[6, -3]$ e. $[1, 2, 3]$ and $[2, 1, -3]$
c. $[3, 6]$ and $[-6, 3]$ f. $[0, 0, 1]$ and $[1, 0, 0]$

3. $[2, 4]$ and $[1, 3]$ are two points on the line $-x + y = 2$. Verify that the vector $[2, 4] - [1, 3]$ is perpendicular to the vector $[-1, 1]$.

4.   Two points which satisfy the equation $x + y + z = 3$ are $[3, 2, -2]$ and $[5, -1, -1]$. Verify that the vector $[3, 2, -2] - [5, -1, -1]$ is perpendicular to the vector $[1, 1, 1]$.

5.   Verify that the direction in each instance below is the direction of the vector $[1, 1]$.

a.   the direction from $[2, -1]$ to $[3, 0]$
b.   the direction from $[2, -1]$ to $[5, 2]$
c.   the direction from $[\frac{1}{2}, \frac{1}{2}]$ to $[1, 1]$
d.   the direction from $[\frac{1}{2}, \frac{1}{2}]$ to $[c, c]$ where $c > \frac{1}{2}$

6.   Sketch the graph of the equation $x + 3y = 3$ and the vector $[1, 3]$. Note the perpendicularity of the vector $[1, 3]$ to the line.

7.   Prove that any two vectors $\mathbf{v}'$, $\mathbf{v}''$ which satisfy the equation $\mathbf{a} \cdot \mathbf{v} = b$ determine a vector $\mathbf{v}'' - \mathbf{v}'$ which is perpendicular to $\mathbf{a}$. (*Hint*: Write out the general expression of the dot product for $\mathbf{v}'$, $\mathbf{v}''$ and $\mathbf{v}'' - \mathbf{v}'$, and use the fact that $\mathbf{a} \cdot \mathbf{v}' = b = \mathbf{a} \cdot \mathbf{v}''$.)

3.3   CONVEX SETS

Our introductory example in Sec. 3.1 dealt with linear equations, but we were more concerned with inequalities which described all the possible solutions for that problem. Linear inequalities can also be written using dot products of vectors. In particular

$$\mathbf{a} \cdot \mathbf{v} \geq b$$

describes those vectors $\mathbf{v}$ for which the dot product is a number greater than or equal to b. We know that $\mathbf{a} \cdot \mathbf{v} = b$ describes a hyperplane, and we will now examine the condition $\mathbf{a} \cdot \mathbf{v} > b$. In special cases we have dealt with this inequality in Sec. 3.1.

As the value of $b > 0$ increases, the hyperplane $\mathbf{a} \cdot \mathbf{v} = b$ lies further from the origin in the direction of $\mathbf{a}$; while if b is a negative number, the hyperplane is in the "opposite" direction of $\mathbf{a}$ from the origin. We claim that the set of vectors for which $\mathbf{a} \cdot \mathbf{v} > b$ is the set of *all vectors on one side of the hyperplane* $\mathbf{a} \cdot \mathbf{v} = b$. These vectors will be in the direction of $\mathbf{a}$ from the hyperplane. By the same reasoning the inequality $\mathbf{a} \cdot \mathbf{v} < b$ is satisfied by all vectors on the opposite side of the hyperplane $\mathbf{a} \cdot \mathbf{v} = b$; these vectors will be in the direction of $-\mathbf{a}$ from the hyperplane.

Example 3.12   The graph of the line $[3, 2] \cdot [x, y] = 6$ is shown in Fig. 3.14. The points $[x, y]$ for which $[3, 2] \cdot [x, y] > 6$ lie in the direction of $[3, 2]$ from that

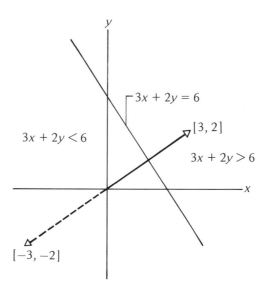

Figure 3.14 The hyperplane $3x + 2y = 6$ and two half-spaces

line, and the points $[x, y]$ for which $[3, 2] \cdot [x, y] < 6$ lie in the direction of $[-3, -2]$ from that line. We call the set of vectors determined by the inequality $\mathbf{a} \cdot \mathbf{v} \geq b$ a *half-space*, and the hyperplane $\mathbf{a} \cdot \mathbf{v} = b$ is called the *bounding hyperplane* of the half-space.

CONVEXITY A half-space has a useful geometric property which at first thought hardly seems worth mentioning. But we will soon discover that this property is a key factor in our ability to solve the general linear programing problem. The property we are talking about is stated in the following definition.

DEFINITION 3.3 A set is called *convex* if for any two points **u** and **v** in the set every point on the line segment joining **u** to **v** is also in the set.

A set is called *polyhedral convex* if it is the intersection of a finite number of half-spaces.

We will only deal with polyhedral convex sets. Several examples of polyhedral convex sets are pictured in Fig. 3.15. In each case we have illustrated the line segments joining pairs of points in those sets. Notice that in each case the entire line segment is completely contained in the set as is required by Definition 3.3.

To help clarify the idea of a convex set, Fig. 3.16 demonstrates some sets which are not convex. In Fig. 3.16a and b we see that many pairs of points can be joined by a line segment contained in the set, but in each case there

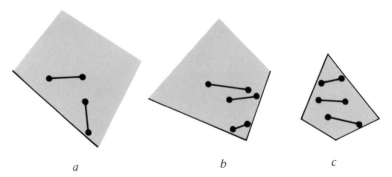

Figure 3.15   Several polyhedral convex sets

are also pairs of points for which the line segment joining them does not lie entirely within the set, thus violating the condition stated in Definition 3.3.

We have not illustrated these ideas in three dimensions simply because the ideas are no different nor more complicated there, yet the graphs are harder to draw and interpret. See Exercise 3.3.6 for several three-dimensional examples.

It is an easy exercise to show that a polyhedral convex set is in fact a convex set. Exercise 3.3.7 asks you to prove that the intersection of any finite number of half-spaces is a convex set.

EXTREME POINTS The bounding hyperplanes of the half-spaces which are intersected to form a polyhedral convex set usually intersect each other. Whenever several of the bounding hyperplanes intersect in just one point of the convex set, this point is a "corner point" of the polyhedral convex set and is of special interest. In Fig. 3.15 there are several such corner points. The formal description of these points is the purpose of our next definition. The intuitive description of corner points is not used in this definition; yet they really describe the same thing.

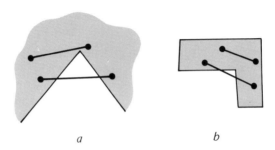

Figure 3.16   Sets which are not convex

DEFINITION 3.4 A point **v** is an *extreme point* of a convex set if every line segment containing **v** and contained in the set has **v** as an endpoint.

In Fig. 3.15 the extreme points are exactly the "corners" of those sets. Notice that any line segment which is contained in the convex set and which contains a corner point must have that corner point as an endpoint.

In actual practice we will locate these extreme points by finding the points of intersection of the bounding hyperplanes of the polyhedral convex set. With that purpose in mind notice that in Fig. 3.15b the pair of bounding hyperplanes determines an extreme point.

In Fig. 3.15c there are four corner points, but not every pair of bounding hyperplanes determines a corner point of the polyhedral convex set. Whenever the intersection of the bounding hyperplanes is a point which is not in the set, this point is *not* an extreme point of the set. Thus the polyhedral convex set in Fig. 3.15c has just four extreme points, since two pairs of bounding hyperplanes intersect at points which are not in the convex set.

We end this section with illustrations in three dimensions. In three dimensions, three planes will typically intersect in one point; this situation is depicted in Fig. 3.17.

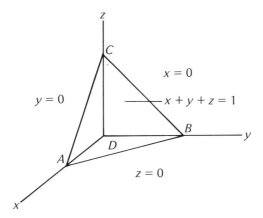

Figure 3.17 The intersections of four planes

The planes $y = 0$ and $z = 0$ intersect in the x axis, and that line intersects the plane $x + y + z = 1$ at the point A. Similarly the point B is the point of intersection of the three planes $x = 0$, $z = 0$, and $x + y + z = 1$. The half-spaces $x \geq 0$, $y \geq 0$, and $z \geq 0$ intersect in the set of all points with only nonnegative coordinates. The intersection of that set with the half-space $x + y + z \leq 1$ is the polyhedral convex set of all points with nonnegative coordinates which lie on or below the plane $x + y + z = 1$ in Fig. 3.17. This polyhedral convex set has the four extreme points A, B, C, and D.

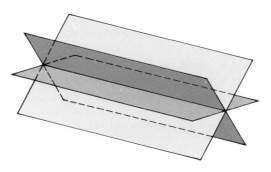

Figure 3.18   Three planes intersecting in a line

Finally in Fig. 3.18 we demonstrate a special case which can arise when three planes are intersected. This example illustrates that the intersection of three planes can be a line.

Exercise 3.3   1.   On a coordinate system shade the half-space $[3, -2] \cdot [x, y] \geq 0$.
2.   Write a vector inequality for the half-space bounded by the hyperplane $[1, -3] \cdot [x, y] = 2$ and containing the point $(0, 0)$.
3.   Graph the polyhedral convex set determined by the intersection of the half-spaces

$$[1, 1] \cdot [x, y] \geq 0 \qquad [1, 2] \cdot [x, y] \leq 5 \qquad [1, -1] \cdot [x, y] \leq 2$$

4.   Find the extreme points of the convex set determined in Exercise 3.3.3.
5.   For the diagram given in Fig. 3.17 name the hyperplanes which intersect at the point C.
6.   Determine which of the sets below are convex.

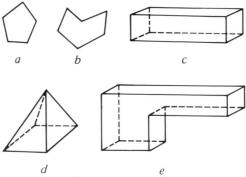

7.   Prove that the intersection of finitely many half-spaces is a convex set. (*Hint:* If two points are in the intersection of all the half-spaces, then those two points must be in each half-space. Use the convexity of the half-space to draw the necessary conclusion about the line segment joining these two points.)

3.4 LINEAR FUNCTIONS

The general linear programing problem concerns the values of a function at the various points of a polyhedral convex set. Specifically, the function in each case is a linear function—a definition is given below. We are going to study some special properties of linear functions which make it possible to solve linear programing problems.

DEFINITION 3.5 A function f defined for n-dimensional vectors $\mathbf{v}$ is a *linear function* if there is an n-dimensional vector $\mathbf{a}$ so that $f(\mathbf{v}) = \mathbf{a} \cdot \mathbf{v}$.

Example 3.13 Suppose f is a linear function such that $f(\mathbf{v}) = [3, -1] \cdot \mathbf{v}$, here $\mathbf{a} = [3, -1]$. Then for the vector $[2, 3]$, f has the value

$$f[2, 3] = [3, -1] \cdot [2, 3] = 6 - 3 = 3$$

For the general vector $[x, y]$ the value of f is given by the formula

$$f[x, y] = [3, -1] \cdot [x, y] = 3x - y$$

Example 3.14 Suppose the linear function f is determined by the vector $\mathbf{a} = [1, -2, 1]$. At the point $[0, 1, 1]$ the function has the value

$$f[0, 1, 1] = [1, -2, 1] \cdot [0, 1, 1] = 0 - 2 + 1 = -1$$

For an arbitrary point in three dimensions, $[x, y, z]$, the linear function f has the value given by the formula

$$f[x, y, z] = [1, -2, 1] \cdot [x, y, z] = x - 2y + z$$

We are specifically interested in computing the values of a linear function. Our next lemma states an important fact about those values and gives us a method for computing certain values.

LEMMA 3.1 If f is a linear function defined for the vectors $\mathbf{u}$ and $\mathbf{v}$, then for the vector $t\mathbf{u} + (1 - t)\mathbf{v}$, f has the value

$$f[t\mathbf{u} + (1 - t)\mathbf{v}] = tf(\mathbf{u}) + (1 - t)f(\mathbf{v})$$

No proof will be given for Lemma 3.1. Instead we will present the essential idea of the proof in Example 3.15. The key to understanding the statement of Lemma 3.1 is to recall (refer to Sec. 2.3) that for two points $\mathbf{u}$ and $\mathbf{v}$ and any number t the point $t\mathbf{u} + (1 - t)\mathbf{v}$ is on the line through $\mathbf{u}$ and $\mathbf{v}$, and when $0 \le t \le 1$ this point is on the line segment joining $\mathbf{u}$ and $\mathbf{v}$.

Example 3.15 Consider the linear function $f(\mathbf{v}) = [3, -1] \cdot \mathbf{v}$ and the two points $[1, 1]$ and $[3, 2]$. Any point with coordinates given by $t[1, 1] + (1 - t)[3, 2] = [t + 3 - 3t, t + 2 - 2t] = [3 - 2t, 2 - t]$ lies on the line through $[1, 1]$ and

[3, 2]. When $t = \frac{1}{2}$, this point is the midpoint of the line segment joining [1, 1] and [3, 2], namely [2, $\frac{3}{2}$]. Furthermore, the values of the function f at these points are

$$f[1, 1] = [3, -1]\cdot[1, 1] = 3 - 1 = 2$$
$$f[2, \tfrac{3}{2}] = [3, -1]\cdot[2, \tfrac{3}{2}] = 6 - \tfrac{3}{2} = \tfrac{9}{2}$$
$$f[3, 2] = [3, -1]\cdot[3, 2] = 9 - 2 = 7$$

and as the lemma states

$$f[2, \tfrac{3}{2}] = \tfrac{1}{2}f[1, 1] + (1 - \tfrac{1}{2})f[3, 2]$$
$$= \tfrac{1}{2}(2) + \tfrac{1}{2}(7) = 1 + \tfrac{7}{2} = \tfrac{9}{2}$$

We can also verify that for any point on the line through [1, 1] and [3, 2], that is, any point which is of the form

$$t[1, 1] + (1 - t)[3, 2] = [3 - 2t, 2 - t]$$

the value of f is

$$[3, -1]\cdot[3 - 2t, 2 - t] = 3(3 - 2t) - 1(2 - t)$$
$$= 9 - 6t - 2 + t$$
$$= 7 - 5t$$

Next we verify that the lemma is correct by computing

$$tf[1, 1] + (1 - t)f[3, 2] = t(2) + (1 - t)7$$
$$= 7 - 5t$$

The main fact about linear functions which we need to know is stated as Lemma 3.2.

LEMMA 3.2   Let $\mathbf{u}$ and $\mathbf{v}$ be two points and f a linear function for which $f(\mathbf{u}) \leq f(\mathbf{v})$. Then

$$f(\mathbf{u}) \leq f(\mathbf{w}) \leq f(\mathbf{v})$$

for any point $\mathbf{w}$ on the line segment joining $\mathbf{u}$ and $\mathbf{v}$.

PROOF   Any point $\mathbf{w}$ on the line segment joining $\mathbf{u}$ and $\mathbf{v}$ can be written as $\mathbf{w} = t\mathbf{v} + (1 - t)\mathbf{u}$ for some number t such that $0 \leq t \leq 1$. When $t = 0$, $\mathbf{w} = \mathbf{u}$, and when $t = 1$, $\mathbf{w} = \mathbf{v}$; while for $0 < t < 1$ the point $\mathbf{w}$ is strictly between $\mathbf{u}$ and $\mathbf{v}$. Lemma 3.1 tells us that

$$f(\mathbf{w}) = tf(\mathbf{v}) + (1 - t)f(\mathbf{u})$$

or written another way

$$f(\mathbf{w}) = f(\mathbf{u}) + t[f(\mathbf{v}) - f(\mathbf{u})]$$

Since $f(\mathbf{v}) \geq f(\mathbf{u})$, the difference $f(\mathbf{v}) - f(\mathbf{u})$ is nonnegative, and for $t \geq 0$, the expression $t[f(\mathbf{v}) - f(\mathbf{u})]$ is nonnegative. This means that $f(\mathbf{w})$ equals $f(\mathbf{u})$ plus a nonnegative quantity, so that

$$f(\mathbf{w}) \geq f(\mathbf{u})$$

Next, because $t \leq 1$ and $f(\mathbf{v}) - f(\mathbf{u})$ is the difference between the values of f at these two points, $t[f(\mathbf{v}) - f(\mathbf{u})]$ is less than or equal to this difference. We conclude that $f(\mathbf{u}) + t[f(\mathbf{v}) - f(\mathbf{u})]$ is less than or equal to $f(\mathbf{v})$.

An example will indicate the relevance of Lemma 3.2 to the problem of finding the largest and/or smallest value of a linear function for all points of a polyhedral convex set.

Example 3.16 Consider the linear function $f(\mathbf{v}) = [4, -3] \cdot \mathbf{v}$. The polyhedral convex set shown in Fig. 3.19 has the three extreme points $[0, 0]$, $[-2, 3]$, and $[6, 4]$. Lemma 3.2 tells us that the value $f(\mathbf{w}) = [4, -3] \cdot \mathbf{w}$ is a number between the values $[4, -3][-2, 3] = -17$ and $[4, -3] \cdot [6, 4] = 12$, since $\mathbf{w}$ lies on the line segment joining $[-2, 3]$ and $[6, 4]$.

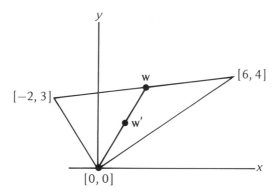

Figure 3.19 A convex set with extreme points $[0, 0]$, $[-2, 3]$, and $[6, 4]$

Now the point $\mathbf{w}'$ lies on the line segment joining $[0, 0]$ and $\mathbf{w}$, so the value $f(\mathbf{w}')$ must be between the values $f[0, 0] = 0$ and $f(\mathbf{w})$. While we have not found an explicit value for $f(\mathbf{w})$, we do know that it is between -17 and 12. Since $f[0, 0]$ is between -17 and 12, $f(\mathbf{w}')$ must also be between the values -17 and 12.

We conclude from this example—since $\mathbf{w}'$ could be any point in the convex set—that the largest and smallest values of the function f are attained at $[6, 4]$ and $[-2, 3]$, respectively, two extreme points of the convex set.

Our next theorem tells us the extent to which the conclusion of Example 3.16 is true. This important result is at the heart of any solution to a linear programing problem.

THEOREM 3.1 Let f be a linear function defined on a polyhedral convex set C. Then

a. If f has a maximum value for points in C, there is an extreme point of C where the value of f is maximum.
b. If f has a minimum value on C, there is an extreme point of C where the value of f is minimum.

The proof of Theorem 3.1 uses essentially the same argument as Example 3.16, and that proof is omitted.

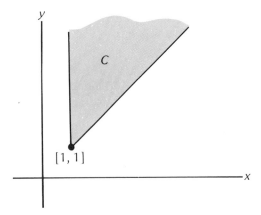

Figure 3.20 The polyhedral convex set formed by the intersection of $x \geq 1$ and $x - y \leq 0$

The essential fact in Theorem 3.1 which we must take note of is that to locate the maximum and/or minimum value of a linear function on a polyhedral convex set (if these values exist) it is only necessary to find the values at the extreme points.

One final example will be given to show you that a linear function need not have a maximum value on a polyhedral convex set. This example also explains why Theorem 3.1a begins with the phrase "if f has a maximum value for."

Example 3.17 Let the linear function be $f(\mathbf{v}) = [1, 1] \cdot \mathbf{v}$ and the polyhedral convex set C be as shown in Fig. 3.20.

The bounding hyperplanes of C are $x = 1$ and $x - y = 0$. To see that f has no maximum value on C, we need only observe that on the hyperplane $x - y = 0$ or $x = y$ the value of f, $x + y$, is as large as we wish. For any value we might give, no matter how large, it is possible to find a point of C where the value of f is even larger.

On the other hand, the value of f is smaller at $[1, 1]$ than at any other point of C. Since f has a smallest value, Theorem 3.1b applies, and this smallest value occurs at an extreme point.

A similar example could be given to show that a linear function need not have a minimum value on a polyhedral convex set.

Exercise 3.4

1. Consider the linear function $f(\mathbf{v}) = [1, -1] \cdot \mathbf{v}$. Find the value of f at $[0, 1]$, at $[2, 1]$, and at each indicated point. Use Lemma 3.1 to find the values in each case.

a. at the point $\frac{1}{4}[0, 1] + \frac{3}{4}[2, 1]$
b. at the point $2[0, 1] - [2, 1]$
c. at the midpoint of the line segment joining $[0, 1]$ and $[2, 1]$
d. at the point $\frac{1}{3}$ of the distance along the line segment joining $[0, 1]$ and $[2, 1]$ and nearest to $[0, 1]$

2. For the linear function $f(\mathbf{v}) = [1, 2, -1] \cdot \mathbf{v}$ and the two points $[1, 1, 1]$ and $[-1, 2, -1]$ verify Lemma 3.2 for each of the points below:

a. $\frac{1}{2}[1, 1, 1] + \frac{1}{2}[-1, 2, -1]$ c. $\frac{1}{3}[1, 1, 1] + \frac{2}{3}[-1, 2, -1]$
b. $\frac{1}{4}[1, 1, 1] + \frac{3}{4}[-1, 2, -1]$ d. $\frac{2}{3}[1, 1, 1] + \frac{1}{3}[-1, 2, -1]$

3. The linear function $f(\mathbf{v}) = [-1, 2] \cdot \mathbf{v}$ has a maximum and a minimum for the points of the polyhedral convex sets pictured below. Find the minimum and maximum values in each case and the points where each of those values is attained.

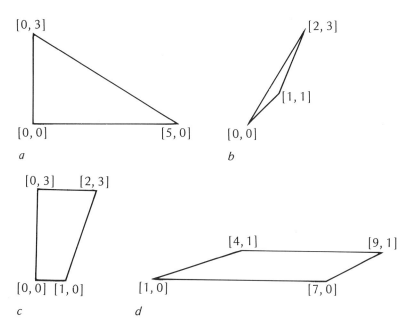

4.   The linear function $f(\mathbf{v}) = [3, -1, -4] \cdot \mathbf{v}$ has a maximum and a minimum value on the polyhedral convex sets C given below. Find the maximum and the minimum value in each case and the points where those values are attained.

a.   The extreme points are $[0, 1, 0]$, $[5, 6, 0]$, $[4, 2, 6]$, $[1, 1, 3]$.

b.   The set is the one pictured

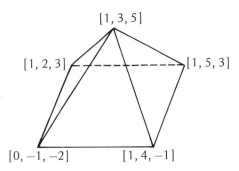

*3.5   SOME THEORY—SUPPORTING AND SEPARATING HYPERPLANES

The points which are in a polyhedral convex set and belong to one or more of the bounding hyperplanes of the set are called *boundary points*. The extreme points of a convex set are boundary points, but many boundary

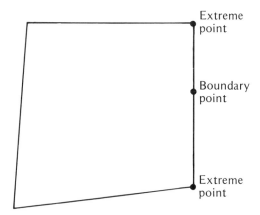

Figure 3.21   A convex set and its boundary

points are not extreme points. Figure 3.21 shows a typical polyhedral convex set for which the set of boundary points is indicated by the black line on the right.

If one of the two half-spaces determined by a hyperplane contains a convex set *A*, and the hyperplane itself contains at least one point of *A*, then the hyperplane is called a *supporting hyperplane of A*. Clearly every boundary point of a polyhedral convex set is contained in some supporting hyperplane, since the bounding hyperplane containing the point is also a supporting hyperplane. There are many supporting hyperplanes at an extreme point of a polyhedral convex set. Figure 3.22 gives an illustration of a convex set and some supporting hyperplanes.

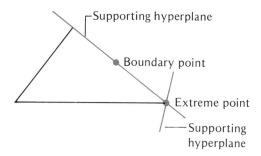

Figure 3.22   Supporting hyperplanes of a polyhedral convex set

An extension of this idea is needed for some of the theory discussed in Chapter 5. What is needed later is the existence of a hyperplane which lies between two convex sets. To make this more precise we state the following formal result.

THEOREM 3.2   Given any two polyhedral convex sets which do not intersect there is a hyperplane, defined by an equation $\mathbf{a} \cdot \mathbf{v} = b$, so that the half-space $\mathbf{a} \cdot \mathbf{v} \leq b$ contains one of the convex sets and the half-space $\mathbf{a} \cdot \mathbf{v} \geq b$ contains the other convex set.

The hyperplane $\mathbf{a} \cdot \mathbf{v} = b$ is called a *separating hyperplane*.

While we will not give a proof of this theorem, it is easy to state why the result is plausible. Since the sets do not intersect, we should be able to find two points, one in each of the two sets, which are nearest to each other among all choices of the points. In Fig. 3.23 these two nearest points are labeled **u** and **w**.

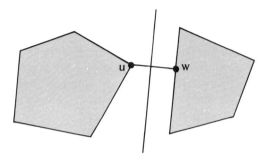

Figure 3.23   Two convex sets and a separating hyperplane

For the two points **u** and **w** in Fig. 3.23 there is a direction from **u** to **w** determined by the vector **w** − **u**. Now the line passing through the midpoint of and perpendicular to the line segment joining **u** and **w** is a separating hyperplane given by the equation

$$(\mathbf{w} - \mathbf{u}) \cdot \mathbf{v} = b$$

Example 3.18   The convex sets shown in Fig. 3.24 are nearest to each other for the two points $[0, 1]$ and $[1, 1]$. The direction from $[0, 1]$ to $[1, 1]$ is given by the vector $[1, 1] - [0, 1] = [1, 0]$. The midpoint of the line segment joining those points is the point $[0, 1] + \frac{1}{2}[1, 0] = [\frac{1}{2}, 1]$. Thus the hyperplane through the point $[\frac{1}{2}, 1]$ which is perpendicular to $[1, 0]$ has the equation $[1, 0] \cdot [x, y] = \frac{1}{2}$.

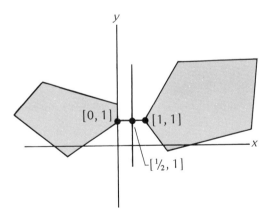

Figure 3.24   Nearest points and a separating hyperplane

Exercise 3.5   1.   For the convex set illustrated write the equations for a supporting hyperplane at the point **a** and for a supporting hyperplane at **b**. For which of these points is there a freedom of choice?

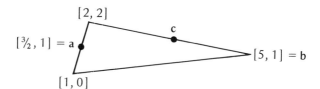

2.   Verify that $[1, 3] \cdot \mathbf{v} = 8$ determines a supporting hyperplane at the point **c** of the convex set in Exercise. 3.5.1.

3.   Find a separating hyperplane for the two convex sets illustrated in each case.

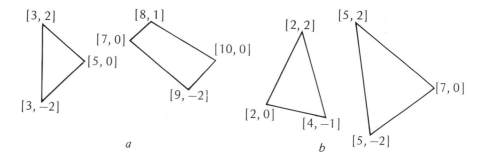

4.   Find a separating hyperplane for the two convex sets given by the intersection of the half-spaces

$$[1, 3] \cdot [x, y] \geq 0$$
$$[-1, 3] \cdot [x, y] \leq 0 \qquad \text{and}$$
$$[1, 0] \cdot [x, y] \leq 5$$

$$[1, 3] \cdot [x, y] \leq -1$$
$$[-1, 3] \cdot [x, y] \geq 0$$
$$[-1, 0] \cdot [x, y] \geq 3$$

3.6 MATHEMATICAL FORMULATION OF PROBLEMS

Most problems in mathematics originate as word problems which must be given a mathematical formulation. You have had previous experience with word problems in your study of algebra and in Sec. 3.1, where the introduction to linear programing problems was given. In this section convex sets and linear functions are used to give mathematical formulations for linear programing problems.

You will recall that in Sec. 3.1 we dealt with a problem for which we wanted to find the minimum cost to produce a fortified cereal. We will begin this section with an example in which a maximum value for a function produces the desired solution.

Example 3.19   The Relax furniture company manufactures two products, living room and dining room suites, at its Memphis plant. Management schedules the number of each product to be manufactured each week so that the profit earned from operations is a maximum. Certain limitations are imposed on the scheduling by the availability of workers. The production employees at this plant are 40 finishers, 25 upholsterers, and 30 carpenters, each of whom is employed 40 hr/wk. A solution to the scheduling problem may not involve overtime for any employees nor any change in the work force. The manhours of labor required to produce each product are given in Table 3.2 for each category of workers.

	Finishers	Upholsterers	Carpenters
Living room suite	3	12	4
Dining room suite	20	2	10

Table 3.2   Manhours of labor required for each unit produced

Finally, the profit to the manufacturer is $80 for each living room suite and $50 for each dining room suite. Assuming that any number of living room and dining room suites can be sold, what number of each does the Relax management schedule for production in a week?

THE MATHEMATICAL FORMULATION   Management can vary the numbers of living room suites and dining room suites to be produced so these will become variables for our formulation. Accordingly we define:

x to be the number of living room suites scheduled per week

y to be the number of dining room suites scheduled per week

Now we can express the weekly profit for any production schedule (x, y) as a function $f(x, y)$. Since each living room suite produced sells for an $80 profit, the x suites result in $80x$ of profit. The y dining room suites contribute $50y$ of profit. Thus the total profit is given by the linear function

$$f(x, y) = 80x + 50y$$

The linear function $f(x, y)$ is to be as large as possible.

Next we determine the set of all feasible solutions to the problem. From Table 3.2 we determine several restrictions on the possible values of x and y. The 40 finishers can provide at most $40(40) = 1600$ hr/wk, while x living room suites and y dining room suites require $3x$ and $20y$ manhours of labor from the finishers so we must have

$$3x + 20y \leq 1600$$

From Table 3.2 a second restriction on x and y results from the limited number of manhours the 25 upholsterers can provide. Since the total hours available are $(25)(40) = 1000$ and $12x + 2y$ hr of upholsterers' labor are needed, we must have

$$12x + 2y \leq 1000$$

The third restriction is determined from the information given for the 30 carpenters who provide at most $(30)(40) = 1200$ hr of labor. From Table 3.2 we find that $4x + 10y$ hr of carpenters' labor are required, thus

$$4x + 10y \leq 1200$$

Finally we note that both x and y must be nonnegative for a meaningful solution to this problem.

For Example 3.19 the polyhedral convex set of feasible solutions can be expressed in a concise notation with a vector-matrix product. The three half-spaces

$$3x + 20y \leq 1600$$
$$12x + 2y \leq 1000$$
$$4x + 10y \leq 1200$$

can be simultaneously specified by writing

$$\begin{bmatrix} 3 & 20 \\ 12 & 2 \\ 4 & 10 \end{bmatrix} \begin{bmatrix} x \\ y \end{bmatrix} \leq \begin{bmatrix} 1600 \\ 1000 \\ 1200 \end{bmatrix}$$

if we understand the inequality to mean that each component of the vector on the left

$$\begin{bmatrix} 3 & 20 \\ 12 & 2 \\ 4 & 10 \end{bmatrix} \begin{bmatrix} x \\ y \end{bmatrix} = \begin{bmatrix} 3x + 20y \\ 12x + 2y \\ 4x + 10y \end{bmatrix}$$

must be less than or equal to the corresponding component of the vector on the right. We can also express the two half-spaces $x \geq 0$, $y \geq 0$ by this scheme. We will first write them as

$$-x \leq 0 \qquad -y \leq 0$$

in order to have the inequality stated as less than or equal. The complete description of the set of feasible solutions is then

$$\begin{bmatrix} 3 & 20 \\ 12 & 2 \\ 4 & 10 \\ -1 & 0 \\ 0 & -1 \end{bmatrix} \begin{bmatrix} x \\ y \end{bmatrix} \leq \begin{bmatrix} 1600 \\ 1000 \\ 1200 \\ 0 \\ 0 \end{bmatrix}$$

Example 3.20 For Example 3.1 the set of feasible solutions was given by the intersection of the half-spaces

$$25x + 15y \geq 1$$
$$20x + 20y \geq 1$$
$$15x + 30y \geq 1$$
$$x \geq 0$$
$$y \geq 0$$

This polyhedral convex set is specified by the inequality

$$\begin{bmatrix} 25 & 15 \\ 20 & 20 \\ 15 & 30 \\ 1 & 0 \\ 0 & 1 \end{bmatrix} \begin{bmatrix} x \\ y \end{bmatrix} \geq \begin{bmatrix} 1 \\ 1 \\ 1 \\ 0 \\ 0 \end{bmatrix}$$

To summarize we specify the format of the mathematical formulation of a linear programing problem.

Summary

1. A linear function $f(\mathbf{v}) = \mathbf{a} \cdot \mathbf{v}$ called the *objective function* is given. The object of the problem is to find either the maximum value or the minimum value of the objective function.
2. A polyhedral convex set of *feasible solutions* specified by an inequality of the form $\mathbf{Av} \leq \mathbf{b}$ or $\mathbf{Av} \geq \mathbf{b}$ is given. Here $\mathbf{A}$ is a matrix of numbers, $\mathbf{v}$ a vector of variables, and $\mathbf{b}$ a vector of numbers. A problem stated in the format outlined above is called a *linear program*.

Exercise 3.6 1. In the problem of Example 3.19 suppose there are 30 finishers, 25 upholsterers, and 20 carpenters employed at the plant. Give a formulation of the problem with this change in the numbers of workers.
2. In Example 3.19 suppose that the profit on each living room suite is reduced to $50, the profit on dining room suites increases to $55, and all other aspects of the problem are as stated originally. Formulate the linear programing problem to maximize the profit under these new conditions.
3. In Example 3.19 suppose the manhours of labor required for each unit produced changes to the values given in the table below:

	Finishers	Upholsterers	Carpenters
Living room suites	2	9	5
Dining room suites	15	3	9

Formulate the problem to maximize total profit using the profits given in Example 3.19.
4. A steel company sells 200,000 tons of top-grade, 50,000 tons of

grade B, and 100,000 tons of mill-grade cold-rolled sheet each week. The company operates two roller mills. The larger has an average output of 300,000 tons of cold-rolled sheet per week. It produces 50 percent top-grade, 25 percent grade B, and 25 percent mill-grade cold-rolled sheet per week, while the smaller mill has an average output of 150,000 tons of which 60 percent is top-grade, 10 percent grade B, and 30 percent mill-grade. The cost per week to operate the smaller mill is two-thirds that of the larger mill. Formulate a linear program to determine the proportion of a week each mill should operate so that the sales demand is met at a minimum cost.

5.   A dietician wishes to prepare a meal which has a minimum caloric value but which supplies at least 4 oz of protein, 2 oz of fat, and 1 oz of carbohydrate. The meal consists of a salad which has 30 cal/oz and is 50 percent protein, 10 percent fat, and 40 percent carbohydrate; meat which has 80 cal/oz, is 60 percent protein, 30 percent fat, and 10 percent carbohydrate; vegetable with 15 cal/oz, is 15 percent protein, 35 percent fat, and 50 percent carbohydrate; and a dessert at 300 cal/oz which is 5 percent protein, 15 percent fat, and 80 percent carbohydrate. Formulate the programing problem determined by these requirements.

6.   In planning a new city the developers have available a 100,000-acre tract of land. The land is to be zoned residential and industrial-business. For aesthetic reasons it is necessary to have at least 70 percent of the area devoted to residential use, but for practical reasons of employment, the residential area cannot exceed 90 percent of the total area.

a.   Formulate a linear program to find the desirable acreage to use in each zone in order to maximize the city tax base. It is estimated that each residential acre will provide $25,000 assessed valuation and each industrial acre $190,000.

b.   Formulate a linear program to find the acreage for each zone so as to maximize net revenue to the city. A real estate tax of $5 per $1000 of assessed value will be imposed. It is estimated that each industrial-business acre will produce pollution which will cost $900 of tax money to combat while each residential acre will produce pollution which will cost $50 of tax money to combat.

7.   An airline has two types of jets. The standard model accommodates 30 first-class and 170 economy-class passengers, while the larger model is fitted for 50 first-class and 150 economy-class passengers. The standard model operates at two-thirds the cost of the larger model. 350 first-class and 650 economy-class passengers must be transported. Formulate a linear program to determine how many of each plane should be used if cost is to be minimized.

8.   Classifying all workers in a city into one of three categories, white collar, blue collar, and executive, we wish to determine the proportion of workers for each category which will yield a maximum average income.

Each executive is paid twice the wage of a blue-collar worker, and each white-collar worker earns nine/tenths as much as a blue-collar worker. At least 3 white-collar workers are needed for each 10 blue-collar workers. The society will support no more than 1 executive for every 50 white- and blue-collar workers.

Formulate a linear program for this problem.

9.   *Compacts or sports cars?* A midwest assembly plant produces both compacts and sports cars. The plant has a limited capacity and must allocate the number of each type of car to assemble. Market research indicates that in this year no more than 50,000 compacts and 2500 sports cars will be sold from this plant. Certain assembly operations such as painting, interior finishing, and inspection limit the plant output. The time requirements and the annual time available for those three operations are given below:

Operation	Painting	Interior finishing	Inspection
Compacts	15 min	20 min	5 min
Sports cars	20 min	20 min	7 min
Annual time available	13,000 hr	17,000 hr	5200 hr

The profit margin on a compact is $100 and on a sports car $135. Find the number of each type of car which should be assembled this year in order to maximize the profit margin for this assembly plant.

Formulate a linear program for this problem.

10.   The Stodgy Fund is an investment firm with several types of securities in its portfolio. Broad rules governing their selection of investments are:

a.   at least 50 percent of the capital must be in blue-chip stocks and AAA bonds

b.   no more than 30 percent of the capital may be in B bonds and municipal bonds

c.   growth stocks may make up no more than 35 percent of the investments

d.   at most 25 percent of the capital may be in bonds of all types

Current yields on investments are given below:

Type of security	Blue-chip stocks	Growth stocks	AAA bonds	B bonds	Municipal bonds
Percentage yield	5.9	3	6	8.25	5.7

Find the current portfolio which maximizes the yield. Formulate a linear program for this problem.

11.   A group of younger investment analysts at Stodgy Fund would like to

see more emphasis placed on growth of investments. They estimate the percentage growth of the various types of securities as:

Type of security	Blue-chip stocks	Growth stocks	AAA bonds	B bonds	Municipal bonds
Percentage growth	3	12	2	7	1

Find the portfolio which maximizes growth of the investments yet adheres to the investment rules a through d of the preceding exercise.

Formulate a linear program for this problem.

12. A compromise is reached at Stodgy Fund so that the portfolio which gives the maximum of three times the yield plus the growth is agreed upon. Find the portfolio which attains that maximum.

Formulate a linear program for this problem.

13. *Purchase or manufacture?* Viable Computers can purchase two major components for their input/output consoles, or they can manufacture them. The first component, an MOS circuit, can be purchased for $12 and manufactured for $8. The second component, a minimemory, can be purchased for $18 and manufactured for $15. In the manufacturing process the MOS circuit requires 3 min of processing time on a circuit printer, the minimemory requires 4 min. The MOS circuit uses 2 min of processing time on the circuit test equipment, and the minimemory 1 min of testing time. The circuit printer is available for 2500 min/wk, the test equipment for 1000. Production requires 300 MOS circuits and 500 minimemories per week. Determine how many MOS circuits and minimemories should be manufactured and how many should be purchased each week.

Formulate a linear program for this problem.

3.7 SOLVING LINEAR PROGRAMS

The facts about linear functions which we considered in Sec. 3.4 can now be applied to find the solution to linear programs. In Sec. 3.4 we saw that *if* the linear function has a maximum (or minimum) value on a polyhedral convex set then there is an extreme point at which the function has its maximum (or minimum) value. For any linear program that is known to have a solution, we only need to locate all extreme points of the polyhedral convex set and then compare the values of the objective function at these points. The maximum (or minimum) of the objective function will be the largest (or smallest) of its values at extreme points.

Example 3.21   The polyhedral convex set from the problem of Sec. 3.4 was given in Example 3.20 as

$$\begin{bmatrix} 25 & 15 \\ 20 & 20 \\ 15 & 30 \\ 1 & 0 \\ 0 & 1 \end{bmatrix} \begin{bmatrix} x \\ y \end{bmatrix} \geq \begin{bmatrix} 1 \\ 1 \\ 1 \\ 0 \\ 0 \end{bmatrix}$$

This problem was solved in Sec. 3.1, so we know it has a solution and that we can determine that solution by finding the extreme points of the convex set.

A systematic method for locating *all* extreme points of this set involves locating all points of intersection for the bounding hyperplanes. In this problem the hyperplanes are lines in the plane, and so any pair of them can intersect at an extreme point. Thus we will systematically list *all pairs* of equations for the hyperplanes and then solve them for the points of intersection. One systematic way of listing the pairs is to first write all pairs which use the first equation $25x + 15y = 1$ (those are the pairs numbered 1 to 4 below), then for the second equation write all pairs which have not been written yet (those pairs are 5 through 7), and continue in this manner to obtain

1. $25x + 15y = 1$
   $20x + 20y = 1$
2. $25x + 15y = 1$
   $15x + 30y = 1$
3. $25x + 15y = 1$
   $x = 0$
4. $25x + 15y = 1$
   $y = 0$
5. $20x + 20y = 1$
   $15x + 30y = 1$
6. $20x + 20y = 1$
   $x = 0$
7. $20x + 20y = 1$
   $y = 0$
8. $15x + 30y = 1$
   $x = 0$
9. $15x + 30y = 1$
   $y = 0$
10. $x = 0$
    $y = 0$

Solving all 10 pairs of equations locates the following 10 points of intersection:

1. $[1/40, 1/40]$
2. $[1/35, 2/105]$
3. $[0, 1/15]$
4. $[1/25, 0]$
5. $[1/30, 1/60]$
6. $[0, 1/20]$
7. $[1/20, 0]$
8. $[0, 1/30]$
9. $[1/15, 0]$
10. $[0, 0]$

To determine which of these are actually extreme points, we must determine if the point lies in the intersection of all the half-spaces. Thus we check to see that all the inequalities are satisfied for each of the 10 points. In each case the point was determined by two of the equations; so we do not have to check the two corresponding inequalities.

For example, the point $[1/40, 1/40]$ is the point of intersection of the first two equations, thus we need only check the inequalities $15x + 30y \geq 1$, $x \geq 0$, and $y \geq 0$ to verify that $[1/40, 1/40]$ lies in all the half-spaces which intersect to give the convex set. Next we check the point $[\frac{1}{35}, \frac{2}{105}]$ which is the point of intersection of the first and third equations. This point does not satisfy the inequality $20x + 20y \geq 1$, so $[\frac{1}{35}, \frac{2}{105}]$ lies on the wrong side of that hyperplane and cannot be an extreme point. The results for all 10 points are now given:

1. $[\frac{1}{40}, \frac{1}{40}]$, yes
2. $[\frac{1}{35}, \frac{2}{105}]$, no since $20(\frac{1}{35}) + 20(\frac{2}{105}) < 1$
3. $[0, \frac{1}{15}]$, yes
4. $[\frac{1}{25}, 0]$, no since $20(\frac{1}{25}) < 1$
5. $[\frac{1}{30}, \frac{1}{60}]$, yes

6. $[0, \frac{1}{20}]$, no since $15(\frac{1}{20}) < 1$
7. $[\frac{1}{20}, 0]$, no since $15(\frac{1}{20}) < 1$
8. $[0, \frac{1}{30}]$, no since $15(\frac{1}{30}) < 1$
9. $[\frac{1}{15}, 0]$, yes
10. $[0, 0]$, no since none of the first three are satisfied

The extreme points are the four points $[\frac{1}{40}, \frac{1}{40}]$, $[0, \frac{1}{15}]$, $[\frac{1}{30}, \frac{1}{60}]$, $[\frac{1}{15}, 0]$. All 10 of the points should be located in Fig. 3.4 but note especially the extreme points. We next compute the value of the objective function $f(x,y) = (1/1000)x + (1.5/1000)y$ (see Sec. 3.1). The values are given in Table 3.3 below.

	x	$\frac{1}{40}$	0	$\frac{1}{30}$	$\frac{1}{15}$
	y	$\frac{1}{40}$	$\frac{1}{15}$	$\frac{1}{60}$	0
$(\frac{1}{1000})x + (\frac{1.5}{1000})y$		$\frac{1}{16,000}$	$(\frac{1}{10,000})$	$\frac{7}{120,000}$	$\frac{1}{15,000}$

Table 3.3 Values of the objective function at the extreme points

The number $7/120,000$ is the smallest value of $f(x, y)$ for the four extreme points (compare $7/120,000$ and $1/16,000 = 7/112,000$). We conclude that the minimum value of the objective function on this convex set is $7/120,000$ and this minimum value is attained at the point where $x = \frac{1}{30}$, $y = \frac{1}{60}$.

Example 3.22 The problem of Example 3.19 was to find the maximum of the function $f(x, y) = 80x + 50y$ for points in the convex set given by

$$\begin{bmatrix} 3 & 20 \\ 12 & 2 \\ 4 & 10 \\ -1 & 0 \\ 0 & -1 \end{bmatrix} \begin{bmatrix} x \\ y \end{bmatrix} \leq \begin{bmatrix} 1600 \\ 1000 \\ 1200 \\ 0 \\ 0 \end{bmatrix}$$

We know that this problem has a solution because the profit cannot be arbitrarily large when the resources are limited. One of the extreme points must give the solution. The possible extreme points are found from the following systems of equations:

1. $\quad 3x + 20y = 1600$ 5. $\quad 12x + 2y = 1000$ 8. $\quad 4x + 10y = 1200$
$\quad\quad 12x + 2y = 1000$ $4x + 10y = 1200$ $-x = 0$

2. $\quad 3x + 20y = 1600$ 6. $\quad 12x + 2y = 1000$ 9. $\quad 4x + 10y = 1200$
$\quad\quad 4x + 10y = 1200$ $-x = 0$ $-y = 0$

3. $\quad 3x + 20y = 1600$ 7. $\quad 12x + 2y = 1000$ 10. $\quad -x = 0$
$\quad\quad -x = 0$ $-y = 0$ $-y = 0$

4. $\quad 3x + 20y = 1600$
$\quad\quad -y = 0$

From the systems of equations we find the 10 points of intersection for the hyperplanes. Those 10 points are:

1. $[2800/39,\ 2700/39]$ 5. $[475/7,\ 650/7]$ 8. $[0, 120]$
2. $[160, 56]$ 6. $[0, 500]$ 9. $[300, 0]$
3. $[0, 80]$ 7. $[250/3, 0]$ 10. $[0, 0]$
4. $[1600/3, 0]$

Checking to determine which of these are actually extreme points, we find the following results:

1. $[2800/39,\ 2700/39]$, yes   6. $[0, 500]$, no since $20(500) > 1600$
2. $[160, 56]$, no since   7. $[250/3, 0]$, yes
$\quad 12(160) + 2(56) > 1000$
3. $[0, 80]$, yes   8. $[0, 120]$, no since $20(120) > 1600$
4. $[1600/3, 0]$, no since   9. $[300, 0]$, no since $12(300) > 1000$
$\quad 12(1600/3) > 1000$
5. $[475/7,\ 650/7]$, no since   10. $[0, 0]$, yes
$\quad 3(475/7) + 20(650/7) > 1600$

The extreme points and the value of the objective function are given in Table 3.4.

x	$2800/39$	0	$250/3$	0
y	$2700/39$	80	0	0
$80x + 50y$	$359,000/39$	4000	$20,000/3$	0

Table 3.4   The values of the objective function at the extreme points

The largest value for the objective function is $20,000/3$ which occurs at the extreme point $x = 250/3$, $y = 0$. Thus the solution is to schedule $250/3$ living room suites and no dining room suites to earn a profit of $20,000/3$.

For problems which have only two variables, it may actually be more work to find all possible extreme points than it is to sketch the convex set and focus only on the obvious extreme points; however, the method of finding all extreme points works in any number of dimensions. Our next example shows an application of this technique to a problem in three dimensions.

Example 3.23   Consider the problem of finding the maximum of the function $f(x, y, z) = x + y + z$ for points of the polyhedral convex set

$$\begin{bmatrix} -2 & 1 & 2 \\ 1 & 1 & -1 \\ -1 & 0 & 0 \\ 0 & -1 & 0 \\ 0 & 0 & 1 \end{bmatrix} \begin{bmatrix} x \\ y \\ z \end{bmatrix} \leq \begin{bmatrix} 1 \\ 3 \\ 0 \\ 0 \\ -1 \end{bmatrix}$$

Since the intersection of at least three hyperplanes is required to determine a point in three dimensions, we must use three equations at a time and in all combinations to locate the possible extreme points. Again a systematic method for listing the equations is needed. We will begin by listing all combinations which use the first equation, listing them according to the scheme we used for listing pairs as we complete each system of three equations by listing a pair of equations chosen from the remaining four. Those combinations give us the first six systems listed; next we list all systems not already listed which use the second equation and so on.

The resulting list is

1. $-2x + y + 2z = 1$ 5. $-2x + y + 2z = 1$ 8. $x + y - z = 3$
    $x + y - z = 3$ $-x = 0$ $-x = 0$
    $-x = 0$ $z = -1$ $z = -1$
2. $-2x + y + 2z = 1$ 6. $-2x + y + 2z = 1$ 9. $x + y - z = 3$
    $x + y - z = 3$ $-y = 0$ $-y = 0$
    $-y = 0$ $z = -1$ $z = -1$
3. $-2x + y + 2z = 1$ 7. $x + y - z = 3$ 10. $-x = 0$
    $x + y - z = 3$ $-x = 0$ $-y = 0$
    $z = -1$ $-y = 0$ $z = -1$
4. $-2x + y + 2z = 1$
    $-x = 0$
    $-y = 0$

From these systems of equations we locate the points of intersection and check to see if they are in the proper half-spaces. The following results are obtained:

1. $[0, 7/3, -2/3]$, no since   6. $[-3/2, 0, -1]$, no since
    $-2/3 > -1$ $-(-3/2) > 0$
2. no point of intersection   7. $[0, 0, -3]$, yes
3. $[-1/3, 7/3, -1]$, no since   8. $[0, 2, -1]$, yes
    $-(-1/3) > 0$
4. $[0, 0, 1/2]$, no since   9. $[2, 0, -1]$, yes
    $1/2 > -1$
5. $[0, 3, -1]$, no since   10. $[0, 0, -1]$, yes
    $3 - (-1) = 4 > 3$

For the extreme points of this polyhedral convex set the values of the objective function are listed in Table 3.5.

x	0	0	2	0
y	0	2	0	0
z	−3	−1	−1	−1
$x + y + z$	−3	1	1	−1

Table 3.5　The values of the objective function at the extreme points

If the objective function has a maximum value, it is 1 and that value occurs at the point $[2, 0, -1]$. Since we have no intuitive interpretation for this problem, we do not know if there is obviously a largest value of $x + y + z$ for points in this convex set. See the remarks below.

To determine if a value M of the objective function is a maximum (or minimum) when it is not obvious to us that the objective function must have a maximum (or minimum) value requires more theory. In practice it is usually clear that the problem has a solution—in such cases we need only examine the extreme points. When a linear program is stated as in Example 3.23 we must verify that the problem has a solution.

If there are no points of the convex set which are arbitrarily far away from the origin, the convex set is called bounded. *For a bounded polyhedral convex set every linear function has a maximum and a minimum value.* Some of the polyhedral convex sets we have considered are bounded, and some are not. In Fig. 3.4, the convex set is not a bounded one for all points above the bounding lines are in that set. The convex set in Fig. 3.19 is a bounded set, the point $[6, 4]$ is obviously the most distant point from the origin; on the other hand, the set shown in Fig. 3.20 is not a bounded set. If you sketch the polyhedral convex set of feasible solutions to the problem of Example 3.19, you will find that the set is bounded (see Exercise 3.7.8).

Exercise 3.7　1.　Find all the extreme points of the polyhedral convex set

$$\begin{bmatrix} -2 & 3 \\ 4 & 7 \\ 6 & -5 \\ -1 & 0 \\ 0 & -1 \end{bmatrix} \begin{bmatrix} x \\ y \end{bmatrix} \leq \begin{bmatrix} 6 \\ 28 \\ 30 \\ 0 \\ 0 \end{bmatrix}$$

2.　Determine the maximum and the minimum values of the function $f(\mathbf{v}) = [6, 1] \cdot \mathbf{v}$ for those points in the polyhedral convex set where

$$\begin{bmatrix} 2 & 1 \\ -1 & 0 \\ 1 & -2 \end{bmatrix} \begin{bmatrix} x \\ y \end{bmatrix} \leq \begin{bmatrix} 4 \\ 0 \\ 2 \end{bmatrix}$$

(*Hint*: You will have to be sure that there really is a maximum and minimum so you should sketch the set to see that it is bounded.)

3.   Exercise 3.6.4 has the formulation:

Find the minimum of $x + \frac{2}{3}y$ for

$$\begin{bmatrix} 150{,}000 & 90{,}000 \\ 75{,}000 & 15{,}000 \\ 75{,}000 & 45{,}000 \\ 1 & 0 \\ 0 & 1 \\ -1 & 0 \\ 0 & -1 \end{bmatrix} \begin{bmatrix} x \\ y \end{bmatrix} \geq \begin{bmatrix} 200{,}000 \\ 50{,}000 \\ 100{,}000 \\ 0 \\ 0 \\ -1 \\ -1 \end{bmatrix}$$

Solve this linear program.

*4.   Exercise 3.6.5 has the formulation:

Find the minimum of $30s + 80m + 15v + 300d$ where

$$\begin{bmatrix} 0.5 & 0.6 & 0.15 & 0.05 \\ 0.1 & 0.3 & 0.35 & 0.15 \\ 0.4 & 0.1 & 0.5 & 0.8 \end{bmatrix} \begin{bmatrix} s \\ m \\ v \\ d \end{bmatrix} \geq \begin{bmatrix} 4 \\ 2 \\ 1 \end{bmatrix}$$

and $s \geq 0$, $m \geq 0$, $v \geq 0$, $d \geq 0$.
  Solve this linear program.

5.   Exercise 3.6.6a has the formulation:

Find the maximum of $25{,}000r + 190{,}000i$ where

$$\begin{bmatrix} 1 & 1 \\ 1 & 0 \\ -1 & 0 \\ 0 & -1 \end{bmatrix} \begin{bmatrix} r \\ i \end{bmatrix} \leq \begin{bmatrix} 100{,}000 \\ 90{,}000 \\ -70{,}000 \\ 0 \end{bmatrix}$$

Solve this linear program.

6.   Exercise 3.6.7 has the formulation:

Find the minimum of $\frac{2}{3}S + L$ where

$$\begin{bmatrix} 30 & 50 \\ 170 & 150 \\ 1 & 0 \\ 0 & 1 \end{bmatrix} \begin{bmatrix} S \\ L \end{bmatrix} \geq \begin{bmatrix} 350 \\ 650 \\ 0 \\ 0 \end{bmatrix}$$

Solve this linear program.

7.   Exercise 3.6.8 has the formulation:

Find the maximum of $2e + b + 0.9w$ for

$$\begin{bmatrix} 0 & -\frac{1}{10} & \frac{1}{3} \\ -1 & \frac{1}{50} & \frac{1}{50} \\ -1 & -1 & -1 \\ 1 & 0 & 0 \\ 0 & 1 & 0 \\ 0 & 0 & 1 \end{bmatrix} \begin{bmatrix} e \\ b \\ w \end{bmatrix} \geq \begin{bmatrix} 0 \\ 0 \\ -1 \\ 0 \\ 0 \\ 0 \end{bmatrix}$$

Solve this linear program.

8.   Sketch the polyhedral convex set specified by

$$\begin{bmatrix} 3 & 20 \\ 12 & 2 \\ 4 & 10 \\ -1 & 0 \\ 0 & -1 \end{bmatrix} \begin{bmatrix} x \\ y \end{bmatrix} \leq \begin{bmatrix} 1600 \\ 1000 \\ 1200 \\ 0 \\ 0 \end{bmatrix}$$

a.   Show that this set is bounded by finding a circle which contains the set.
b.   Notice that both x and y have largest possible values and smallest possible values for all points in the convex set.
c.   Show that $ax + by$ must have a largest possible value and a smallest possible value for all points of the convex set. For simplicity, assume that both a and b are positive numbers.

3.8   DUALITY AND THE SIMPLEX ALGORITHM

Solving a linear program is routine once the program has been formulated. In fact, it is possible to write a list of instructions for an electronic computer to follow in solving a linear program. A list of instructions which can be executed by a computer to solve a problem is called an *algorithm*. The algorithm commonly used to solve linear programs is called the *simplex algorithm* (the term *simplex* refers to a geometric concept rather than the difficulty of the algorithm). Before explaining more of the simplex algorithm, we will define a concept which relates certain linear programs to each other. This concept, called *duality*, makes it easier to describe the simplex algorithm.

In defining duality we are going to give greater attention to row and column vectors for specifying a linear program. As a reminder, we state again that a vector **u** which multiplies a matrix **A** on the left is necessarily a row vector and the product is a row vector. Duality is now formally presented in Definition 3.6.

DEFINITION 3.6   For the linear function $f(\mathbf{v}) = \mathbf{a} \cdot \mathbf{v}$, the problem of finding the *maximum* of f, for those vectors $\mathbf{v}$ where $\mathbf{A} \cdot \mathbf{v} \leq \mathbf{b}$, is called the *dual* to the problem of finding the *minimum* of the linear function $g(\mathbf{u}) = \mathbf{b} \cdot \mathbf{u}$ where $\mathbf{u} \cdot \mathbf{A} \geq \mathbf{a}$, and conversely.

The use of row and column vectors makes it easier to state duality. Notice that $\mathbf{a}$ and $\mathbf{b}$ interchange positions in the dual linear programs, that $\mathbf{u}$ is a row vector of variables, that $\mathbf{v}$ is a column vector of variables, and in general that $\mathbf{u}$ and $\mathbf{v}$ are points from two different coordinate systems. It is also important that we have written the maximum problem by using the inequality less than or equal to, while the minimum problem is stated by an inequality using greater than or equal to. An example will help to establish these distinctions.

Example 3.24   The maximum linear program with objective function $f(\mathbf{v}) = [40, 25] \cdot \mathbf{v}$ where

$$\begin{bmatrix} 2 & 5 \\ 6 & 0 \\ 0 & 10 \end{bmatrix} \begin{bmatrix} x \\ y \end{bmatrix} \leq \begin{bmatrix} 10 \\ 10 \\ 16 \end{bmatrix}$$

is dual to the minimum linear program with objective function

$$g(\mathbf{u}) = \mathbf{u} \cdot \begin{bmatrix} 10 \\ 10 \\ 16 \end{bmatrix}$$

and where

$$[r, s, t] \begin{bmatrix} 2 & 5 \\ 6 & 0 \\ 0 & 10 \end{bmatrix} \geq [40, 25]$$

Theorem 3.3 states the importance of duality to the problem of finding solutions to linear programs.

THEOREM 3.3   If a linear program has a solution, then the dual linear program has a solution. The value of the objective function at the solution is the same for both linear programs.

We will not prove this result for you. We state Theorem 3.3 because it is the basis of our approach to giving a general method for solving linear programs. We will present a method for solving linear programs which gives the solution, simultaneously, for both dual problems.

SIMPLEX ALGORITHM   The simplex algorithm not only allows us to find solutions routinely, but with the aid of a computer, we can solve problems for which it would be humanly impossible to do the necessary amount of computation.

There is a danger in producing solutions routinely, since analysis of information must be done by someone who understands the results. To help you to understand the simplex algorithm step by step, we will present an account of the information produced at each step.

The algorithm we describe carries out each step in an attempt to find the maximum of the objective function; hence we always deal with a maximum problem. When the answer to the problem is found, we will be able to give the solution to the dual minimum problem, as well as to the stated maximum problem. Whenever a minimum problem is posed, we will first formulate the dual maximum problem and then use the simplex algorithm to solve both problems simultaneously.

An example will serve as our primary method of giving the simplex algorithm.

Example 3.25   For purposes of this example we pose the following linear program. Find the maximum of the linear function $f(x, y) = 40x + 25y$ for points of the convex set

$$\begin{bmatrix} 2 & 5 \\ 6 & 0 \\ 0 & 10 \end{bmatrix} \begin{bmatrix} x \\ y \end{bmatrix} \le \begin{bmatrix} 1000 \\ 1000 \\ 1600 \end{bmatrix}$$

and where both $x \ge 0$ and $y \ge 0$. *In the simplex algorithm the condition $x \ge 0$, $y \ge 0$ will always be assumed to hold.* The polyhedral convex set described above is shown in Fig. 3.25.

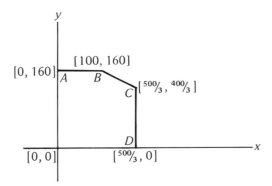

Figure 3.25   The set of feasible solutions

The simplex algorithm begins by specifying the linear program as a table of numbers. Table 3.6 gives the format for this table of numbers using the notation of a general maximum problem in which $f(\mathbf{v}) = \mathbf{a} \cdot \mathbf{v}$, and the convex set is given by $\mathbf{A} \cdot \mathbf{v} \le \mathbf{b}$ with the additional condition $\mathbf{v} \ge \mathbf{0}$ to be understood.

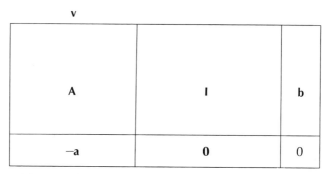

Table 3.6   The initial simplex table

When $\mathbf{A}$ is an $m \times n$ matrix, the vector $\mathbf{a}$ is n-dimensional and $-\mathbf{a} = [-a_1, -a_2, \ldots, -a_n]$; $\mathbf{I}$ indicates an $m \times m$ matrix with i,jth element 1 if $i = j$ and 0 if $i \ne j$.

The vector $\mathbf{0}$ is m-dimensional with all components 0; $\mathbf{b}$ is of course the m-dimensional vector from the linear program; and finally the lower right-hand corner contains the number 0. The vector $\mathbf{v}$ of variable symbols is written above the columns of $\mathbf{A}$.

For our example this is given in Table 3.7.

x	y				
2	5	1	0	0	1000
6	0	0	1	0	1000
0	10	0	0	1	1600
−40	−25	0	0	0	0

Table 3.7   The initial simplex table for Example 3.24

STEP 1   The first step is designed to consider all extreme points adjacent to the origin (A and D in Fig. 3.25); choose one for which the value of the objective function will be larger than zero (the value at the origin), and evaluate the objective function at that point. The actual procedure by which this is carried out is now described.

Choose one of the negative elements from the bottom row of the table. While it is not obvious, a negative element indicates at which of the adjacent extreme points the objective function will have a positive value. In this example either of the first two numbers will suffice; we will choose −25. This choice focuses our attention on the second column. For purposes of reference we will call this the *pivot column*.

We next wish to select a row from the upper section of the table to be the *pivot row*. To determine the pivot row, we examine the quotient of each element of the last column in the table by the element in the corresponding row of the pivot column; the bottom row of the table is not used in this computation. We seek to determine that row for which the quotient is the *smallest positive number*. For our example this involves computing 1000 divided by 5 and 1600 divided by 10 (ignore the second row since division by zero is not defined). The smaller positive quotient occurs in the third row, which is therefore the pivot row. The pivot column and pivot row together determine the *pivot point*. Here the pivot point is the entry in the third row and second column.

GEOMETRIC SIDELIGHT   Selection of the pivot row has a geometric significance. Recall that each bounding hyperplane of the convex set is used to determine a row of the matrix **A** which in turn determines the initial simplex table. When computing the quotients we are searching for the hyperplane which intersects the y axis at a point with positive coordinate and nearest to the origin; that is why we choose the smallest positive quotient.

We circle the number in the pivot point as a means of identifying the pivot point, and we write the variable name appearing above the pivot column as a label to the left of the pivot row. This is all preliminary to the main computations to be carried out; our table now has the appearance shown in Table 3.8.

	x	y				
	2	5	1	0	0	1000
	6	0	0	1	0	1000
y	0	⑽	0	0	1	1600
	−40	−25	0	0	0	0

Table 3.8   The pivot point chosen from Table 3.7

We now systematically add multiples of the pivot row to the other rows including the bottom row in order to obtain zeros in all entries of the pivot column except in the pivot point. We divide the pivot row by the element appearing in the pivot point in order to obtain a 1 in the pivot point.

In Table 3.9, section *a* records the result after we divide the pivot row by the element in the pivot point. Table 3.9*b* and *c* record the completion of the simplex table as we would naturally fill it in while doing the computations.

Table 3.9*c* contains the following information. The point $x = 0$, $y = 160$

	x	y				
y	0	①	0	0	$1/10$	160

a. *The pivot row has been divided by 10.*

	x	y				
	2	0	1	0	$-1/2$	200
y	0	①	0	0	$1/10$	160

b. *Add -5 times the pivot row of Fig. 3.5 to the first row of Table 3.8.*

	x	y				
	2	0	1	0	$-1/2$	200
	6	0	0	1	0	1000
y	0	①	0	0	$1/10$	160
	-40	0	0	0	$5/2$	4000

c. *Copy row 2 unaltered and add 25 times the pivot row to the bottom row of the table in Table 3.8.*

Table 3.9   Completion of the simplex table for step 1

is an extreme point, and the value of the objective function at [0, 160] is 4000. A variable written to the left of a row (here the only such variable is y) has the value given by the element in the last column of that row. Any variable which is not written on the left is given the value zero. The values assigned the variables identify an extreme point, and the value of the objective function at that extreme point appears in the bottom right corner of the table. Notice that the point [0, 160] is the extreme point labeled A in Fig. 3.25.

STEP 2   The procedure is repeated using Table 3.9c. We choose a negative number from the bottom row to determine a new pivot column. We divide the elements of the last column by the elements from the same row which appear in the pivot column. (The result is 200 divided by 2 and 1000 divided by 6.) The smaller quotient is in the top row which becomes the new pivot row, and the pivot point is in the first row and first column. The pivot point is circled, and the variable x written to the left of the first row.

Dividing the first row by the value of the pivot element, adding and subtracting multiples of the pivot row to all other rows transforms Table 3.9c into Table 3.10.

	x	y				
x	①	0	$\frac{1}{2}$	0	$-\frac{1}{4}$	100
	0	0	-3	1	$\frac{3}{2}$	400
y	0	1	0	0	$\frac{1}{10}$	160
	0	0	20	0	$-\frac{15}{2}$	8000

Table 3.10   Completion of the simplex table for step 2

We now read that at the point [100, 160]—the extreme point labeled B in Fig. 3.25—the value of the objective function is 8000 (read in the bottom right corner).

The simplex table still has a negative number in the bottom row indicating that the objective function will have a larger value at the next adjacent extreme point. The new pivot column (column 5) has the two positive quotients $400/(\frac{3}{2})$ and $160/(\frac{1}{10})$ with the smaller value in the second row. This pivot column has *no variable heading so we do not write a variable name to the left of the pivot row.* The remaining steps are as before and result in Table 3.11.

	x	y				
x	1	0	0	$\frac{1}{6}$	0	$\frac{500}{3}$
	0	0	-2	$\frac{2}{3}$	①	$\frac{800}{3}$
y	0	1	$\frac{1}{5}$	$-\frac{1}{15}$	0	$\frac{400}{3}$
	0	0	5	5	0	10,000
			r	s	t	

Table 3.11   The final simplex table

There are now *no negative numbers in the bottom row.* This is the signal that computations terminate. The extreme point $[\frac{500}{3}, \frac{400}{3}]$ determines the solution to this maximum problem with the value 10,000 for the objective function.

THE MINIMUM PROBLEM   According to Theorem 3.3 the dual minimum problem has a solution with the minimum value of the objective function equal to 10,000. You may verify that the dual to the maximum problem we have solved is to find the minimum of the linear function $g(r, s, t) = 1000r + 1000s + 1600t$ where

$$[r, s, t] \begin{bmatrix} 2 & 5 \\ 6 & 0 \\ 0 & 10 \end{bmatrix} \geq [40, 25]$$

To complete the solution for this minimum problem, we must know the extreme point at which the minimum value is attained. The coordinates of that extreme point appear along the bottom row of the table. The values of r, s, t at the solution which we have found are given by the numbers directly above r, s, t in the bottom row of Table 3.11. Thus $r = 5$, $s = 5$, and $t = 0$ provide the value $1000(5) + 1000(5) + 1600(0) = 10,000$ for the objective function. This value is in agreement with the number we claimed to be the minimum, the number appearing in the bottom right corner of the simplex table.

Instructions for Simplex Algorithm

1. Set up the simplex table with **A**, **b**, and −**a** from the maximum problem.
2. Select a column for which the entry in the bottom row is negative; this is the pivot column. If there is no negative number in the bottom row, go to 6.
3. Divide the elements of the upper section of the last column by the elements in the corresponding row of the pivot column. Select a row for which this quotient is the smallest positive number, and denote this row as the pivot row. The pivot row and pivot column together determine the pivot point.
4. Circle the pivot point. Write the variable name, if any, of the pivot column as a label for the pivot row. If a variable name was used to label the pivot row at a previous step, erase that previous label.
5. Divide each entry in the pivot row by the pivot entry, obtaining a 1 in the pivot point. Obtain zeros in all entries of the pivot column except at the pivot point by adding to or subtracting from each of the other rows the appropriate multiple of the pivot row. Go to 2.
6. Computation terminates. The extreme point for the solution to the maximum problem has coordinates read from the last column of all labeled rows, with the label determining which coordinate has that value. All coordinates which are not labeled have the value 0.

The extreme point for the solution to the minimum problem appears along the bottom row of the table in the center section. The coordinates are written in the order that they appear in **u**.

The maximum of the objective function $f(\mathbf{v}) = \mathbf{a} \cdot \mathbf{v}$ equals the minimum of the objective function $g(\mathbf{u}) = \mathbf{u} \cdot \mathbf{b}$ and is the number in the bottom right corner of the final simplex table.

Exercise 3.8   *Use the simplex algorithm to solve these problems.*

1.   Find the maximum of $5x + 3y + 4z$ for the convex set determined by

$$\begin{bmatrix} 2 & 1 & 2 \\ 1 & 3 & 2 \\ 2 & 2 & 3 \end{bmatrix} \begin{bmatrix} x \\ y \\ z \end{bmatrix} \leq \begin{bmatrix} 8 \\ 5 \\ 4 \end{bmatrix}$$

where $x \geq 0$, $y \geq 0$, $z \geq 0$.

2.   Minimize $6r + s + 5t$ when $[r, s, t] \geq [0, 0, 0]$ and

$$[r, s, t] \begin{bmatrix} 2 & 1 \\ 1 & 2 \\ 1 & 3 \end{bmatrix} \geq [3, 5]$$

3.   For a linear program we find $\mathbf{a} = [3, 5, -2]$, $\mathbf{b} = [3, 4]$, and

$$\mathbf{A} = \begin{bmatrix} 2 & -1 & 3 \\ 1 & 2 & 4 \end{bmatrix}$$

Solve both the maximum and the dual minimum problems.

4.   We want to find the minimum of $20s + 20t$ where

$$[s, t] \begin{bmatrix} 1 & 3 & 5 \\ 2 & 2 & 2 \end{bmatrix} \geq [2, 4, 5]$$

and $[s, t] \geq [0, 0]$.

Solve this minimum problem graphically and by the simplex method as a comparison of the work required by the two methods.

5.   Find the maximum of $9x + 5y + z$ when

$$\begin{bmatrix} 2 & -5 & 3 \\ 2 & 3 & -1 \\ 3 & -2 & 1 \end{bmatrix} \begin{bmatrix} x \\ y \\ z \end{bmatrix} \leq \begin{bmatrix} 12 \\ 3 \\ 2 \end{bmatrix}$$

3.9   MORE ON THE SIMPLEX ALGORITHM

We have avoided all discussion of possible complications which might arise when using the simplex algorithm. The steps which were explained in the last section are not adequate to complete the computations in every case, and you should be aware of certain pitfalls. You will recall that the word *algorithm* means a list of instructions for carrying out a computation. The algorithm we presented in Sec. 3.8 will produce the answer in most cases, but in order to work every time, more instructions are necessary. For example, we have not worried about the event that certain steps cannot be carried out. It might happen that the entries in the pivot column are all zero or that we might be unable to locate a pivot row because all quotients are negative. To give an algorithm for solving all linear programs then requires instructions to deal with the special cases. To that extent we have not presented the complete simplex algorithm. If there is an electronic computer or a computing center available to you, there will undoubtedly be a library program called the simplex algorithm which uses a more complete list of instructions than you have been given.

Our purpose in this section is not to extend the list of instructions given in the last section but instead to discuss some of the problems which might arise as well as to give conditions which assure success for the algorithm you now know about.

CONDITIONS FOR THE SIMPLEX ALGORITHM   The algorithm given in Sec. 3.8 will produce solutions for both dual problems if the conditions below are satisfied.

For the maximum problem with objective function

$$f(\mathbf{v}) = \mathbf{a} \cdot \mathbf{v} \qquad \text{where } \mathbf{A} \cdot \mathbf{v} \leq \mathbf{b}$$

the conditions are:

1.   The problem has a solution.
2.   The components of **b** are all nonnegative.
3.   If the convex set is in n-dimensional space, then no more than n of the bounding hyperplanes intersect at one extreme point.

The remainder of this section is devoted to examples of linear programs which are excluded by conditions 1, 2, and 3. Our examples are chosen to illustrate what can happen when carrying out the algorithm if one of these conditions is not met. We will see that in some cases the algorithm provides a solution even when one or more of 1, 2, or 3 is not satisfied.

Of course when condition 1 is not satisfied, no method can produce a solution. Our first example will pose a linear program which we know has no solution, but we will try the algorithm anyway to observe the result.

Example 3.26   The linear function $x + y$ has no maximum value on the polyhedral convex set where $x \geq 0$, $y \geq 0$ and

$$\begin{bmatrix} -1 & 1 \\ 1 & -2 \end{bmatrix} \begin{bmatrix} x \\ y \end{bmatrix} \leq \begin{bmatrix} 1 \\ 2 \end{bmatrix}$$

Figure 3.26 gives a sketch of the polyhedral convex set, which you will note is not bounded and so there are positive values of x and y as large as we want which are contained in this set.

We now give the simplex table after choosing the first column as pivot

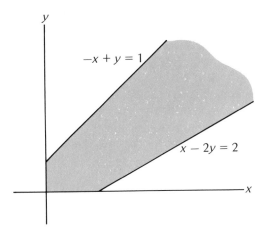

Figure 3.26　The convex set of Example 3.26

column and completing the computations for one step of the simplex algorithm. Table 3.12 contains the initial and final simplex tables.

x	y			
−1	1	1	0	1
1	−2	0	1	2
−1	−1	0	0	0

x

x	y			
0	−1	1	1	3
①	−2	0	1	2
0	−3	0	1	2

Table 3.12　The simplex tables for Example 3.26

The second step of our simplex algorithm names the second column as pivot column, but there are no positive quotients of the elements of this column with those of the last column. What does this mean? It means that there is an edge of the convex set along which the value of the objective function increases, but there is no extreme point along that edge—the function can increase indefinitely. *The linear program has no solution.*

The next example we have chosen will show that condition 2 is not crucial; there can be components of **b** which are zero or even negative yet we find a solution.

Example 3.27　For this example the objective function is $x + y$ and the convex set is given by $x \geq 0$, $y \geq 0$ and

$$\begin{bmatrix} 1 & 0 \\ 0 & 1 \\ -1 & -1 \end{bmatrix} \begin{bmatrix} x \\ y \end{bmatrix} \leq \begin{bmatrix} 2 \\ 3 \\ -2 \end{bmatrix}$$

This polyhedral convex set is shown in Fig. 3.27. Obviously the objective function $x + y$ has its largest value at the extreme point $[2, 3]$ where the value is 5. Let us use the simplex algorithm even though the third component of **b** is -2 in violation of condition 2.

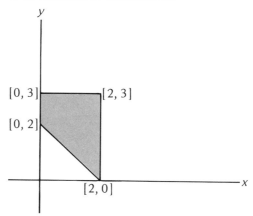

Figure 3.27   The convex set of Example 3.27

Table 3.13 gives the simplex tables which result and shows that we find the solution even though condition 2 is not satisfied.

x	y				
1	0	1	0	0	2
0	1	0	1	0	3
−1	−1	0	0	1	−2
−1	−1	0	0	0	0

a.   *The initial simplex table*

	x	y				
x	①	0	1	0	0	2
	0	1	0	1	0	3
	0	−1	1	0	1	0
	0	−1	1	0	0	2

b.   *The simplex table after one step*

	x	y				
x	1	0	1	0	0	2
y	0	①	0	1	0	3
	0	0	1	1	1	3
	0	0	1	1	0	5

c.   *The final simplex table*

Table 3.13   The simplex tables leading to a solution for Example 3.27

It is possible, when condition 2 is not satisfied, that the algorithm fails to find the solution. This happens when the algorithm fails to determine an extreme point of the convex set. Exercise 3.9.4 provides a situation of this type.

Finally we will show one example of what might happen if condition 3 is not satisfied.

Example 3.28   Consider the linear program of Example 3.27 again. Notice in Table 3.13 that the three lines with equations $y = 0$, $x + y = 2$, and $x = 2$ all intersect at the point $[2, 0]$. Since these three lines, including $y = 0$, are among the bounding hyperplanes of the convex set, and they all three intersect at the point $[2, 0]$ in two dimensions, condition 3 is not satisfied. We have already seen in Example 3.27 that the algorithm gives a solution, but there is one point worth noting which the solution given in Example 3.27 did not bring out. At the first step in the algorithm—the step used to obtain Table 3.13b— there was a choice of pivot row since the first and third quotients are both 2. We will next show what happens if the alternate choice of pivot row is used. Table 3.14 shows the simplex tables which result.

x	y				
1	0	1	0	0	2
0	1	0	1	0	3
−1	−1	0	0	1	−2
−1	−1	0	0	0	0

a. *The initial simplex table*

	x	y				
	0	−1	1	0	1	0
	0	1	0	1	0	3
x	(1)	1	0	0	−1	2
	0	0	0	0	−1	2

b. *The simplex table after one step*

Table 3.14   The simplex tables for Example 3.28

Now the choice of pivot row at the first step leads us to the fifth column as the pivot column for the second step—but none of the quotients are positive so we cannot select a pivot row. Obviously the way out of this dilemma is to make the other choice of pivot row at the first step when the choice was made. Of course, there was no instruction given in Sec. 3.8 to handle this, so that the algorithm fails to find the solution when this difficulty is encountered. An added instruction to handle such cases is part of the complete list of instructions for the simplex algorithm.

Exercise 3.9   1.   Use the simplex algorithm for the maximum problem $f(x, y) = x$ on the convex set $x \geq 0$, $y \geq 0$ and

$$\begin{bmatrix} -2 & 1 \\ 1 & -1 \end{bmatrix} \begin{bmatrix} x \\ y \end{bmatrix} \leq \begin{bmatrix} 1 \\ 1 \end{bmatrix}$$

       a. Note that the convex set is unbounded, and x can be as large as we want to have it.

       b. What happens in the simplex table?

    2.   The convex set given by

$$\begin{bmatrix} 1 & 0 \\ 0 & 1 \\ -1 & -1 \end{bmatrix} \begin{bmatrix} x \\ y \end{bmatrix} \leq \begin{bmatrix} 2 \\ 3 \\ -5/2 \end{bmatrix}$$

has a negative component in the vector **b**. Determine if the simplex algorithm gives the maximum of the objective function $f(x, y) = x - y$.

    3.   For the convex set where $x \geq 0$, $y \geq 0$, $z \geq 0$, and

$$\begin{bmatrix} 0 & 2 & 3 \\ 3 & 2 & 0 \end{bmatrix} \begin{bmatrix} x \\ y \\ z \end{bmatrix} \leq \begin{bmatrix} 3 \\ 3 \end{bmatrix}$$

find the maximum of the objective function $x + y + z$.

       a. Use the second column as the first pivot column.

       b. If you do not reach an answer, go back to the step at which there was a choice for pivot row and try another choice.

       c. Sketch the region in three dimensions, and note the four planes which intersect at one point.

    4.   In Exercise 3.6.11 we have the formulation:

          Maximize $3r + 12s + 2t + 7u + v$ where $[r, s, t, u, v] \geq [0, 0, 0, 0, 0]$ and

$$\begin{bmatrix} -1 & 0 & -1 & 0 & 0 \\ 0 & 0 & 0 & 1 & 1 \\ 0 & 1 & 0 & 0 & 0 \\ 0 & 0 & 1 & 1 & 1 \\ 1 & 1 & 1 & 1 & 1 \end{bmatrix} \begin{bmatrix} r \\ s \\ t \\ u \\ v \end{bmatrix} \leq \begin{bmatrix} -1/2 \\ 3/10 \\ 7/20 \\ 1/4 \\ 1 \end{bmatrix}$$

In the successive steps of the simplex algorithm, it is possible to avoid using the first row as a pivot row yet the algorithm will terminate. The "solution" you find will not satisfy the inequality stated in the first line above; i.e., you will not have $-r - t \leq 1/2$. Verify this statement. (*Hint:* Begin by selecting the second column as the pivot column. Next select the fourth column as the pivot column at the second step. Any method of finishing this "solution" will give you a point which is not in the convex set.)

3.10   APPLICATIONS

Linear programing was invented as a mathematical tool to solve problems in business and industry. Our examples have emphasized the purely practical problems which can be solved with this tool, but there are theoretical applications in economics and management as well. Recently, applications of linear programing have been made in other disciplines, and we report on two such cases here. The first application is to a problem in anthropology, and the second concerns an educational problem.

Application 3.1   Measuring Cultural Intensity

In the article, "A linear programing approach to cultural intensity," by Hans Hoffman (*Game Theory in the Behavioral Sciences*, University of Pittsburgh Press, 1969), the author discusses the problem of measuring the cultural intensity of a society. Hoffman's point of view in this article is that of an anthropologist who, having observed a primitive Indian tribe, wants to have an empirical measure of societal organization.

Hoffman suggests that the "organization of a culture" be determined from the allocation of wealth, time, or tribal resources in general. Specifically, Hoffman compares the time spent by a tribe in each of several economic activities to the optimal allocation of time necessary to accomplish comparable goals. Presumably, then, a well-organized society is one which has worked out a good allocation of resources. In this example, the resources are the manhours of productive labor for one family.

Hoffman gives an interesting interpretation for the meaning of feasible solutions in this context. He says a feasible economic organization in a tribe is one for which the tribe flourishes. In contrast, a nonfeasible organization produces an inadequate economic payoff resulting in an unstable society. This situation is characterized by emigration, infanticide, etc.

For an example of his ideas, Hoffman considered the Shipibo Indian village on the upper Amazon river. The village is supported by three types of economic activity: fishing, farming, and varieties of commerce. Data on the allocation of working hours to each of these economic activities were collected for one family consisting of two adult males, two adult females, and three children.

Table 3.15 gives a summary of one week's typical economic production and allocation of time to various activities for this family.

Further, it was believed that this Shipibo family would thrive on any diet which supplied at least as much protein and carbohydrate as their current diet as given in Table 3.15. That is, other diets are acceptable to this family as long as they supply at least 5130 g of protein and 5440 g of carbohydrate. We will now discuss a model for allocation of labor and from the model determine an optimum allocation.

Activity	Production	Grams of protein	Grams of carbohydrate	Working time, hr
Fishing	50 lb of fish	4900	0	25
Farming	24 lb of yucca, fruit, etc.	138	3264	19
Commerce	16 lb of miscellaneous foodstuff	92	2176	43
Total		5130	5440	87

Table 3.15   The economic production of a Shipibo family

There are five cultural, biological, and natural circumstances affecting the activities of this family which Hoffman mentions as influences on their use of time. We enumerate those circumstances:

1.   Each adult can contribute a maximum of 90 hr of activity per week.
2.   The culture of the Shipibo does not allow women to fish. This of course affects the allocation of time for the family.
3.   It is not always possible to fish since the lake near the village sometimes has waves too high for the shallow canoes used. The waves are too high an average of 1 hr each day.
4.   There is an abundance of land suitable for farming.
5.   There is a lack of opportunity for commercial activity. An estimated 54 hr/wk is the maximum amount of time which could be productively devoted to this activity.

Using variables x, y, z for the number of hours spent in fishing, farming, and commerce respectively, we are able to formulate a linear program to minimize the total time necessary to produce an acceptable diet. This solution is also feasible within the cultural, biological, and natural constraints imposed on this family.

To formulate the linear program which results from the given information, we will first impose the obvious conditions which result from the hours available for various activities.

First the constraint on total working hours available is

$$x + y + z \leq 4(90) = 360$$

Because women are not allowed to fish, and waves prevent fishing 7 hr/wk, we find the restriction on the number of hours of fishing to be

$$x \leq 2(90) - 2(7) = 166$$

The third condition relating to commercial activity is simply

$$z \leq 54$$

Next, the nutritional requirement of 5130 g of protein is expressed as an inequality. We determine the number of grams of protein produced per hour for each activity. Then, using those numbers as coefficients, we obtain the inequalities. For example, 25 hr of fishing produced 4900 g of protein, and so we will use $(4900/25)x$ as the number of grams of protein which are produced through x hr of fishing. In similar fashion we then find the inequality

$$(4900/25)x + (138/19)y + (92/43)z \geq 5130$$

which expresses the protein requirement of the diet. This inequality guarantees that if x hr fishing, y hr farming, and z hr commerce are performed in a week then at least 5130 g of protein will be available in the diet.

Carbohydrate available from the various sources as indicated in Table 3.15 yields a second inequality. Again we use as coefficients the g/hr of carbohydrate which result from each activity. The resulting inequality is

$$(3264/19)y + (2176/43)z \geq 5440$$

The five inequalities which we have derived determine the set of feasible solutions for the allocation of economic activity. Since we wish to find the allocation which requires the minimum number of hours the objective function is $x + y + z$, the number of hours devoted to all three activities. The convex set of feasible solutions to this linear program is given by those points where $x \geq 0$, $y \geq 0$, $z \geq 0$, and

$$[x, y, z] \begin{bmatrix} -1 & -1 & 0 & 196 & 0 \\ -1 & 0 & 0 & 138/19 & 3264/19 \\ -1 & 0 & -1 & 92/43 & 2176/43 \end{bmatrix} \geq [-360, -166, -54, 5130, 5440]$$

The various minus signs which appear in these inequalities are necessary because all the inequalities are stated as greater than or equal to conform to the standard format.

The dual to this problem requires the maximum value of

$$-360s - 166t - 54u + 5130v + 5440w$$

when

$$\begin{bmatrix} -1 & -1 & 0 & 196 & 0 \\ -1 & 0 & 0 & 138/19 & 3264/19 \\ -1 & 0 & -1 & 92/43 & 2176/43 \end{bmatrix} \begin{bmatrix} s \\ t \\ u \\ v \\ w \end{bmatrix} \leq \begin{bmatrix} 1 \\ 1 \\ 1 \end{bmatrix}$$

It is impossible to draw the convex set in five dimensions, and it is a tedious job to find all the extreme points for this set so we will use the simplex algorithm to solve the linear program.

The initial and final simplex tables are given in Table 3.16.

	s	t	u	v	w				
	-1	-1	0	196	0	1	0	0	1
	-1	0	0	$138/19$	$3264/19$	0	1	0	1
	-1	0	-1	$92/43$	$2176/43$	0	0	1	1
	360	166	54	-5130	-5440	0	0	0	0

a. The initial simplex table

	s	t	u	v	w				
v	$-1/196$	$-1/196$	0	1	0	$1/196$	0	0	$1/196$
w	$-1793/319,872$	$69/319,872$	0	0	1	$-69/319,872$	$19/3264$	0	$1793/319,872$
	$-91/129$	$10/2107$	-1	0	0	$-10/2107$	$-38/129$	1	$91/129$
	$910/3$	$6977/49$	54	0	0	25	$95/3$	0	$170/3$

b. The final simplex table

Table 3.16   The simplex tables for the linear program of Application 3.1

Remember that we are solving the minimum problem so that the solution we are interested in is read from the bottom row of the final simplex table. The extreme point $x = 25$, $y = 95/3$, and $z = 0$ corresponds to the solution requiring a minimum number of hours of labor. The solution is to spend 25 hr fishing, $95/3$ hr farming, and to allot no time for commerce. This solution involves $170/3$ hr of labor, a saving of about 30 hr over the allocation of time used by the Shipibo Indians.

Hoffman goes on in his paper to discuss the possible ways for using the observed allocation of time as a measure of cultural intensity when compared to the minimum time solution of the linear program. We must mention here that Hoffman incorrectly formulated the linear program for this problem. While Hoffman was not interested in an explicit solution to the linear program, he would have been surprised to discover that with his formulation the minimum was attained at the origin, $[0, 0, 0]$, a solution which produces no food whatsoever! Again we want to emphasize the importance of being able to give a correct formulation to a linear programing problem.

Exercise 3.10 1.   Formulate the problem of this example ignoring the constraints 1 through 5 which were called cultural, biological, and natural considerations; i.e., use only the dietary conditions. Solve this formulation of the problem, and compare your solution to the one found in the text.
2.   Suppose the total hours available for fishing were 15 hr/wk. Impose this constraint and the dietary ones. Solve the resulting problem. How does this solution differ from your solution in Exercise 3.10.1?
3.   If the Brazilian government completes their program of road development in the upper Amazon valley, it is estimated that an hour spent on

commerce will produce 3 times as much food value as at present. Will road development create circumstances for which a different optimum solution results?

4.   Criticize the basic assumptions of the Hoffman paper. (Should all forms of work be counted equally as hours of labor? Is diversity of labor valuable in itself?)

a. Find the minimum value for the objective function which counts an hour of fishing as twice that of an hour of either farming or commerce, $f(x, y, z) = 2x + y + z$.

b. Suppose the Shipibo Indians spend 43 hr each week bartering in the market square because of the social value. Reformulate the problem using the requirement that $z \geq 43$, and find the solution to this new linear program. How many hours of labor are saved by the minimum solution over the observed allocation with this new requirement added?

Application 3.2   Allocation of Class Time

Hershkowitz, Jensen, and Mills writing in the *5th Annual Urban Symposium* for the application of computers to problems of urban society report on an application of linear programing to allocate class time to differing methods for teaching developmental reading. At Drew University each student in the developmental reading program is tested for reading speed and comprehension at the beginning of each class period. The student's reading speed and comprehension level coupled with the instructor's goals for the class period determine a linear program. The solution of the linear program specifies the fraction of class time that each student should be exposed to each of six teaching methods in order to attain the instructor's goals at a minimum cost for the teaching methods used. The result is an individualized plan of instruction for each student during each class period.

A computer is used to solve the linear program and to give the student an assignment to devote a certain fraction of class time to each of the teaching methods. The six teaching methods used in this course are:

1. pacer
2. exhortation
3. T-scope and films
4. note taking
5. question recall
6. flow of thought

The fraction of class time devoted to teaching method i will be denoted by x_i, $i = 1, 2, \ldots, 6$.

Using standard statistical methods to study the effectiveness of each teaching method to increase reading speed and comprehension level makes it possible to predict how much improvement to expect for a chosen allocation of class time to each of the six methods. More specifically, if a student has a reading speed between 200 and 220 wpm, and if the allocation of class time to each of the six teaching methods is given by the fractions indicated in the vector $[x_1, x_2, x_3, x_4, x_5, x_6]$, then

$$0.24x_1 + 0.15x_2 + 0.25x_3 + 0.06x_4 + 0.05x_5 + 0.07x_6$$

is the predicted percentage increase in reading speed which the authors reported. For example if the entire class period is devoted to method 1 (the pacer)—i.e., we allocate $[1, 0, 0, 0, 0, 0]$—then the percentage increase in reading speed is predicted to be $(0.24)(1)$ or a 24 percent increase.

It should be noted that the coefficients in the expression above will change with the reading speed. Thus for a reading speed of 600 wpm, we probably would not predict such a large increase in reading speed by using a pacer for the entire period.

A similar prediction can be made for the change in comprehension level. After a class period devoted to the allocation $[x_1, x_2, \ldots, x_6]$, the prediction for a student whose comprehension level is between 81 and 85 percent is a percentage change given by the expression

$$-0.41x_1 - 0.19x_2 + 0.07x_3 + 0.00x_4 - 0.15x_5 - 0.20x_6$$

Notice that it is possible to have a negative percentage increase, meaning that it is possible to lower the comprehension level by the use of certain teaching methods (hopefully those methods improve the speed).

At Drew University the costs of the six teaching methods are proportional to the corresponding components of the vector $[10, 20, 75, 25, 20, 25]$; that is to say exhortation, the second method, is twice as costly as the pacer, method 1, while T-scope and films, method 3, is 3 times as costly as note taking and flow of thought, methods 4 and 6. To determine the cost of a particular allocation of class time, we write the function

$$[10, 20, 75, 25, 20, 25] \cdot [x_1, x_2, x_3, x_4, x_5, x_6]$$

which is the function we want to minimize.

A linear programing problem results when the instructor specifies the goal for a class period. For example, a goal of a 10 percent increase in reading speed and no loss of current comprehension level for a student who currently reads 200 to 220 wpm and who comprehends 81 to 85 percent is expressed by the two inequalities

$$0.24x_1 + 0.15x_2 + 0.25x_3 + 0.06x_4 + 0.05x_5 + 0.07x_6 \geq 0.1$$

and

$$-0.14x_1 - 0.19x_2 + 0.07x_3 + 0x_4 - 0.15x_5 - 0.20x_6 \geq 0$$

It is obvious that we also require $x_1 \geq 0$, $x_2 \geq 0$, $\ldots$, $x_6 \geq 0$, and $x_1 + x_2 + \cdots + x_6 = 1$ for a meaningful solution.

Remember that the simplex algorithm produces a solution which satisfies inequality constraints. To accommodate an equality constraint such as $x_1 + x_2 + \cdots + x_6 = 1$, we can use two inequalities, namely,

$$x_1 + x_2 + x_3 + x_4 + x_5 + x_6 \geq 1$$

and

$$x_1 + x_2 + x_3 + x_4 + x_5 + x_6 \leq 1$$

The formulation of this problem is then to find the minimum of the cost function

$$[10, 20, 75, 25, 20, 25] \cdot [x_1, x_2, x_3, x_4, x_5, x_6]$$

subject to those allocations $[x_1, x_2, x_3, x_4, x_5, x_6]$ for which

$$[x_1, x_2, x_3, x_4, x_5, x_6] \begin{bmatrix} 1 & -1 & 0.24 & -0.14 \\ 1 & -1 & 0.15 & -0.19 \\ 1 & -1 & 0.25 & 0.07 \\ 1 & -1 & 0.06 & 0 \\ 1 & -1 & 0.05 & -0.15 \\ 1 & -1 & 0.07 & -0.20 \end{bmatrix} \geq [1, -1, 0.1, 0]$$

The simplex table for the dual maximum problem is given in Table 3.17.

s	t	u	v							
1	−1	0.24	−0.14	1	0	0	0	0	0	10
1	−1	0.15	−0.19	0	1	0	0	0	0	20
1	−1	0.25	0.07	0	0	1	0	0	0	75
1	−1	0.06	0	0	0	0	1	0	0	25
1	−1	0.05	−0.15	0	0	0	0	1	0	20
1	−1	0.07	−0.20	0	0	0	0	0	1	25
−1	1	−0.1	0	0	0	0	0	0	0	0

Table 3.17   The initial simplex table for Application 3.2

Solving the problem by the simplex algorithm, we find that the solution is to allocate 7 percent of the class time to method 1, 14 percent to method 3, and 79 percent of the class time to method 4. Thus the pacer will be used for about $3\frac{1}{2}$ min, the T-scope and films will be used for about 7 min, and note taking will occupy the remainder of a 50-min class period.

Exercise 3.10   5.   Consider only the first, third, and fourth teaching methods. Formulate the problem in terms of those three variables only. Solve your formulation, and check the solution against the solution given in the text.
6.   Change the goal of the class period to: attain a 10 percent propor-

tionate increase in reading speed and a 5 percent increase in comprehension. Only consider methods 1, 3, and 4. Does this change in goal yield a different solution?

7. Can you determine what will happen if the instructor sets an unattainable goal for the class period? Will the linear program detect that an unattainable goal has been specified?

3.11 FOOD FOR THOUGHT— AN ALGORITHM FOR TRANSPORTATION PROBLEMS

The simplex algorithm provides a basic computational method for solving linear programing problems, but there is a danger in always relying on one algorithm. This danger lies in the fact that one method of solution will generally not be efficient for all problems. Certain types of problems are solved more efficiently by algorithms which are specifically designed for those problems. When a problem type is encountered frequently in practice, it becomes important to have efficient algorithms. (An interesting research activity is to devise such algorithms.)

We are going to discuss an algorithm which provides an efficient method for solving "transportation problems." This problem type involves two groups of objects, say "warehouses" and "stores," and a product which is transported from the warehouses to the stores. Each store is to receive and each warehouse is to send a specified quantity of the product. The problem arises when we are required to determine a routing which will minimize (say) total transportation cost.

We consider an example which illustrates the use of this algorithm and which also fixes the transportation-problem concept.

A Reallocation Problem

A car rental company, due to its one-way rental service, must reallocate its rental automobiles among agencies in various cities. On a particular occasion the company had a surplus of 48 autos in Cleveland, 31 in Albany, and 16 in Buffalo. These surplus rental units had to be redistributed so that New York received 43, Chicago 22, and Philadelphia 30.

The company wanted to accomplish the reallocation of these rental cars keeping the total distance traveled at a minimum. The three cities Albany, Buffalo, and Cleveland are called "warehouses," while Chicago, New York, and Philadelphia are called "stores." The rental cars are to be transported from "warehouses" to "stores."

In order to formulate the mathematics we will write A for Albany, B for Buffalo, and C for Cleveland while using lowercase c for Chicago, n for New York, and p for Philadelphia. Variables in this problem will be Ac for the number of automobiles sent from Albany to Chicago, with An, Ap, Bc, etc., having similar meanings.

From the information given we derive six equations which must be satisfied by these variables. The first three equations are determined by the number of automobiles at A, B, and C respectively; and the next three equations are determined by the number of automobiles to be sent to c, n, and p.

$$Ac + An + Ap = 31 \tag{3.1}$$
$$Bc + Bn + Bp = 16 \tag{3.2}$$
$$Cc + Cn + Cp = 48 \tag{3.3}$$
$$Ac + Bc + Cc = 22 \tag{3.4}$$
$$An + Bn + Cn = 43 \tag{3.5}$$
$$Ap + Bp + Cp = 30 \tag{3.6}$$

Table 3.18 gives the distances in miles between all warehouse-store pairs.

	Chicago	New York	Philadelphia
Albany	800	150	230
Buffalo	550	380	365
Cleveland	350	500	425

Table 3.18   Mileages between cities, computed using interstate highways and stated to the nearest 5 mi

Using the mileages from Table 3.18, the function which expresses the total number of miles traveled to accomplish the reallocation for a particular feasible solution is

$$800Ac + 150An + 230Ap + 550Bc + 380Bn + 365Bp + \\ 350Cc + 500Cn + 425Cp$$

At this point we could of course apply the simplex algorithm to find the minimum distance.

The alternative algorithm we wish to demonstrate requires that we find some feasible solution to the problem. We can find a feasible solution by stating any assignment to the variables so that the six equations are satisfied. For example we may arbitrarily begin by setting $Ac = 2$, then specifying $An = 29$, now $Ac + An = 31$ and so $Ap = 0$. Next perhaps we take $Bn = 14$ so that $An + Bn = 43$ and so $Cn = 0$. We let $Bc = 2$, $Cc = 18$, and $Cp = 30$. This assignment is shown in Fig. 3.28; note that all six equations are satisfied by this assignment.

The feasible solution of Fig. 3.28 was chosen to illustrate a basic fact about transportation problems and to show the basis on which the algorithm operates. Note that both A and B are assigned to send automobiles to both n and c. We will show that a better solution results when a reassignment is made, and one of these four routes is canceled; this forms the basis for our algorithm.

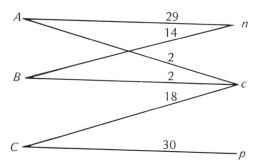

Figure 3.28   A feasible solution to the transportation problem

To find this better solution, we examine the mileages between the four cities. Figure 3.29 shows the relevant mileages for A, B, c, and n.

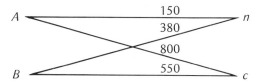

Figure 3.29   Relevant mileages

Note that, if an automobile from A to c (see Fig. 3.28) were reassigned to n while one auto assigned from B to n is reassigned to c as a replacement, there results a saving in total mileage. The automobile from A travels 150 mi instead of 800, and the automobile from B travels 550 rather than 380 mi.

The change in mileage is computed by the sum

$$(150 - 800) + (550 - 380) = -480$$

The negative value indicates 480 mi is saved by this reassignment. Of course we get the same saving for each auto reassigned, and we should therefore reassign both of the cars scheduled from Albany to Chicago, sending them instead to New York. As part of this reassignment we also divert two cars scheduled from Buffalo to New York, sending them to Chicago. Note that the new feasible solution which results no longer has an Albany-to-Chicago route. The total mileage required by the new solution is 2(480) = 960 mi less than the solution given in Fig. 3.28. Figure 3.30 shows the new assignments and the routes for this solution.

The essential fact is that the solution proposed in Fig. 3.28 can be improved because the routes used by that solution contain a *circuit*. A circuit is a path which begins and ends at the same location, follows only the routes used by the solution, although not necessarily in the same direction, and follows each route at most once (it need not follow every route). In Fig. 3.28

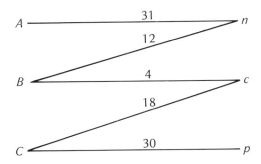

Figure 3.30   A new feasible solution

there is a circuit starting at B using the route B to c, then c to A, next A to n, and finally n to B. We showed above that there is a better solution (the one given by Fig. 3.30) which eliminates one of the routes in the circuit (the route c to A). This better solution does not have a circuit. A formal statement which we will not prove asserts this fact.

THEOREM 3.4   If a feasible solution to a transportation problem has a circuit, then there is a reassignment which gives at least as good a solution and for which there is no circuit.

From Theorem 3.4 and our example we see that there is no point in proposing a feasible solution (such as was proposed in Fig. 3.28) which has a circuit. We now describe the algorithm in detail.

1. Begin with a feasible solution which has no circuits.
2. Select a route, not used in the solution, which if added to the routes of the solution creates a circuit.
3. Reassign cars so that one car uses this new route and assignments along other routes of the circuit are adjusted so that Eqs. (3.1) to (3.6) are satisfied. (This reassignment is uniquely determined; it affects each route in the circuit and only those routes.)
4. Analyze the reassignment to determine if a saving in total mileage resulted from the reassignment.
5. If there is no saving in total mileage, delete the route which was tried, then try a route which has not been previously considered to improve on this solution. If every route has been previously considered, go to step 7; otherwise return to step 3.
6. If a saving in total mileage resulted by the reassignment, then reassign as many automobiles as possible to use this route. It is possible to reassign automobiles until one of the routes in the circuit no longer has any automobiles assigned to it. That route can be deleted. A new feasible

solution (one with no circuits) results from this reassignment. Return to step 2.

7. The algorithm terminates; it is not possible to improve the current feasible solution.

These steps of the algorithm are now illustrated as we follow them in solving the example. Step 1 uses the feasible solution given in Fig. 3.30. At step 2 we select the route B to p, creating the circuit involving B, C, c, and p. In step 3 we wish to also note that a unique reassignment is determined. Assigning an automobile from B to the route B to p sends one too many cars to p. To have a feasible solution we thus divert a car which was assigned from C to p, sending it to c, and finally one fewer car from B to c. This reassignment is then a feasible solution and the only one we could make which uses the route B to p. (Notice that each route of the circuit is affected by either an increase or a decrease in the number of cars assigned to use the route.)

In step 4 we must determine if this reassignment is advantageous.

To analyze this question we introduce a useful notation. In Fig. 3.31 we indicate by arrows directed away from the "warehouse" the reassignments and by arrows directed toward the warehouse the assignments which are cancelled. Thus arrows directed away from a warehouse indicate a positive (added) mileage and arrows directed toward a warehouse are negative (canceled) mileage. The sum of these positive and negative mileages indicates the net change in mileage introduced by the reassignment. A positive sum means that a larger mileage is required for the reassignment, and a negative sum indicates a saving in total mileage. (Again we note that the arrows use every route in the circuit.)

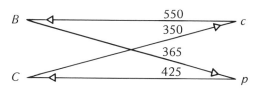

Figure 3.31   The directed mileages involved by adding B to p

We now show the relevant directed mileages for adding the route B to p in Fig. 3.31. We compute the net change in mileage traveled by the sum

$$(365 - 550) + (350 - 425) = -260$$

and we conclude that the route B to p is advantageous. We therefore go to step 6, and we reassign all four cars assigned from B to c (see the solution in Fig. 3.30) to the route B to p. Figure 3.32 shows the new feasible solution.

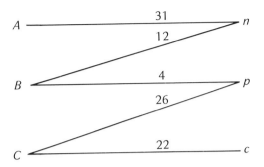

Figure 3.32   A new feasible solution

In Fig. 3.32 we have reversed the order of p and c to avoid lines which cross. We then return to step 2.

The route A to p is chosen first and the reassignment made. Figure 3.33 is used for the analysis at step 4. The analysis of Fig. 3.33 results in a net mileage change of

$$(230 - 150) + (380 - 365) = 95$$

thus the route A to p is not advantageous; we go to step 5 and select the route A to c.

Adding the route A to c gives a more complicated reassignment than our previous cases. Remember that we must compensate for adding A to c by using *only* routes already used in the feasible solution of Fig. 3.32. Figure 3.34 gives the resulting reassignment at step 3 and is used for the analysis at step 4.

The net change in mileage which results is

$$(800 - 150) + (380 - 365) + (425 - 350) = 740$$

This is no saving so we are again at step 5.

The next possible route is B to c. Figure 3.35 gives the relevant informa-

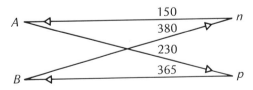

Figure 3.33   The directed mileages involved by adding the route A to p

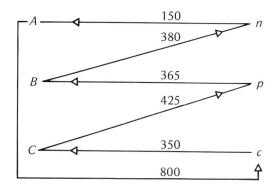

Figure 3.34   The directed mileages involved by adding A to c

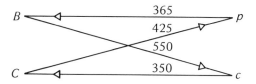

Figure 3.35   The directed mileages involved by adding the route B to c

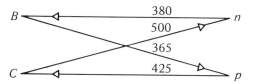

Figure 3.36   The directed mileages involved by adding the route C to n

tion about steps 3 and 4. The net change introduced by a B-to-c route is

$$(550 - 365) + (425 - 350) = 260$$

so it offers no improvement.

The final possible route not considered is C to n. From Fig. 3.36 the change incurred by adding the route C to n is

$$(365 - 380) + (500 - 425) = 60$$

We are at step 5, and we have exhausted all routes which could be used but are not used by the feasible solution of Fig. 3.32. None of these possible routes leads to an improvement over the existing solution. Thus we go to step 7, the algorithm terminates, and we conclude that the minimum distance required to reallocate the automobiles results from the assignment shown in Fig. 3.32.

The distance required is computed from the sum

$$150An + 380Bn + 365Bp + 350Cc + 425Cp$$

When the values for these variables are supplied from Fig. 3.32, the total mileage is found to be 29,420 mi.

DEGENERACY When some subset of warehouses can exactly supply the needs of some subset of stores, then the algorithm will not always work. The problem which arises in that case is that the addition of a route *may* not introduce a circuit since there is a feasible solution which involves at least two separate groups of warehouses and stores. You may wish to try such an example to see what problems arise when you use the transportation algorithm to search for a solution.

Exercise 3.11

1. Set up the example of the text for solution by the simplex algorithm (do not solve it). It should be clear from the size of the table that the algorithm you have just learned is shorter.
2. Suppose Philadelphia requires 52 cars and Chicago none. Solve the transportation problem of this section using this change in the conditions.
3. If locations A, B, and C are warehouses for 10, 7, and 18 units of goods, respectively, and location d requires 15 units while location e needs the remainder, find the least expensive method of distribution when the costs of shipping each unit are given by the table below.

	A	B	C
d	7	8	12
e	3	6	15

4. Solve the transportation problem which corresponds to the feasible solution shown below. The transportation costs per unit are given by

	A	B	C
d	5	4	2
e	4	5	3
f	2	4	5

COMPUTER PROJECTS

It has been pointed out in the text that an algorithm is a scheme for systematic computation. The simplex algorithm is commonly used to produce computer routines for solving linear programing problems and is found in many computing center libraries of programs. All that is required to use such library programs is a complete knowledge of "input" and "output."

Project 1: Simplex Library Program

a. Visit the computing center library and learn the input/output for the simplex algorithm program which the computing center has available.
b. Use the computing center library program to solve the linear programing problem, "Allocation of class time," found as Application 3.2.

Project 2: Special Algorithm

Read the Sec. 3.11 if you have not already done so. For this project you are to write a computer program to carry out the computations for the special algorithm for transportation problems.

Some suggestions for notation may be helpful to you. By listing the warehouses numerically $i = 1, 2, \ldots, I$ and the stores $j = 1, 2, \ldots, J$ a pair of numbers (i, j) can be used to represent the route from i to j. Then each possible route is given by a distinct pair (i, j), and all possible routes can be easily listed by the use of two "DO loops," $i = 1$ to I and $j = 1$ to J.

A function $q(i, j)$ will give the quantity of goods shipped from i to j along the route (i, j), and the function $C(i, j)$ will denote the cost of shipping a unit of goods along the route i to j. [$C(i, j)$ may be just the distance from i to j in some cases.]

$$\sum_{j=1}^{J} q(i, j) = \text{amount of goods at warehouse } i$$

$$\sum_{i=1}^{I} q(i, j) = \text{amount of goods shipped to store } j$$

The printout from your program should include the matrix of quantities $q(i, j)$ for each feasible solution considered and the total cost for that solution, which is

$$\sum_{i=1}^{I} \sum_{j=1}^{J} q(i, j)C(i, j)$$

4   GAME THEORY

4.1   INTRODUCTION

The theory of games has the distinction of being one of the few branches of mathematics invented with an eye toward applications in the social sciences. Although the word *game* sounds less than serious, the goal of game theory is to analyze rational behavior in many types of real-life competitive situations. Applications of special importance include the analysis of economic behavior and military strategy, but social scientists have gone so far as to regard a West Indies fisherman attempting to catch as many fish as possible as competing in a game with nature as his opponent. Psychologists have shown interest in experiments comparing the actual behavior of people in competitive situations with the "rational" behavior recommended by game theory. But just as simple gambling situations motivate the basic ideas of probability theory, game theory is most easily introduced by studying simple recreational games.

We will consider only games involving two competing players. This is a great restriction, as many competitive situations involve more than two "sides." What is more, we interpret "competing" strictly—the formal definition we will give in Sec. 4.2 implies that cooperation is never advantageous. The theory therefore does not apply very well to many competitive situations in which both sides may gain from cooperation, such as labor-management bargaining. But studying only games between two strictly competing players yields a reasonably simple theory which still has many applications. You should read Sec. 4.11 to appreciate the complications that are introduced when cooperation can be advantageous.

The first basic concept of game theory is that of a *strategy*. Recreational games such as chess or checkers often consist of a long sequence of moves made alternately by the two players, and other competitive situations such as war or negotiation have a similar pattern. The players usually do not consider in advance what moves they will make at each stage of the game, but respond to moves made by their opponent. Yet it is possible to conceive that a player might decide in advance what move he would make at every step of the entire game, including planned responses to every possible move of his opponent. If he does this, his side of the game could be played by a clerk who knows nothing about the game, but simply follows the detailed plan. *Such a complete description of how to play in all possible circumstances is called a strategy.* If both players write down strategies, a referee can play the complete game following these strategies and decide the outcome.

A strategy (at least a strategy with any chance of winning) for a game such as chess or checkers is very complicated. But it is possible to write down quite good strategies, for computer programs have been written which can defeat better-than-average human players in chess or checkers. Indeed, thinking of a strategy as a computer program will help you grasp the idea— it is a detailed set of instructions providing for every possible contingency which can be followed in a mechanical way by a machine incapable of independent judgment.

Example 4.1 Matching Fingers

In this very simple game each player chooses to present either one or two fingers. If the numbers presented match, Player 1 wins; if they differ, Player 2 wins. This is a one-step game, so a strategy is just a choice of which possible move to make. Each player has only two possible strategies: "one finger" and "two fingers." If you know, for example, that Player 1 has chosen "one" and Player 2 has chosen "one," then you know that Player 1 will win. (This simple game has great social utility in determining which team will bat first in sandlot baseball games.)

Example 4.2 Tictactoe

Player 1 plays first in this familiar game and uses X's, while Player 2 uses O's. For convenience in describing strategies we number the nine spaces to be filled as shown below.

1	2	3
4	5	6
7	8	9

Tictactoe layout

Here is a strategy for Player 1:

1. On the first move, place an X in 1.

2. On the second move, place an X in 9 unless it is filled. If 9 is filled, place an X in 7.
3. On the third move, if 1 and 9 have X's and 5 is empty, place an X in 5. If 1 and 7 have X's and 4 is empty, place an X in 4. Otherwise place an X in the highest-numbered empty space.
4. Complete the game by placing an X in the highest-numbered empty space on each move.

It is possible to give strategies for tictactoe which will always win or draw against any strategy of your opponent.

The players in a game usually think of the game in move-by-move form, which we call the *extensive form* of the game. The idea of a strategy is a considerable conceptual simplification, for we now think of each player as choosing a single strategy rather than going through a complicated series of moves. To formulate a mathematical model for a game we therefore begin by specifying the set of all possible strategies for each of the players. Of course, it is not possible in practice to list all possible strategies for a game such as chess or checkers. But once you understand what a strategy is, it is not hard to conceive of the set of all strategies a player could possibly use.

Exercise 4.1   1.   Here is a strategy for Player 2 in tictactoe:

a. Place an O in 5 unless it is filled. If 5 is filled, place an O in 1.
b. Complete the game by placing an O in the lowest-numbered empty space on each move.

You as referee have been given the strategy of Example 4.2 by Player 1 and this strategy by Player 2. Play the game using these strategies and determine the outcome.

2.   Explain carefully why the following is *not* a strategy for Player 1 in tictactoe.

a. Place an X in 5 on the first move.
b. Place an X in 1 on the second move.
c. Place an X in 9, at which point the game ends with a win for Player 1.

3.   You are Player 2 in tictactoe. You know that Player 1 will use the strategy of Example 4.2. Write a strategy that will defeat him.

*4.   Write a strategy for Player 1 in tictactoe which guarantees that he will always win or draw.

5.   Each of two players holds a hand of three cards numbered 1, 2, and 3. On the first move each player shows a card, and the higher card wins 2. If the cards are the same, each wins 1. On the second move each player shows one of his two remaining cards with the same scoring arrangement. On the third move the last card is shown by each player with the same scoring

again. List all possible strategies for a player in this game. (*Hint*: Not all strategies consist of a fixed order in which to show the cards, ignoring the opponent's moves.)

4.2 ZERO SUM TWO-PERSON GAMES

We have seen that it is possible to think of each player in a two-person game as choosing a particular strategy from the set of all possible strategies available to him. The outcome of the game is determined by the strategies chosen. In recreational games the outcome is usually win, lose, or draw. But the outcome of an economic competition may be given in dollars, the outcome of a military competition in area lost, or casualties incurred, and so on. We insist that three basic properties hold:

1. The outcome depends only on the strategies used by the two players.
2. The outcome can be measured in numerical units.
3. The amount won by one player is lost by the other.

Each of these three properties is worth thinking about. The first means that the problem facing you as a player in a game is to choose a strategy which will do well even though the outcome also depends on what your opponent does. The decision problem here is therefore quite different from the case in which (as in Application 1.4) the outcome depends only on your own action.

The second property seems generally true at first, but careful thought shows that it is not obvious. Even when the outcome is measured in dollars, considerations of pride and prestige often enter as well. What is more, the "value" or "utility" of money to a player may decrease when large amounts are involved. Most of us do not regard $1 million as being worth 10 times as much to us as $100,000. To be precise, we do not think that a sure $100,000 has the same value as a gamble which has probability of $1/10$ of yielding $1 million and probability of $9/10$ of yielding nothing. Now it turns out that if a player can express a preference for one of each pair of gambles, then it is possible to assign a numerical *utility* to each outcome such that all the player's preferences are summed up by saying that he always chooses to maximize the expected utility. All that is required is that the player's preferences not be self-contradictory. We do not want to be more precise about utility theory, but only to point out that it *is* reasonable to measure outcomes numerically after all.

The third property implies that cooperation between the opposing players can never occur if both are rational. For any amount won by Player 1 is lost by Player 2 and vice versa, so that a player can assist his opponent only by directly injuring himself. Such games are called *zero sum* games, for in such a game the sum of the amounts won by the two players is always zero if we count an amount lost as a negative amount won. Zero sum games are the only games we will study. Notice that the players must agree about the utility of the outcomes—if Player 1 feels he has won 10, Player 2 must feel that he has lost 10.

We will always describe the result of a zero sum game by giving the amount won by Player 1. If Player 2 wins 10, this means that Player 1 has lost 10, and the payoff is -10. Property 2 says that the payoff is a number, and property 1 says that it depends only on the strategies used. That is to say, the amount won by Player 1 is a *function* of the strategies employed—not only Player 1's strategy but Player 2's as well. This discussion has led to the following formal definition of a game.

DEFINITION 4.1   A *game* (two-person zero sum game) consists of

1.   the set X of all possible strategies for Player 1
2.   the set Y of all possible strategies for Player 2
3.   a *payoff function* $M(x, y)$ which gives for every x in X and y in Y the amount won by Player 1 (which is equal to the amount lost by Player 2)

We will often refer to Player 1 simply as 1 and to Player 2 simply as 2. The context should make clear when 1 refers to Player 1 and when it is a number. When each player has only finitely many strategies to choose from, the payoff function can be displayed in a simple form. In this case there are only finitely many possible payoffs, and we present them as entries of a matrix. In this payoff matrix the *i,j*th entry is the payoff which results when 1 chooses his *i*th strategy and 2 chooses his *j*th strategy. Thus the payoffs for a strategy of Player 1 against all strategies of Player 2 form a row of the matrix. Similarly, the payoffs of a strategy for 2 against all strategies of 1 form a column of the matrix. Such *finite games* are the only games we will study in detail.

DEFINITION 4.2   If $X = \{x_1, x_2, \ldots, x_n\}$ and $Y = \{y_1, y_2, \ldots, y_m\}$, the *payoff matrix* of the game is the $n \times m$ matrix whose *i,j*th entry is $M(x_i, y_j)$, the amount won by 1 when 1 uses strategy x_i and 2 uses strategy y_j. In this case the game is called an $n \times m$ game.

Example 4.3   Matching Fingers

This game is described in Example 4.1. Since the only outcomes are "win" and "lose," we will take the payoff (amount won by 1) to be 1 if he wins and -1 if he loses. The only possible strategies for 1 are

x_1:   show 1 finger
x_2:   show 2 fingers

and 2's possible strategies are

y_1:   show 1 finger
y_2:   show 2 fingers

This is a 2×2 game. The payoff matrix is given in Table 4.1.

$$
\begin{array}{c c}
 & 2 \\
 & \begin{array}{cc} y_1 & y_2 \end{array} \\
\begin{array}{c} x_1 \\ x_2 \end{array} &
\begin{array}{|cc|}
\hline
1 & -1 \\
-1 & 1 \\
\hline
\end{array}
\end{array}
$$

Table 4.1   Payoff matrix for matching fingers

This is just a convenient way to display the fact that the payoff function is

$$M(x_1, y_1) = M(x_2, y_2) = 1$$
$$M(x_1, y_2) = M(x_2, y_1) = -1$$

Example 4.4   Duopolistic Competition

Suppose that there are only two sellers of a certain commodity. By analogy with "monopoly," this situation is called a *duopoly*. Assume for simplicity that each of the two sellers (1 and 2) has available only three possible promotional and pricing strategies. Assume further that total sales of the commodity are constant at $10 million/mo, and that the way in which this amount is divided between 1 and 2 is known for each combination of strategies. Then the payoff can be specified by giving only 1's sales, and we have a zero sum game (see Exercise 4.2.1 for details). The payoff matrix in millions of dollars of monthly sales for 1 might be as in Table 4.2. Player 1 wants to make the payoff as large as possible, while Player 2 wants to make the payoff as small as possible.

$$
\begin{array}{c c}
 & 2 \\
 & \begin{array}{ccc} y_1 & y_2 & y_3 \end{array} \\
\begin{array}{c} x_1 \\ x_2 \\ x_3 \end{array} &
\begin{array}{|ccc|}
\hline
1 & 7 & 2 \\
3 & 0 & 8 \\
4 & 5 & 6 \\
\hline
\end{array}
\end{array}
$$

Table 4.2   Payoff matrix for duopolistic competition

Example 4.5   Rommel and Montgomery

Rommel is retreating from Mersa Matruh to Sidi Barrani with three companies. There are two possible routes (route 1 and route 2); and he can send zero, one, two, or three companies along route 1, the remaining companies taking route 2. Montgomery has three companies in position to block either route, and an additional company already in place along route 1 which cannot move for lack of fuel. So Montgomery can place one, two, three, or four companies along route 1, the remaining companies covering route 2. Now the terrain is such that on route 1 the superior force will destroy or capture the entire inferior force. On route 2 the inferior force will lose only the difference in numbers of companies between the two forces (but of course the inferior force cannot lose more companies than it has on that route.) If the forces on either route are equal, the British will withdraw, allowing the Germans to pass along that route with no losses to either side.

Rommel's possible strategies are y_0, y_1, y_2, and y_3 where y_i means that he sends i companies by route 1 and $3 - i$ by route 2. Montgomery's possible strategies are x_1, x_2, x_3, and x_4 where the subscript again denotes the number of companies sent to route 1. We will let the payoff be the number of companies lost by the Germans minus the number lost by the British. Montgomery is therefore Player 1, since he wants to maximize this payoff while Rommel wants to minimize it. We can now compute the payoff matrix. If the strategies chosen are x_3 and y_0, the payoff is -1, for the British have three companies on route 1 where the Germans have none (no losses) and one company on route 2 where the Germans have three (that British company is lost). If the strategies chosen are x_1 and y_1 the payoff is 1, for each side has one company on route 1 (no losses) and on route 2 the British outnumber the Germans 3 to 2 (so the Germans lose the difference, one company). The result of such reasoning is the payoff matrix of Table 4.3. This is a 4×4 game.

		y_0	y_1	y_2	y_3
	x_1	0	1	0	-1
1	x_2	-1	1	1	-2
	x_3	-1	0	2	0
	x_4	0	1	2	3

Table 4.3   Payoff matrix for Rommel and Montgomery

The payoff function essentially specifies the rules of the game by showing what outcome any choice of strategies will have. This process occurred in Example 4.5 where we translated a set of statements about the competitive situation into a payoff matrix.

You must infallibly remember the convention used to label the players. Player 1 chooses a *row* of the payoff matrix, and is the *maximizing* player who wants to make the payoff as large as possible. Player 2 chooses a *column* of the payoff matrix, and is the *minimizing* player who wants to make the payoff as small as possible. The actual payoff is the entry of the payoff matrix falling in the row chosen by Player 1 and the column chosen by Player 2.

Exercise 4.2

1.   For the duopoly of Example 4.4 write the matrix whose entries are Player 2's sales volumes. Show that Example 4.4 is a zero sum game if we take Player 1's "winnings" to be his actual sales and Player 2's "loss" to be the *difference* between his potential sales ($10 million) and his actual sales.

2.   Derive all the entries of Table 4.3 not already derived in Example 4.5.

3.   Suppose Example 4.5 is modified so that if the forces on either route are equal, they fight to the finish and all units on that route are lost. Does this game have a payoff matrix different from that of Table 4.3?

4.   Suppose that Example 4.5 is modified so that the inferior force on either route is completely lost. Find the new payoff matrix.

5.   In the game called Morra, each player shows one or two fingers and simultaneously calls one of the numbers 2, 3, or 4. A player who calls out the total number of fingers shown by both players wins that amount from his opponent. (That means that if both are correct or neither is correct no money changes hands.) List all the strategies for both players and give the payoff matrix for Morra.

6.   Two fair dice are rolled independently. Players 1 and 2 may bet 1 or 2 units each. If 7 or 11 come up, 1 wins 2's bet; otherwise 2 wins 1's bet. For example, if 1 bets 2 and 2 bets 1 and 7 comes up, then 1 wins 1. Since the amount paid to 1 by 2 is random, take the payoff function to be the expected value of the amount paid. Find the payoff matrix for this game. [*Remark*:   This shows how games containing a random element not under the players' control fit into the framework of game theory. In this case the payoff is a random variable since it depends on the randomness present as well as on the strategies employed by 1 and 2. We take the payoff function $M(x, y)$ to be the expected payoff when 1 uses strategy x and 2 uses strategy y. Both players can compute $M(x, y)$ since the probability distribution of the payoff is public knowledge. So poker, as well as chess, can in principle be analyzed by game theory.]

7.   The Northwest Company and the Hudson's Bay Company together control the entire trade of a certain region in the Canadian north in 1820. They are each planning to build one post in an area containing three settlements, and for security reasons must build the post in one of the settlements. Settlement 2 has twice the annual trade volume of settlement 1, and settlement 3 has three times the trade volume of 1. If both companies build in the same place they divide the total trade evenly. If they build in different settlements, each gets the total trade of all settlements to which it is closer than its competitor. The locations of the settlements are given below. Find the payoff matrix for this duopoly problem.

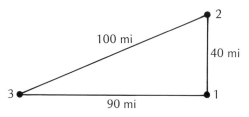

8.   Give the payoff matrix (Player 1's winnings) for the game of Exercise 4.1.5.

9.   Which of the following competitive situations would you be willing to describe (at least roughly) by a zero sum two-person game? Explain your reasons.

a.   incoming ICBMs versus antimissile missiles
b.   United Auto Workers versus General Motors in contract negotiations
c.   two auto dealers trying to sell a car to the same customer
d.   nation A positioning submarines versus nation B choosing routes for convoys
e.   lobbyists attempting to influence federal pollution-control legislation
f.   two vice presidents of a company, one of whom will be promoted to president

4.3 STRICTLY DETERMINED GAMES

We have thus far concentrated on formulating games and have said nothing about how a player should decide which strategy to use in actually playing a game. Since the payoff depends on the actions taken by *both* players, it is not at all obvious what either player should do. To introduce the idea of the solution of a game, we return to a previous example.

Example 4.6   The duopoly situation of Example 4.4 has the payoff matrix (the entries are the sales of Player 1) given in Table 4.4.

		2		
		y_1	y_2	y_3
	x_1	1	7	2
1	x_2	3	0	8
	x_3	4	5	6

Table 4.4   Payoff matrix for the duopoly example

Suppose that each company suspects that the other has a "spy" and can therefore learn which strategy will be used. To what extent can they protect themselves?

Consider first Player 1's point of view. He wishes to make the payoff (his sales) as large as possible, but he would be foolish to simply search for the maximum possible payoff (which is 8) and accordingly choose strategy x_2. For the outcome depends on his competitor's strategy as well, and if 2 should choose strategy y_2, then Player 1 will have no sales at all if he uses x_2! Since Player 2 has a spy, he will choose strategy y_2 if Player 1 uses x_2. To protect himself, Player 1 must be conservative and ask, "How large a payoff can I *guarantee* myself, regardless of what 2 does?" When he chooses a strategy, he knows that he will receive one of the payoffs in the corresponding row of the payoff matrix, but which one depends on 2. So he is certain to receive at least the minimum payoff in that row. Thus choosing x_1 guarantees him at least 1, choosing x_2 guarantees him no less than 0, and x_3 guarantees him at least 4. So Player 1's "conservative" strategy is x_3—this choice guarantees him at least 4 whatever 2 may do, and no other strategy can guarantee him this much. Notice what 1 has done analytically: *He has computed the minimum entry in each row of the payoff matrix and chosen the row for which this minimum is largest.* He is certain to gain 4 even though Player 2 has a spy.

What about Player 2? He wants to make the payoff as *small* as possible, since it represents sales lost to his competitor. But if he chooses strategy y_2 because it offers a chance to hold Player 1 to 0 sales, Player 1 will use information from his spy to choose x_1. Then Player 1 will achieve sales of 7. To protect himself against the spy, Player 2 can ask, "How small can I force the payoff to be, regardless of what Player 1 does?" If Player 2 chooses y_1, he cannot lose more than 4; while if he chooses y_2, he can lose as much as 7; and if he chooses y_3, he can lose as much as 8. Player 2's "conservative" strategy is therefore y_1—this choice guarantees that he will lose no more than 4, and no other strategy can guarantee to hold his losses this low. Analytically, *Player 2 has computed the maximum entry in each column of the payoff matrix and chosen the column for which this maximum is smallest.*

Example 4.6 illustrates how a player can protect himself from a spy as far as possible. The resulting strategy has an additional justification—it is a conservative strategy which guarantees the player as much as can be guaranteed without knowing what his opponent will do. Player 1 acts conservatively by maximizing his minimum possible gain, so his conservative strategy is called maximin. Player 2's conservative strategy is called minimax, for it minimizes his maximum possible loss. Here is a formal definition for a finite game having payoff matrix **M** with entries $m_{i,j}$ representing the payoff when Player 1 chooses strategy x_i and Player 2 chooses strategy y_j.

DEFINITION 4.3   A *maximin strategy* for Player 1 is a strategy x_k such that the minimum element in row k of **M** is at least as large as the minimum element of any other row. The *lower value* of the game is the value of the minimum element in row k.

A *minimax strategy* for Player 2 is a strategy y_i such that the maximum element in column i of **M** is no larger than the maximum element of any other column. The *upper value* of the game is the value of the maximum element in column i.

There may be more than one minimax or maximin strategy in a game, but all such strategies give the same upper and lower values. As the names suggest, the upper value is always at least as large as the lower value. This is reasonable, since if Player 2 has a spy then Player 1 can guarantee himself the lower value; while if Player 2 has the spy, he can be sure of getting the upper value. Surely Player 1 can do better when he has the spy. We will give a proof of this result in Sec. 4.10.

Now in the game of Example 4.6 we notice something remarkable. The maximin strategy is x_3, and the lower value is 4; while the minimax strategy is y_1, and the upper value is also 4. That is, the same number is the maximum of the row minima and the minimum of the column maxima. The floor which Player 1 can put under his gain is the same as the ceiling which Player 2 can put on his loss. Player 1 can guarantee that he will gain at least 4, and Player 2 can guarantee that Player 1 will not gain more than 4. So if both players have spies or both play conservatively, we can predict that Player 1's sales volume will be \$4 million.

DEFINITION 4.4   A finite game is *strictly determined* if its upper and lower values are equal. In this case the common value of the upper and lower values is called the *value* of the game. A *solution* to a strictly determined game is given by stating the value, a minimax strategy and a maximin strategy.

Example 4.7   To review the treatment of the game of Table 4.4:

a. The row minima are 1, 0, and 4 in that order. The largest of these is 4, so the maximin strategy for Player 1 is x_3 and the lower value is 4.
b. The column maxima are 4, 7, and 8 in that order. The smallest of these is 4, so the minimax strategy for Player 2 is y_1 and the upper value is 4.
c. Since the upper and lower values are equal, this game is strictly determined and the value is 4.

The game of Table 4.5 is a minor modification of that of Table 4.4. To treat this game,

a. The row minima are 1, 0, 4 again. So the maximin strategy is x_3 and the lower value is 4.
b. The column maxima are 6, 7, and 8 in that order. The smallest of these is 6, so the minimax strategy for Player 2 is y_1 and the upper value is 6.
c. Since the upper value is greater than the lower value, this game is not strictly determined.

		y_1	y_2	y_3
	x_1	1	7	2
1	x_2	3	0	8
	x_3	6	5	4

Table 4.5   A game which is not strictly determined

Minimax and maximin strategies are often called *optimal strategies* for Players 2 and 1 respectively. They are "best" strategies in a strictly determined game *if* the other player has a spy, or *if* the other player uses his optimal strategy, or *if* a player decides to base his choice on what a strategy can guarantee him. Neither player can profit from a unilateral decision not to use his optimal strategy, though he may want to use a different strategy if he has reason to think that his opponent will not play his optimal strategy. In the absence of such information, a player may do very badly if he does not use his optimal strategy. So we argue that in many situations, both players in a strictly determined game will use their optimal strategies, and the actual outcome of the game will be equal to its value. By contrast, notice that if Player 1 plays his maximin strategy x_3 in the game of Table 4.5, the *worst* thing that 2 can do is to play his minimax strategy y_1. So games which are not strictly determined do not have a predictable outcome. Here is a final example.

Example 4.8   Return to Rommel and Montgomery (see Table 4.3). Montgomery's maximin strategy is x_4, which guarantees him at least 0 (no losses) while any other strategy allows Rommel to inflict a net loss (negative payoff) on the British. Rommel's minimax strategy is y_0, which guarantees him no losses while any other strategy allows Montgomery to inflict a net loss on the Germans. So this game is strictly determined, and the solution consists of strategies x_4 and y_0 with value 0—the two forces avoid each other and neither suffers any casualties. Notice that if Montgomery *knows* Rommel will use y_0, he can do no better than x_4 (though x_1 is just as good). Similarly, even if Rommel knows Montgomery will use x_4, his only rational choice is y_0. (If you think this outcome is not the one Montgomery would want, reread the statement of the example. Rommel can send his entire force by route 2, knowing that Montgomery cannot block him with a superior force there since one British unit is stuck on route 1. Montgomery can inflict no losses on Rommel in this situation, so he might as well wait to fight another day. Rommel would have solved this problem informally—but did you see this "obvious" solution when you read Example 4.5?)

Exercise 4.3   1.   Find all minimax and maximin strategies for the following games and solve them if they are strictly determined.

a.

	y_1	y_2	y_3
x_1	4	-1	6
x_2	3	7	-2
x_3	8	0	4

b.

	y_1	y_2	y_3	y_4
x_1	2	1	-4	-3
x_2	0	0	-1	1
x_3	-1	3	-2	-5

2.   Find the maximin and minimax strategies for the following games and solve them if they are strictly determined:

a. the game of Exercise 4.2.4
b. the game of Exercise 4.2.5
c. the game of Exercise 4.2.6
d. the game of Exercise 4.2.7

3.   Give an example of a strictly determined game for which the maximin strategy is not unique. Give an example of a strictly determined game for which neither the maximin nor the minimax strategy is unique.

4.   Suppose that the payoff in Rommel and Montgomery is simply the number of German companies lost. Write the new payoff matrix, and show that the solution to the game is just as it was with the original payoff function.

5.   . A science fiction story has the following plot. A war between the humans and the aliens is in progress in which both are following "best" strategies as computed by great computers. The aliens have a higher technology, so the predictable result is that they will win. A brilliant human individual decides to abandon the "best" strategy and fight unpredictably. The result is that the aliens are defeated. Discuss this in the light of what you know about game theory if

a. the game in question is strictly determined
b. the game is not strictly determined

6.   Make up a simple 3×3 game with a value and play it five times with each of several of your friends who do not know game theory. You play Player 2's minimax strategy each time. Do your friends play maximin at the beginning? Do they learn to play maximin when you repeatedly play the same (minimax) strategy against them? [For a survey of experiments of this type done by psychologists, read "Experimental games: A review" by Anatol Rapoport and Carol Orwant *Behavioral Science* **7**, 1962, 1–37].

7.   A manufacturer of special-purpose electronic tubes has agreed to replace any which are defective. His gain on a tube sold is $50. If a defective tube is sold, it must be replaced at a cost of $100, so that the manufacturer loses $50 if he sells a defective tube. It is possible to detect defective tubes by testing, which costs $10 per tube. A defective tube has a scrap value of $20. The manufacturer has the following strategies available:

x_1:   sell without testing

x_2:   scrap without testing

x_3:   test: scrap if defective and sell if not

x_4:   test: sell if defective and scrap if not

x_5:   test and sell (ignore test result)

x_6:   test and scrap (ignore test result)

Consider the manager to be Player 1 in a game in which Player 2 (Nature) has two strategies:

y_1: tube is defective

y_2: tube is not defective

Find the payoff matrix of this game and solve it. (Nature is assumed to receive any amount lost by the manufacturer. We will comment on Nature as an opponent in Chapter 5.) Several of the manufacturer's possible strategies are "obviously absurd." Which are they? Does your result change if these strategies are omitted?

8. A manufacturer of snowmobiles (ugh) must decide in advance how many to produce for the coming winter season. The winter will be either severe (y_1), normal (y_2), or moderate (y_3). The manufacturer knows the appropriate production strategies (x_1, x_2, and x_3 respectively) which will maximize his profit in each type of winter, and can compute the result of each of these strategies for the other types of winter. The table of his profit is:

	y_1	y_2	y_3
x_1	120	50	−10
x_2	100	100	50
x_3	80	80	80

Solve this game against Nature. Do you think that in practice the manufacturer should assume that Nature is a rational Player 2 who will play minimax against him?

9. Suppose that in Rommel and Montgomery, Montgomery is able to provide fuel to the British company on route 1. He now has an additional pure strategy x_0—send all four companies to route 2. Solve this larger version of the game.

10. In an article "Military decision and game theory" (Journal of the Operations Research Society of America, **2**, 1954, 365−385), O. G. Haywood gives the following factual example. In February of 1943 a Japanese convoy is seen about to leave Rabaul to reinforce Lae. The convoy can travel north or south of New Britain Island, each route requiring 3 days. The American general, Kinney, wants to bomb the convoy. He can send most of his reconnaissance aircraft either north or south of the island, with only routine patrols on the other route. Poor visibility and rain are forecast for the northern route, fair weather for the southern. General Kinney (in his memoirs) estimated that if the Japanese and he both chose the northern route, he would find the convoy on the second day and have 2 days of bombing. If he sent his planes south, and the Japanese went north, the convoy would be near Lae before being discovered and only 1 day of bombing would be possible. If the Japanese went south, and Kinney sent his planes north, he would sight the convoy on the second day anyway and have 2 days of bombing. If both went south, he would find the convoy immediately

and have 3 days of bombing. What should General Kinney and the Japanese commander do? (They both did the right thing, by the way.)

11.   You might suppose that if a game is strictly determined with value v and x_i and y_j are strategies with $M(x_i, y_j) = v$, then x_i must be maximin and y_j must be minimax. Explain why the game

$$\begin{array}{|cc|}\hline 0 & 1 \\ -1 & 0 \\ \hline \end{array}$$

shows that this is not true.

4.4   MIXED STRATEGIES

We have seen that strictly determined games have a predictable outcome given by a solution of the game. We have also seen that many games are not strictly determined. The central result of game theory states that if we add strategies of a new type to each player's set of possible strategies, then the enlarged game is strictly determined and has a solution. The added strategies are derived from the strategies already present in a way best described by an example.

Example 4.9   The payoff matrix for the game of Matching Fingers (Example 4.3) is repeated in Table 4.6. You can see that the upper value of this game is 1, and that the lower value is −1, so that the game has no solution. Although there is no clearly defined "best" strategy for either player in a single play of this game, it *is* clear that a player who intends to play the game repeatedly should *not* use the same strategy each time. For if he did, his opponent would soon learn what to expect and could then win every time. [Notice that in games with a value, each player should continue to use his maximin (or minimax) strategy no matter how often the game is repeated.] If a player alternates strategies according to some simple pattern, his opponent might also detect this and use this knowledge to win repeatedly. So a player who matches fingers repeatedly may well be led to *randomly* choose the number of fingers he will present.

$$
\begin{array}{c c}
 & \begin{array}{cc} 2 \\ y_1 \quad y_2 \end{array} \\
1 \quad \begin{array}{c} x_1 \\ x_2 \end{array} & \begin{array}{|cc|} \hline 1 & -1 \\ -1 & 1 \\ \hline \end{array}
\end{array}
$$

Table 4.6   Payoff matrix for matching fingers

Let us concentrate on Player 1. Suppose he decides to choose the number of fingers he will present by tossing a fair coin (which he hides from

Player 2) and presenting one finger if heads, two if tails. His payoff against either of Player 2's strategies is now a random variable. The expected value of that random variable tells him (by the law of large numbers) the long-term average payoff he will receive. If Player 2 plays y_1, the expected payoff is, from Table 4.6,

$$1 \cdot P(1 \text{ uses } x_1) + (-1) \cdot P(1 \text{ uses } x_2) = 1 \cdot \tfrac{1}{2} - 1 \cdot \tfrac{1}{2} = 0$$

and if Player 2 plays y_2, the expected payoff is

$$(-1) \cdot P(1 \text{ uses } x_1) + 1 \cdot P(1 \text{ uses } x_2) = -1 \cdot \tfrac{1}{2} + 1 \cdot \tfrac{1}{2} = 0$$

This means that by randomly choosing x_1 with probability $\tfrac{1}{2}$ and x_2 with probability $\tfrac{1}{2}$ Player 1 can guarantee an expected payoff of 0, whichever strategy Player 2 chooses. Since he could previously guarantee only -1, the use of randomly chosen strategies is advantageous.

In Example 4.9, Player 1 discovered a new approach to playing a game. Rather than choosing a particular strategy, he chose x_1 with probability $\tfrac{1}{2}$ and x_2 with probability $\tfrac{1}{2}$. He might instead have chosen x_1 with probability $\tfrac{1}{3}$ and x_2 with probability $\tfrac{2}{3}$. In general, his new approach specifies the probabilities

$$p_1 = P(\text{strategy } x_1 \text{ is chosen})$$
$$p_2 = P(\text{strategy } x_2 \text{ is chosen})$$

rather than specifying a particular strategy. In a game in which Player 1's set of possible strategies is $X = \{x_1, \ldots, x_n\}$, he can play the game by giving a probability vector $\mathbf{p} = [p_1, p_2, \ldots, p_n]$ rather than stating which specific strategy x_i he will use. To choose a specific strategy, Player 1 performs a random experiment with n possible outcomes having probabilities $p_1, \ldots, p_n$. If the ith outcome occurs, he uses strategy x_i. We say that Player 1 is using a mixed strategy. Here is a formal definition.

DEFINITION 4.5 Suppose that the set of strategies for Player 1 is $X = \{x_1, \ldots, x_n\}$. A *mixed strategy* for Player 1 is a probability vector $\mathbf{p} = [p_1, \ldots, p_n]$ where for each $i = 1, 2, \ldots, n$

$$p_i = P(\text{strategy } x_i \text{ is used})$$

X^* will denote the set of all mixed strategies for Player 1.

The definition of the set Y^* of all mixed strategies for Player 2 is similar. Recall from Sec. 2.3 that the components p_i of a probability vector are all nonnegative and that their sum is 1. Any n-dimensional probability vector specifies a mixed strategy for Player 1. Similarly, if $Y = \{y_1, \ldots, y_m\}$ is the set of possible strategies for Player 2, any m-dimensional probability vector $\mathbf{q} = [q_1, \ldots, q_m]$ specifies a mixed strategy for Player 2. The original strategies x_i and y_j are called *pure strategies* when we wish to distinguish them from mixed strategies. A pure strategy can be described as a mixed strategy which always makes the same choice, so for convenience we can speak of all strategies, pure or mixed, as members of X^* or Y^*.

If both players use mixed strategies, the payoff which results is a random variable. We will use the expected payoff (an ordinary number) to measure the outcome of the game. This can be justified in two ways. First, the law of large numbers tells us that the average of the actual payoffs of a large number of repetitions of the game will be close to the expected payoff. Second, if the payoff is a utility measuring Player 1's preferences among the pure strategies, then Player 1's preferences among different mixed strategies are described by expected utilities or expected payoffs. This is a fact about utilities which we will not discuss in detail.

Example 4.10   In Matching Fingers, suppose Player 1 uses the mixed strategy

$$\mathbf{p} = [\tfrac{1}{2}, \tfrac{1}{2}]$$

and Player 2 uses the mixed strategy

$$\mathbf{q} = [\tfrac{1}{3}, \tfrac{2}{3}]$$

The actual payoff is a random variable which (from Table 4.6) is 1 if Player 1 uses x_1 and Player 2 uses y_1, and so on. What is

$$P(\text{Player 1 uses } x_1 \text{ and Player 2 uses } y_1)$$

Surely each player conceals his random experiment from the other, so we can assume that the two players operate their mixed strategies *independently* of each other. This means that

$$P(1 \text{ uses } x_1 \text{ and 2 uses } y_1) = P(1 \text{ uses } x_1) \cdot P(2 \text{ uses } y_1)$$
$$= \tfrac{1}{2} \cdot \tfrac{1}{3} = \tfrac{1}{6}$$

The expected payoff when Player 1 uses the mixed strategy $\mathbf{p}$ and Player 2 uses the mixed strategy $\mathbf{q}$ is therefore

$$\sum M(x_i, y_j)P(1 \text{ uses } x_i \text{ and 2 uses } y_j) = (1 \cdot \tfrac{1}{2} \cdot \tfrac{1}{3}) + (-1 \cdot \tfrac{1}{2} \cdot \tfrac{2}{3})$$
$$+ (-1 \cdot \tfrac{1}{2} \cdot \tfrac{1}{3}) + (1 \cdot \tfrac{1}{2} \cdot \tfrac{2}{3})$$
$$= 0$$

We have now completed the process of enlarging a finite game by allowing the players to use mixed strategies and by defining the outcome of the game when mixed strategies are used to be the expected payoff. The result of this process is a new game which is called the mixed extension of the original game.

DEFINITION 4.6   The *mixed extension* of a game with strategy sets $X = \{x_1, \ldots, x_n\}$ and $Y = \{y_1, \ldots, y_m\}$ and payoff function $M(x, y)$ is the game with strategy sets X^* and Y^* and payoff function for $\mathbf{p}$ in X^* and $\mathbf{q}$ in Y^* given by

$$M^*(\mathbf{p}, \mathbf{q}) = \sum_{i=1}^{n} \sum_{j=1}^{m} M(x_i, y_j)p_i q_j$$

Several comments about Definition 4.6 are in order. First, if $\mathbf{p}$ chooses x_i always and $\mathbf{q}$ chooses y_j always, then

$$M^*(\mathbf{p}, \mathbf{q}) = M(x_i, y_j)$$

so that the payoff for those mixed strategies which are really pure strategies is not changed (see Exercise 4.4.3). Second, the double sum used to define M^* is just an instruction to sum over all nm possible pairs of pure strategies x_i and y_j. You may find it convenient to use the vector-matrix formulation for this sum given in Exercise 4.4.7.

Third, there are infinitely many possible mixed strategies, so that the mixed extension of a finite game is no longer a finite game. The payoff function M^* cannot be summarized by a payoff matrix as could M.

Example 4.11 Let us describe completely the mixed extension of Matching Fingers. Any mixed strategy for Player 1 is a probability vector $\mathbf{p} = [\alpha, 1 - \alpha]$ where α is the probability of choosing x_1 and $1 - \alpha$ is the probability of choosing x_2. So X^* is the set of all probability vectors $\mathbf{p} = [\alpha, 1 - \alpha]$ for $0 \leq \alpha \leq 1$. Similarly, Y^* is the set of all probability vectors $\mathbf{q} = [\beta, 1 - \beta]$ for $0 \leq \beta \leq 1$. The expected payoff when Player 1 uses $\mathbf{p}$ and Player 2 uses $\mathbf{q}$ is

$$M^*(\mathbf{p}, \mathbf{q}) = \alpha\beta M(x_1, y_1) + \alpha(1 - \beta)M(x_1, y_2) + (1 - \alpha)\beta M(x_2, y_1)$$
$$+ (1 - \alpha)(1 - \beta)M(x_2, y_2)$$
$$= \alpha\beta - \alpha(1 - \beta) - (1 - \alpha)\beta + (1 - \alpha)(1 - \beta)$$
$$= 4\alpha\beta - 2\alpha - 2\beta + 1$$
$$= (2\alpha - 1)(2\beta - 1)$$

Exercise 4.4 1. Suppose that in the game of Matching Fingers Player 1 uses the mixed strategy $\mathbf{p} = [1/4, 3/4]$ and Player 2 uses the mixed strategy $\mathbf{q} = [3/4, 1/4]$. Compute the expected payoff $M^*(\mathbf{p}, \mathbf{q})$ in two ways: (1) from first principles, as in Example 4.10, and (2) using the result of Example 4.11.

2. Suppose that in Matching Fingers Player 1 uses the mixed strategy $\mathbf{p} = [1/2, 1/2]$. Show that for *any* mixed strategy $\mathbf{q}$ for Player 2, $M^*(\mathbf{p}, \mathbf{q}) = 0$. Can you explain why this result is not surprising since Example 4.9 showed that $\mathbf{p}$ had expected payoff 0 against both y_1 and y_2?

3. In Matching Fingers suppose that Player 1 uses the mixed strategy $\mathbf{p} = [1, 0]$. Find (as in Example 4.9) the expected payoff of this $\mathbf{p}$ when Player 2 plays y_1 and when Player 2 plays y_2. Explain why the mixed strategy $\mathbf{p} = [1, 0]$ is "essentially the same" as the pure strategy x_1. Are the payoffs against y_1 and y_2 the same?

4.   A game has payoff matrix

$$
\begin{array}{cc}
 & \begin{array}{ccc} y_1 & y_2 & y_3 \end{array} \\
\begin{array}{c} x_1 \\ x_2 \end{array} &
\left.\begin{array}{|ccc|} \hline 0 & 1 & 2 \\ 2 & 1 & 0 \\ \hline \end{array}\right.
\end{array}
$$

Player 1 chooses a strategy by tossing a fair coin and using x_1 if heads, x_2 if tails. What is **p** for this strategy? What is the expected payoff if Player 1 uses this **p** and Player 2 uses y_1?

Player 2 chooses a strategy by rolling a fair die and using y_1 if the outcome is 1, y_2 if it is 2, and y_3 if it is 3, 4, 5, or 6. What is **q** for this strategy? What is the expected payoff if Player 2 uses this **q** and Player 1 uses x_1?

What is the expected payoff if Player 1 uses **p** and Player 2 uses **q**, where **p** and **q** are as in the earlier parts of the problem?

5.   Mixed strategies for Player 1 in the game of Exercise 4.4.4 have the form $\mathbf{p} = [\alpha, 1 - \alpha]$ for some $0 \le \alpha \le 1$. Mixed strategies for Player 2 have the form $\mathbf{q} = [\beta, \gamma, 1 - \beta - \gamma]$ for some $\beta \ge 0$, $\gamma \ge 0$ with $\beta + \gamma \le 1$. Compute the expected payoff $M^*(\mathbf{p}, \mathbf{q})$ in terms of α, β, and γ.

6.   Toss a coin to carry out the mixed strategy $\mathbf{p} = [1/2, 1/2]$ for Player 1 in Matching Fingers, and roll a die to carry out the mixed strategy $\mathbf{q} = [1/3, 2/3]$ for Player 2. (Explain how you used the die.)

Play these mixed strategies against each other 20 times, and record the payoff each time. Compute the average payoff and compare it with the expected payoff computed in Example 4.10.

7.   A finite game has payoff matrix **M** with entries $m_{ij} = M(x_i, y_j)$ for $i = 1, \ldots, n$ and $j = 1, \ldots, m$. A mixed strategy **p** for Player 1 will be represented by an n-dimensional *row vector*; a mixed strategy **q** for Player 2 will be represented by an m-dimensional *column vector*. Show that the expected payoff as given by Definition 4.6 can be expressed as $M^*(\mathbf{p}, \mathbf{q}) = \mathbf{pMq}$.

8.   Use Exercise 4.4.7 to express the expected payoff for mixed strategies

$$
\mathbf{p} = [\alpha, 1 - \alpha] \qquad \mathbf{q} = \begin{bmatrix} \beta \\ 1 - \beta \end{bmatrix}
$$

for Matching Pennies as

$$
M^*(\mathbf{p}, \mathbf{q}) = [\alpha, 1 - \alpha] \begin{bmatrix} 1 & -1 \\ -1 & 1 \end{bmatrix} \begin{bmatrix} \beta \\ 1 - \beta \end{bmatrix}
$$

Show that this product gives the same result obtained in Example 4.11.

9.   Repeat Exercise 4.4.4 using the vector-matrix expression for M^* given in Exercise 4.4.7.

4.5   MIXED EXTENSIONS HAVE SOLUTIONS

Our purpose in introducing mixed strategies in games which are not strictly determined is to enlarge the set of strategies so that the extended game has a value. *In that case the value, a maximin mixed strategy and a minimax mixed strategy together will be called a solution of the original game.* Since the mixed extension of a finite game is no longer finite, the details of finding a solution are now complicated. We therefore introduce solutions of non-strictly determined, finite games by studying Matching Fingers once more.

Example 4.12   We know from Example 4.11 that if Player 1 uses a mixed strategy $\mathbf{p} = [\alpha, 1 - \alpha]$ in Matching Fingers and Player 2 uses a mixed strategy $\mathbf{q} = [\beta, 1 - \beta]$, the expected payoff is

$$M^*(\mathbf{p}, \mathbf{q}) = (2\alpha - 1)(2\beta - 1)$$

Let us take Player 1's point of view. For each of his mixed strategies $\mathbf{p}$, he wants to know the smallest expected payoff he can possibly receive. This is the amount that $\mathbf{p}$ can guarantee him. Formally, Player 1 must be able to find the smallest of the numbers $M^*(\mathbf{p}, \mathbf{q})$ when $\mathbf{p}$ is fixed and $\mathbf{q}$ ranges over all Player 2's mixed strategies. Let us take for example $\mathbf{p} = [1/4, 3/4]$. Then if $\mathbf{q} = [\beta, 1 - \beta]$,

$$M^*(\mathbf{p}, \mathbf{q}) = (2 \cdot 1/4 - 1)(2\beta - 1) = 1/2 - \beta$$

Figure 4.1 is a graph of the function $h(\beta) = 1/2 - \beta$ for $0 \le \beta \le 1$. From that graph we can see that $1/2 - \beta$ is smallest when $\beta = 1$, and that the value when $\beta = 1$ is $1/2 - 1 = -1/2$. So if Player 1 uses this $\mathbf{p}$, the smallest expected payoff he can receive is $-1/2$, and this happens when Player 2 uses $\mathbf{q} = [1, 0]$. This $\mathbf{q}$ is of course just another name for the pure strategy y_1.

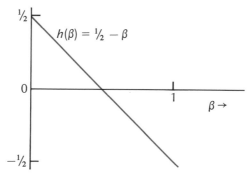

Figure 4.1   Graph of the function $h(\beta) = 1/2 - \beta$

We need a new notation which will enable us to abbreviate "the smallest of the numbers $M^*(\mathbf{p}, \mathbf{q})$ when $\mathbf{p}$ is fixed and $\mathbf{q}$ ranges over all Player 2's mixed strategies." We will abbreviate this as

$$\min_{\mathbf{q}} M^*(\mathbf{p}, \mathbf{q})$$

Similarly

$$\max_{\mathbf{p}} M^*(\mathbf{p}, \mathbf{q})$$

is an abbreviation for "the largest of the numbers $M^*(\mathbf{p}, \mathbf{q})$ when $\mathbf{q}$ is fixed and $\mathbf{p}$ ranges over all Player 1's mixed strategies." If we have a function $h(\beta)$, we use

$$\min_{0 \leq \beta \leq 1} h(\beta)$$

to abbreviate "the smallest of the numbers $h(\beta)$ as β ranges over all values between 0 and 1." Figure 4.1 shows that

$$\min_{0 \leq \beta \leq 1} (1/2 - \beta) = -1/2$$

Using this new notation, it is easy to restate the ideas of Definition 4.3 in the context of the mixed extension of a finite game.

DEFINITION 4.7   A mixed strategy $\mathbf{p}_0$ is a *maximin strategy* for Player 1 if for any $\mathbf{p}$ in X^*

$$\min_{\mathbf{q}} M^*(\mathbf{p}_0, \mathbf{q}) \geq \min_{\mathbf{q}} M^*(\mathbf{p}, \mathbf{q})$$

The *lower value* of the mixed extension is the number

$$\min_{\mathbf{q}} M^*(\mathbf{p}_0, \mathbf{q})$$

A mixed strategy $\mathbf{q}_0$ is a *minimax strategy* for Player 2 if for any $\mathbf{q}$ in Y^*

$$\max_{\mathbf{p}} M^*(\mathbf{p}, \mathbf{q}_0) \leq \max_{\mathbf{p}} M^*(\mathbf{p}, \mathbf{q})$$

The *upper value* of the mixed extension is the number

$$\max_{\mathbf{p}} M^*(\mathbf{p}, \mathbf{q}_0)$$

If Player 1 is to find a maximin strategy, he must find $\min_{\mathbf{q}} M^*(\mathbf{p}, \mathbf{q})$ (the smallest expected payoff possible when he uses $\mathbf{p}$) for each $\mathbf{p}$, then choose the $\mathbf{p}$ for which this minimum is largest. This task is greatly simplified by the fact that for *any* $\mathbf{p}$, $\min_{\mathbf{q}} M^*(\mathbf{p}, \mathbf{q})$ is *always* attained when $\mathbf{q}$ is one of Player 2's pure strategies. In Matching Fingers these are $\mathbf{q} = [1, 0]$ (another name for y_1) and $\mathbf{q} = [0, 1]$ (another name for y_2). Let us check that this is true.

Example 4.13   In Example 4.12 we saw that when $\mathbf{p} = [1/4, 3/4]$,

$$\min_{\mathbf{q}} M^*(\mathbf{p}, \mathbf{q}) = -1/2$$

and that the $\mathbf{q}$ which attained that minimum was $\mathbf{q} = [1, 0]$. Now in fact, no matter what **p** Player 2 uses,

$$\min_{\mathbf{q}} M^*(\mathbf{p}, \mathbf{q})$$

is always attained by $\mathbf{q} = [1, 0]$ or $\mathbf{q} = [0, 1]$. To see this, notice that if we fix $\mathbf{p} = [\alpha, 1 - \alpha]$, then $2\alpha - 1$ is some fixed number, say c. So

$$\min_{\mathbf{q}} M^*(\mathbf{p}, \mathbf{q}) = \min_{0 \le \beta \le 1} c(2\beta - 1)$$

Now $c(2\beta - 1)$ is always of the form $a\beta + b$, so its graph is always a straight line (see Exercise 4.5.1). This means that the smallest value of $a\beta + b$ for β between 0 and 1 is attained either at $\beta = 0$ or at $\beta = 1$.

Example 4.13 illustrates the fact that, to find the worst that can happen when he uses any specified mixed strategy, a player need only look at his opponent's pure strategies. That is the content of the following theorem. The proof appears in Sec. 4.10.

THEOREM 4.1   If **p** in X^* is any mixed strategy for Player 1, then

$$\min_{\mathbf{q}} M^*(\mathbf{p}, \mathbf{q}) = \min_{1 \le j \le m} M^*(\mathbf{p}, y_j)$$

If **q** in Y^* is any mixed strategy for Player 2, then

$$\max_{\mathbf{p}} M^*(\mathbf{p}, \mathbf{q}) = \max_{1 \le i \le n} M^*(x_i, \mathbf{q})$$

Combining Theorem 4.1 with Definition 4.7, we see that to find a maximin mixed strategy for Player 1 we need only investigate the minimum payoff against Player 2's *pure* strategies,

$$\min_{1 \le j \le m} M^*(\mathbf{p}, y_j)$$

The **p** that maximizes this is maximin. Similarly, the **q** that minimizes

$$\max_{1 \le i \le n} M^*(x_i, \mathbf{q})$$

is a minimax strategy for Player 2.

Example 4.14   We know that Matching Fingers is not strictly determined and that the payoff function for the mixed extension of this game is

$$M^*(\mathbf{p}, \mathbf{q}) = (2\alpha - 1)(2\beta - 1)$$

when

$$\mathbf{p} = [\alpha, 1 - \alpha] \qquad \mathbf{q} = [\beta, 1 - \beta]$$

We will try to find a maximin strategy $\mathbf{p_0}$ for Player 1. We now know that a maximin strategy $\mathbf{p_0}$ is a mixed strategy which maximizes

$$\min_{j=1,2} M^*(\mathbf{p}, y_j) = \min \{M^*(\mathbf{p}, y_1), M^*(\mathbf{p}, y_2)\}$$

Since y_1 as a mixed strategy is given by $\beta = 1$ and y_2 is given by $\beta = 0$,

$$M^*(\mathbf{p}, y_1) = 2\alpha - 1$$
$$M^*(\mathbf{p}, y_2) = 1 - 2\alpha$$

The maximin $\mathbf{p}_0$ is therefore obtained by choosing the α between 0 and 1 for which the smaller of these two numbers, $\min \{2\alpha - 1, 1 - 2\alpha\}$, is largest. This is most easily done by drawing the graphs of $2\alpha - 1$ and $1 - 2\alpha$, as in Fig. 4.2. The blue line is the minimum of the two. Figure 4.2 shows that this minimum is largest when $\alpha = \frac{1}{2}$, where the minimum is 0. So the maximin strategy is $\mathbf{p}_0 = [\frac{1}{2}, \frac{1}{2}]$. The lower value is, using Theorem 4.1 once more,

$$\min_{\mathbf{q}} M^*(\mathbf{p}_0, \mathbf{q}) = \min_{j=1,2} M^*(\mathbf{p}_0, y_j)$$
$$= \min \{2 \cdot \frac{1}{2} - 1, 1 - 2 \cdot \frac{1}{2}\} = 0$$

An argument much like that of Example 4.14 (see Exercise 4.5.3) shows that $\mathbf{q}_0 = [\frac{1}{2}, \frac{1}{2}]$ is a minimax strategy for Player 2 in Matching Fingers and that the upper value of the mixed extension of this game is also 0. All our troubles with Matching Fingers are now at an end: The upper and lower values of the mixed extension are equal, so the mixed extension has value 0 and has a solution given by the value 0 and the maximin strategy $\mathbf{p}_0 = [\frac{1}{2}, \frac{1}{2}]$ and minimax strategy $\mathbf{q}_0 = [\frac{1}{2}, \frac{1}{2}]$. This result should seem reasonable. It says that the best a player can do is to randomly choose one or two fingers, choosing each half the time in the long run. If both players do this, then each will win half the time in the long run and the expected payoff is 0.

The good news of game theory is that even very complicated finite games can always be solved if we allow the use of mixed strategies. This is the

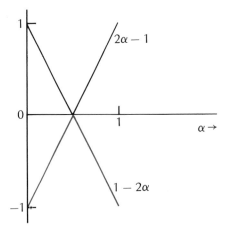

Figure 4.2   Graph of $2\alpha - 1$ and $1 - 2\alpha$ for $0 \le \alpha \le 1$. The blue line is the minimum of these functions.

most important theorem in the subject, discovered in 1927 by John von Neumann at the age of 24.

THEOREM 4.2 Fundamental Theorem of Game Theory The mixed extension of any finite game has a value.

Theorem 4.2 means that if the players can use mixed strategies, any finite game has a predictable outcome in the sense that each player has a mixed strategy which guarantees him as much as any strategy can, and that neither player can profit by not using this optimal strategy unless the other player also abandons his optimal strategy. As we might expect, nothing is gained by introducing mixed strategies in games which are strictly determined. In this case the value of the mixed extension is the same as the value of the original game, and the optimal pure strategies in the original game are optimal strategies in the mixed extension as well. These facts, as well as the fundamental theorem, will be proved in Sec. 4.10.

Exercise 4.5   1.   Draw a graph of the function $h(\beta) = c\beta + d$, for β taking values between 0 and 1 and c and d fixed numbers, in each of the two cases

a. $c > 0$ b. $c < 0$

(*Hint*: Mark your vertical scale from d to $c + d$.)

2.   Find and graph as a function of β the expected payoff $M^*(\mathbf{p}, \mathbf{q})$ in Matching Fingers for $\mathbf{q} = [\beta, 1 - \beta]$ against $\mathbf{p} = [\sqrt[3]{4}, \sqrt[1]{4}]$. For what value of β does this function attain its minimum?

3.   Show by an argument similar to that given in Example 4.14 that $\mathbf{q_0} = [\sqrt[1]{2}, \sqrt[1]{2}]$ is a minimax strategy for Player 2 in the mixed extension of Matching Fingers and that the upper value is 0.

4.   The game with payoff matrix

$$
\begin{array}{cc}
 & \begin{array}{cc} 2 \\ y_1 \quad\; y_2 \end{array} \\
1 \;\; \begin{array}{c} x_1 \\ x_2 \end{array} & \begin{array}{|cc|} \hline 1 & 2 \\ 0 & -1 \\ \hline \end{array}
\end{array}
$$

has the solution (x_1, y_1) with value 1. What is the mixed extension of this game? Show that $\mathbf{p_0} = [1, 0]$ is the maximin strategy for the mixed extension by showing

a. $\min\limits_{j=1,2} M^*(\mathbf{p_0}, y_j) = 1$

b.   if $\mathbf{p} \neq \mathbf{p_0}$, then $M^*(\mathbf{p}, y_1) < 1$

Show similarly that $\mathbf{q_0} = [1, 0]$ is the minimax strategy for the mixed extension and that the value of the mixed extension is 1. (As claimed in the text, the solution of the mixed extension is the same as the solution of the original game.)

5.   Solve the game with matrix

$$\begin{array}{|cc|} \hline 4 & 0 \\ 1 & 3 \\ \hline \end{array}$$

Hint: Show that if $\mathbf{p} = [\alpha,\ 1 - \alpha]$ and $\mathbf{q} = [\beta,\ 1 - \beta]$, then

$$M^*(\mathbf{p}, \mathbf{q}) = (3\alpha - 1)(2\beta - 1) + 2$$

6.   Solve the game with matrix

$$\begin{array}{|cc|} \hline 2 & -1 \\ -2 & 1 \\ \hline \end{array}$$

*7.   Suppose that v is the value of the mixed extension of a finite game, $\mathbf{p_0}$ is a maximin mixed strategy, and $\mathbf{q_0}$ is a minimax mixed strategy. Use Theorem 4.1 and Definition 4.7 to show that

a. $M^*(\mathbf{p_0},\ y_j) \geq v$ for each pure strategy y_j of Player 2
b. $M^*(x_i,\ \mathbf{q_0}) \leq v$ for each pure strategy x_i of Player 1

*8.   We claim that if you can find a mixed strategy $\mathbf{p_0}$ for Player 1, a mixed strategy $\mathbf{q_0}$ for Player 2, and a number v such that

a. $M^*(\mathbf{p_0},\ y_j) \geq v$ for every pure strategy y_j
b. $M^*(x_i,\ \mathbf{q_0}) \leq v$ for every pure strategy x_i

then the mixed extension of the game has value v, $\mathbf{p_0}$ is a maximin strategy, and $\mathbf{q_0}$ is a minimax strategy. To prove this, procede as follows.

c.   Show that item a implies that the lower value of the mixed extension is at least v.
d.   Show that item b implies that the upper value of the mixed extension is no more than v.
e.   Conclude from items c and d that v is the value and that $\mathbf{p_0}$ and $\mathbf{q_0}$ are optimal strategies.

9.   We claim that any 2×2 finite game

$$\begin{array}{|cc|} \hline a & b \\ c & d \\ \hline \end{array}$$

which is not strictly determined has mixed extension with the value

$$v = \frac{ad - bc}{a + d - b - c}$$

maximin strategy for Player 1

$$\mathbf{p_0} = \left[\frac{d - c}{a + d - b - c},\ \frac{a - b}{a + d - b - c} \right]$$

and minimax strategy for Player 2

$$\mathbf{q}_0 = \left[\frac{d - b}{a + d - b - c}, \frac{a - c}{a + d - b - c} \right]$$

Prove that these formulas are correct, using the result of Exercise 4.5.8.
10.   Use the formula of Exercise 4.5.9 to solve the following 2×2 games:

a.   Matching Fingers
b.   the game of Exercise 4.5.5
c.   the game of Exercise 4.5.6

4.6   DOMINATED STRATEGIES

We now know that any finite game has a solution if mixed strategies are allowed, and we have solved a few 2×2 games which were not strictly determined. Exercise 4.5.9 gave formulas for solving any 2×2 game. The next three sections will discuss systematic methods of solving larger games. As you might guess, the difficulty of solution increases when the players have many pure strategies. It is therefore to our advantage to reduce the size of the game as much as possible before attempting to solve it. That is the purpose of this section.

Example 4.15   Let us study the 3×3 game of Table 4.7 from Player 2's point of view. As Player 2 studies the payoffs for strategies y_1 and y_3 with his goal of minimizing the payoff in mind, he notices that *no matter what his opponent does, y_3 is preferable to y_1.*

		2		
		y_1	y_2	y_3
	x_1	1	5	0
1	x_2	5	1	2
	x_3	4	3	1

Table 4.7   Strategy y_3 dominates y_1

This is true because every entry in the third column of the payoff matrix is less than the corresponding entry in the first column. So Player 2 would be foolish to use y_1 when y_3 is available. He would also be foolish to use any mixed strategy which chooses y_1 with positive probability, since again choosing y_3 instead is always better (see Exercise 4.6.5).

So in effect we can eliminate the first column from the payoff matrix. The original 3×3 game has been reduced to a 3×2 game. The reduced game should have the same solution as the original game, since Player 2 would never choose y_1, and hence both players can ignore that strategy.

Example 4.15 illustrates how games can often be reduced in size by eliminating strategies which do no better than another strategy against all the opponent's pure strategies. For Player 2 this means eliminating *columns* with all entries *greater* than or equal to the corresponding entries in another column, and some entries strictly greater. Player 1 eliminates a *row* if all its entries are *less* than or equal to the corresponding entries of another row, and some are strictly less. These ideas lead to a definition.

DEFINITION 4.8   Strategy x_i for Player 1 *dominates* strategy x_j if

$$M(x_i, y_k) \geq M(x_j, y_k)$$

for all Player 2's pure strategies y_k and strict inequality holds for at least one y_k. Strategy y_i for Player 2 *dominates* strategy y_j if

$$M(x_k, y_i) \leq M(x_k, y_j)$$

for all Player 1's pure strategies x_k and strict inequality holds for at least one x_k.

Eliminating dominated strategies never changes the solution of a game, whether or not the game is strictly determined. The first step in solving a game is therefore to reduce its size by eliminating all dominated strategies. The next step is to see if the reduced game has dominated strategies—some may appear which were not dominated in the original game, and they too can be eliminated. In fact, if the original game is strictly determined, you can find the solution by continuing to eliminate dominated strategies until only a 1×1 game remains.

Example 4.16   Two discount chains are planning to open stores in adjacent shopping malls. Each chain has three store types (layout, size, stock, etc.) of different cost and attractiveness. The manner in which sales divide between the stores depends on the choice of types. The estimated monthly profit for chain 1 in all nine possible circumstances is given in Table 4.8 in units of $100,000.

		2 y_1	y_3	y_3
	x_1	1	−1	0
1	x_2	2	−1	3
	x_3	1	3	2

Table 4.8   Payoff matrix for a chain store decision problem

Studying Table 4.8, we see that x_1 is dominated by both x_2 and x_3. Eliminating the first row reduces this 3×3 game to the 2×3 game of Table 4.9a.

$$
\begin{array}{cc}
 & 2 \\
\end{array}
$$

		y_1	y_2	y_3
1	x_2	2	−1	3
	x_3	1	3	2

a.

		y_1	y_2
1	x_2	2	−1
	x_3	1	3

b.

Table 4.9   Reduction of the game of Table 4.8

Now in this 2 × 3 game, y_3 is dominated by y_1. (Notice that y_3 was *not* dominated by y_1 in the original·game.) Eliminating the third column reduces the game to the 2 × 2 game of Table 4.9b. No further reduction is possible. The game is not strictly determined, but the formulas of Exercise 4.5.9 apply. The value of the mixed extension is $v = 7/5$, the maximin mixed strategy is $\mathbf{p_0} = [2/5, 3/5]$, and the minimax mixed strategy is $\mathbf{q_0} = [4/5, 1/5]$. So the optimal strategies for the original game are $\mathbf{p_1} = [0, 2/5, 3/5]$ and $\mathbf{q_1} = [4/5, 1/5, 0]$. (*Comment:* It is hard to believe that mixed strategies would actually be used in such a case in practice, unless either the confrontation was to be repeated in other locations or the chains feared that "spies" might disclose their plans. This indicates that the true utility of various outcomes does not consist only of expected profits, but includes aversion to risk and other factors.)

Exercise 4.6   1.   Reduce each of the following games as far as possible by eliminating dominated strategies. If the resulting game is strictly determined, or if it is a 2 × 2 game which is not strictly determined, solve it.

a.

−1	2	5	7
4	−1	0	−2
4	3	2	−1

d.

100	41	56
99	48	82
80	26	75

b.

−1	3	−1	−3
2	2	2	1
−3	3	−2	−2
1	2	3	−1

e.

0	−2	−1
1	0	2
−1	2	1

c.

1	−1	2
0	3	−2
2	6	−3

f.

1	−1	−1	2
4	3	0	1
−2	1	−1	2

2.   Solve the game of Exercise 4.6.1*f* in the usual way (finding row minima and column maxima). Show that the solution is the same as that obtained in Exercise 4.6.1 by eliminating dominated strategies.

3.   Find and eliminate all dominated strategies in the game Rommel and

Montgomery (Example 4.5). Then solve the reduced game by eliminating strategies which are now dominated.

4.   Solve the duopoly game of Example 4.4 by repeated elimination of dominated strategies.

5.   Suppose that in the game of Example 4.15 Player 1 uses $\mathbf{p} = [\frac{1}{6}, \frac{1}{3}, \frac{1}{2}]$ and Player 2 uses $\mathbf{q} = [\frac{1}{3}, \frac{1}{3}, \frac{1}{3}]$. We claimed that Player 2 can do better by always choosing y_3 instead of y_1. That would transfer probability $\frac{1}{3}$ from y_1 to y_3 and result in using $\mathbf{q}^* = [0, \frac{1}{3}, \frac{2}{3}]$. Show that

$$M^*(\mathbf{p}, \mathbf{q}^*) < M^*(\mathbf{p}, \mathbf{q})$$

by computing expected payoffs. So $\mathbf{q}^*$ does better than $\mathbf{q}$ against this $\mathbf{p}$, and you should be able to see that this is true no matter what $\mathbf{p}$ is.

4.7 GRAPHICAL SOLUTION OF GAMES

The first step in the solution of a game which is not strictly determined is to reduce the size of the payoff matrix as far as possible by eliminating dominated strategies. If the reduced game is $2 \times m$ or $m \times 2$, there is a fairly easy graphical method of solving the game. That is the subject of this section. If both players have more than two pure strategies in the reduced game, more elaborate procedures are needed. These are discussed in the next section.

Example 4.17   Consider the 2×4 game with payoff matrix

	y_1	y_2	y_3	y_4
x_1	1	12	4	0
x_2	3	0	1	9

The column maxima are 3, 12, 4, and 9 so that the upper value is 3. The row minima are both 0, so that the lower value is 0. The game is therefore not strictly determined. There are no dominated rows or columns, so no reduction is possible.

We begin by taking the point of view of the player with only two pure strategies, Player 1 in this case. If Player 1 uses a mixed strategy $\mathbf{p} = [\alpha, 1 - \alpha]$ for some α between 0 and 1, his expected payoffs against each of Player 2's pure strategies are

$$
\begin{aligned}
\text{against } y_1: \quad & (1)(\alpha) + (3)(1 - \alpha) = 3 - 2\alpha \\
\text{against } y_2: \quad & (12)(\alpha) + (0)(1 - \alpha) = 12\alpha \\
\text{against } y_3: \quad & (4)(\alpha) + (1)(1 - \alpha) = 3\alpha + 1 \\
\text{against } y_4: \quad & (0)(\alpha) + (9)(1 - \alpha) = 9 - 9\alpha
\end{aligned}
$$

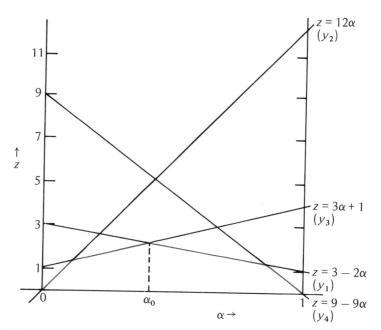

Figure 4.3 Payoffs of $\mathbf{p} = [\alpha, 1 - \alpha]$ against each of Player 2's pure strategies graphed as functions of α

To find a maximin mixed strategy $\mathbf{p}_0$, Player 1 must choose the α which maximizes the minimum of these four numbers. This is difficult algebraically but easy geometrically. The graph of each of the four quantities in question is a straight line. These four straight lines are drawn on the same scale in Fig. 4.3. (It is very easy to graph these lines by marking the values at $\alpha = 0$ and $\alpha = 1$ and connecting them.)

For any α between 0 and 1, the expected payoffs of $\mathbf{p} = [\alpha, 1 - \alpha]$ against Player 2's four mixed strategies are the z coordinates of the points on these four lines lying above α. The blue line in Fig. 4.3 shows the minimum payoff for each α. This minimum is largest when α takes the value α_0 indicated in Fig. 4.3. So $\mathbf{p}_0 = [\alpha_0, 1 - \alpha_0]$ is Player 1's maximin strategy.

What is α_0 and what is the solution to the game? The easiest way to find these results is to notice from Fig. 4.3 that when Player 1 uses $\mathbf{p}_0 = [\alpha_0, 1 - \alpha_0]$, the expected payoffs against y_1 and y_3 are equal and the expected payoffs against y_2 and y_4 are larger. Since Player 2 wants to keep the payoff as small as possible when Player 1 uses $\mathbf{p}_0$, he will never choose y_2 or y_4. Ignoring y_2 and y_4 reduces the game to the 2×2 game

	y_1	y_3
x_1	1	4
x_2	3	1

By the formulas of Exercise 4.5.9 the solution to this 2×2 game is

$$v = \frac{(1)(1) - (4)(3)}{1 + 1 - 4 - 3} = \frac{-11}{-5} = \frac{11}{5}$$

$$\mathbf{p}_0 = \left[\frac{1-3}{-5}, \frac{1-4}{-5}\right] = \left[\frac{2}{5}, \frac{3}{5}\right]$$

$$\mathbf{q}_0 = \left[\frac{1-4}{-5}, \frac{1-3}{-5}\right] = \left[\frac{3}{5}, \frac{2}{5}\right]$$

The solution to the original game is therefore $v = {}^{11}/_5$, $\mathbf{p}_0 = [{}^2/_5, {}^3/_5]$ and $\mathbf{q}_0 = [{}^3/_5, 0, {}^2/_5, 0]$. (A longer way of verifying this is outlined in Exercise 4.7.7.)

The routine illustrated in Example 4.17 can be used to solve $2 \times n$ games in all cases except when the minimum expected payoff against pure strategies is largest at a point in which three or more lines intersect. Such a situation can be treated by more elaborate geometric arguments, but we leave it to the general methods presented in the next section. This case is very rare; usually the situation is as in Fig. 4.3, where only two payoff lines intersect at α_0. In $m \times 2$ games the procedure is similar. Begin with Player 2 (who has only two pure strategies), and compute the expected payoff of any of his mixed strategies $\mathbf{q}$ against each of Player 1's m pure strategies. Search for the point at which the maximum of these is smallest. Here is an example, which also illustrates a complication which may arise.

Example 4.18   Consider the 3×2 game with payoff matrix

	y_1	y_2
x_1	2	−1
x_2	1	1
x_3	−1	2

You should check that this game is not strictly determined and that there are no dominated rows or columns. Since Player 2 has only two pure strategies, we compute the expected payoff of any of his mixed strategies $\mathbf{q} = [\beta, 1 - \beta]$ against each of Player 1's pure strategies,

against x_1: $(2)(\beta) + (-1)(1 - \beta) = 3\beta - 1$
against x_2: $(1)(\beta) + (1)(1 - \beta) = 1$
against x_3: $(-1)(\beta) + (2)(1 - \beta) = 2 - 3\beta$

These three straight lines are graphed in Fig. 4.4. Player 2 wants to choose β to minimize the maximum payoff. The maximum payoff for each β is represented by the blue line in Fig. 4.4.

Note that the maximum payoff is smallest not at one point alone, but for

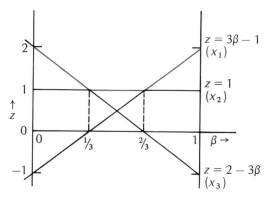

Figure 4.4   Payoffs of $\mathbf{q} = [\beta, 1 - \beta]$ against each of Player 1's pure strategies in Example 4.18

an entire interval of values of β. In this case any β_0 with $\tfrac{1}{3} \le \beta_0 \le \tfrac{2}{3}$ will give a minimax strategy for Player 2. The maximin strategy for Player 1 in this case is the pure strategy x_2 (or $\mathbf{p_0} = [0, 1, 0]$) since that gives a larger expected payoff than either x_1 or x_3 against any of Player 2's minimax strategies. The value of the game is 1, the payoff that results when both players use their optimal strategies.

Example 4.19   Solve the 2 × 4 game with payoff matrix

	y_1	y_2	y_3	y_4
x_1	4	1	−1	1
x_2	0	2	3	4

The game is not strictly determined. Since y_4 is dominated by y_3, we eliminate the fourth column. If $\mathbf{p} = [\alpha, 1 - \alpha]$ is any mixed strategy for Player 1, the expected payoffs against y_1, y_2, and y_3 are

against y_1: $(4)(\alpha) + (0)(1 - \alpha) = 4\alpha$
against y_2: $(1)(\alpha) + (2)(1 - \alpha) = 2 - \alpha$
against y_3: $(-1)(\alpha) + (3)(1 - \alpha) = 3 - 4\alpha$

Graphing these straight lines (Exercise 4.7.1) shows that the minimum of these lines is greatest at the point at which 4α and $2 - \alpha$ intersect. These correspond to y_1 and y_2, so the solution of the game is the same as the solution of

	y_1	y_2
x_1	4	1
x_2	0	2

Using the formulas of Exercise 4.5.9, the solution of this 2×2 game is $v = \frac{8}{5}$, $\mathbf{p_0} = [\frac{2}{5}, \frac{3}{5}]$, and $\mathbf{q_0} = [\frac{1}{5}, \frac{4}{5}]$. So the solution of the original 2×4 game has the same value v and maximin $\mathbf{p_0}$ and minimax strategy $\mathbf{q_0} = [\frac{1}{5}, \frac{4}{5}, 0, 0]$.

Exercise 4.7   1.   Graph the three expected payoffs of Example 4.19 for α between 0 and 1. (Plot the values for $\alpha = 0$ and $\alpha = 1$ and connect with a straight line.) Mark the minimum of these three functions with a heavy line. Mark the α_0 at which the minimum is largest. Which lines intersect at that point?

2.   Add to the graph of Exercise 4.7.1 a plot of the expected payoff of $\mathbf{p} = [\alpha, 1 - \alpha]$ against the dominated strategy y_4, $(1)(\alpha) + (4)(1 - \alpha) = 4 - 3\alpha$. Explain from the graph why eliminating y_4 does not affect the solution of the game.

3.   Solve each of the following $2 \times n$ games:

a.
3	0
0	6

c.
3	1	0
0	1	6

e.
3	2	0	−1
0	1	6	0

b.
3	2	0
0	1	6

d.
3	2	0	4
0	1	6	6

f.
3	2	0
0	2	6

4.   Solve the following $m \times 2$ games:

a.
3	0
2	1
0	6
4	6

b.
3	0
2	4
0	6

5.   In playing Matching Fingers Player 2 notices that Player 1 has a mannerism which tends to give away his play. When Player 1 plays one finger, he clenches his fist in advance three-fourths of the time; when he plays two fingers, his fist is open three-fourths of the time. Player 2 now has four pure strategies. He can ignore Player 1's mannerism and play one finger (y_1) or two fingers (y_4). He can "believe" the clue—playing two fingers when Player 1 clenches his fist and one when he does not (y_2). Or he can "disbelieve" the clue and reverse his responses (y_3).

There is a random element in this game, so we use expected payoffs. For example, suppose Player 1 uses x_1 (shows one finger) and Player 2 uses y_2. Player 1 clenches his fist with probability $\frac{3}{4}$, in which case Player 2 shows two fingers and the payoff is −1. With probability $\frac{1}{4}$ the fist is open, Player 2 shows one finger, and the payoff is 1. So the expected payoff is

$$M(x_1, y_2) = (-1)(\frac{3}{4}) + (1)(\frac{1}{4}) = -\frac{1}{2}$$

a. Show that the payoff matrix for this game is

$$2$$

		y_1	y_2	y_3	y_4
1	x_1	1	$-1/2$	$1/2$	-1
	x_2	-1	$-1/2$	$1/2$	1

b. Solve this game.

6.   A convoy can follow any of four routes. An opposing submarine commander can send a wolfpack to any one of the first three routes, but the fourth route is out of range. The payoff function is the total cost of the convoy. This is the sum of a fixed cost for each route (resulting mainly from the length of the route) and an estimated cost of ships sunk by submarines. The four routes have fixed costs (in arbitrary units) 1, 2, 3, and 4 respectively. The cost of sinkings is 0, unless the wolfpack is sent to the route chosen by the convoy. If the submarines and the convoy choose the same route, the cost of sinkings is 7 on route 1, and 1 on each of routes 2 and 3. Solve this game.

7.   Return to Example 4.17 and to Fig. 4.3. This exercise outlines a way of solving the game without using the formulas for solution of a 2×2 game.

a. Find α_0 by solving $3 - 2\alpha = 3\alpha + 1$, as can be done since α_0 is the value for which these lines intersect. Check that the maximin p_0 is the same as that obtained in Example 4.17.
b. The lower value of the mixed extension is the maximum of the minimum payoff. This is the z coordinate of the minimum graph at the value $\alpha = \alpha_0$. It is the value of the mixed extension by the fundamental theorem. Find this value v and check that it is the same as that obtained in Example 4.17.
c. Since a minimax strategy for Player 2 must have expected payoff v against $p_0 = [\alpha_0, 1 - \alpha_0]$, the minimax q_0 must be a mixture of y_1 and y_3 alone [because those payoff lines pass through the point (α_0, v) and the other two strategies have larger payoffs at α_0]. So $q_0 = [\beta, 0, 1 - \beta, 0]$ for some β between 0 and 1.

Draw a graph of the expected payoff of such a q_0 against x_1 and x_2 as a function of β. Mark the maximum of these two lines with a heavy line, and note that the maximum is smallest at their intersection. So the β_0 at which the lines intersect gives the minimax q_0. Find this β_0 and check your result against that of Example 4.17.

4.8 SOLUTION OF GAMES BY LINEAR PROGRAMING

Our purpose in this section is to develop a method for solving any two-person zero sum game. The methods you have already used to solve strictly determined games, the geometric methods of Sec. 4.7, and the elimination of dominated strategies are all viable methods which are used when applicable. As you know, not all games can be solved by those methods so other techniques for solving games are needed. The theory of linear programing as developed in Chapter 3 provides another technique. We will show in this section how to formulate a linear program from a game. The method is demonstrated by our first example.

Example 4.20   The game with payoff matrix

$$\begin{bmatrix} 1 & 7 & 2 \\ 3 & 0 & 8 \\ 6 & 5 & 4 \end{bmatrix}$$

was cited in Table 4.5 as an example of a game which is not strictly determined. The fundamental theorem guarantees us that the mixed extension has a value, and we write v for this unknown value.

Consider the plight of Player 1, who seeks a maximin mixed strategy for the game. One consequence of Theorem 4.1 is that the maximin strategy $\mathbf{p}^* = [p_1^*, p_2^*, p_3^*]$ must satisfy three inequalities, one determined from each pure strategy of Player 2 (see Exercise 4.5.7a):

$$p_1^* + 3p_2^* + 6p_3^* = M^*(\mathbf{p}^*, y_1) \geq v$$
$$7p_1^* + 0p_2^* + 5p_3^* = M^*(\mathbf{p}^*, y_2) \geq v$$
$$2p_1^* + 8p_2^* + 4p_3^* = M^*(\mathbf{p}^*, y_3) \geq v$$

From this system of inequalities we pass to an equivalent system of inequalities when we divide by v. For this example we will assume that $v > 0$ so that division by v will preserve each inequality. After dividing by v we obtain:

$$\frac{p_1^*}{v} + \frac{3p_2^*}{v} + \frac{6p_3^*}{v} \geq 1$$
$$\frac{7p_1^*}{v} + \frac{0p_2^*}{v} + \frac{5p_3^*}{v} \geq 1$$
$$\frac{2p_1^*}{v} + \frac{8p_2^*}{v} + \frac{4p_3^*}{v} \geq 1$$

It is convenient to treat each of p_1^*/v, p_2^*/v, and p_3^*/v as only one variable. Because $p_1^* + p_2^* + p_3^* = 1$, $p_1^*/v + p_2^*/v + p_3^*/v = 1/v$ and we can determine the value v if p_1^*/v, p_2^*/v, and p_3^*/v are known.

Because Player 1 wants v to be as large as possible, he tries to *minimize* the quantity

$$\frac{1}{v} = \frac{p_1^*}{v} + \frac{p_2^*}{v} + \frac{p_3^*}{v} = \begin{bmatrix} \dfrac{p_1^*}{v}, & \dfrac{p_2^*}{v}, & \dfrac{p_3^*}{v} \end{bmatrix} \cdot \begin{bmatrix} 1 \\ 1 \\ 1 \end{bmatrix}$$

These facts now determine a linear programing problem:

$$\text{Minimize} \quad \left[\frac{p_1^*}{v}, \frac{p_2^*}{v}, \frac{p_3^*}{v}\right] \cdot \begin{bmatrix} 1 \\ 1 \\ 1 \end{bmatrix}$$

over the polyhedral convex set where $[p_1^*/v,\ p_2^*/v,\ p_3^*/v] \geq [0, 0, 0]$ and

$$\left[\frac{p_1^*}{v}, \frac{p_2^*}{v}, \frac{p_3^*}{v}\right] \begin{bmatrix} 1 & 7 & 2 \\ 3 & 0 & 8 \\ 6 & 5 & 4 \end{bmatrix} \geq [1, 1, 1]$$

Using the simplex algorithm to solve this problem, we find that the minimum occurs when (see Exercise 4.8.3)

$$\frac{p_1^*}{v} = 0 \qquad \frac{p_2^*}{v} = \frac{1}{40} \qquad \text{and} \qquad \frac{p_3^*}{v} = \frac{1}{5}$$

Thus

$$\frac{1}{v} = 0 + \frac{1}{40} + \frac{1}{5} = \frac{9}{40}$$

so we find that $v = {}^{40}\!/_9$, $p_1^* = ({}^{40}\!/_9)(0) = 0$, $p_2^* = ({}^{40}\!/_9)(\frac{1}{40}) = \frac{1}{9}$, $p_3^* = ({}^{40}\!/_9)(\frac{1}{5}) = \frac{8}{9}$.

Player 1 can always write the system of inequalities which we express by the vector inequality

$$\mathbf{p}^*\mathbf{M} \geq [v, v, \ldots, v]$$

for any payoff matrix $\mathbf{M}$. At the second step (division by v) in formulating the linear program it is *essential* that v be a positive quantity, for if v is negative all the inequalities are reversed when we divide by v. Since v is unknown to us, we cannot carry out this step for every given payoff matrix $\mathbf{M}$. Fortunately it is possible to consider a related game which we know has a positive value, and the solution of the related game determines the solution to the original game.

If the matrix $\mathbf{M}$ has only positive payoffs, then the game will have a positive value. (Why?) If the lower value of the game is positive, then the mixed extension of the game also has a positive value. Notice that the lower value of the game in Example 4.20 is 4 (playing x_3 guarantees at least a payoff of 4). If the lower value of the game is negative or zero, we can add a positive number c to *all* payoffs in $\mathbf{M}$ to obtain the new payoff matrix $\mathbf{M}^*$. This number c is chosen to be large enough so that the lower value for the game with payoff matrix $\mathbf{M}^*$ is positive and so that the payoffs in $\mathbf{M}^*$ are easy numbers to work with (as many as possible are integers).

The mixed extension of the game with payoff matrix $\mathbf{M}^*$ has a positive value v^*. The important fact is that

$$v = v^* - c$$

(see Exercise 4.8.7) so that v is determined from v^*.

A maximin mixed strategy $\mathbf{p}^*$ for the game with payoff matrix $\mathbf{M}^*$ is also a maximin mixed strategy for the game with payoff matrix $\mathbf{M}$ (see Exercise 4.8.8).

Example 4.21   For the game with payoff matrix

$$\mathbf{M} = \begin{bmatrix} -4 & 2 & -3 \\ -2 & -5 & 3 \\ 1 & 0 & -1 \end{bmatrix}$$

the lower value is -1 (play x_3); so if we add 2 to each payoff, the new matrix

$$\mathbf{M}^* = \begin{bmatrix} -2 & 4 & -1 \\ 0 & -3 & 5 \\ 3 & 2 & 1 \end{bmatrix}$$

has lower value 1, and v^* must be positive. The solution to the game with payoff matrix $\mathbf{M}^*$ is $v^* = {}^{13}\!/_9$, $p_1^* = 0$, $p_2^* = {}^1\!/_9$, $p_3^* = {}^8\!/_9$. The strategy $p^* = [0, \ {}^1\!/_9, \ {}^8\!/_9]$ is therefore also maximin for the game with payoff matrix $\mathbf{M}$ and the value of this game is

$$v = v^* - c = {}^{13}\!/_9 - 2 = -{}^5\!/_9$$

To complete the solution of the game, we must also consider Player 2. For that purpose we again consider an example.

Example 4.22   For the game of Example 4.20, Player 2 is also assured by Theorem 4.1 (see Exercise 4.5.7b) that his minimax mixed strategy $\mathbf{q}^*$ will satisfy the three inequalities which result from each pure strategy for Player 1:

$$q_1^* + 7q_2^* + 2q_3^* = M^*(x_1, \mathbf{q}^*) \leq v$$
$$3q_1^* + 0q_2^* + 8q_3^* = M^*(x_2, \mathbf{q}^*) \leq v$$
$$6q_1^* + 5q_2^* + 4q_3^* = M^*(x_3, \mathbf{q}^*) \leq v$$

Division by v results in an equivalent system of inequalities in the variables q_1^*/v, q_2^*/v, and q_3^*/v for which

$$\frac{q_1^*}{v} + \frac{q_2^*}{v} + \frac{q_3^*}{v} = \frac{1}{v}$$

Since Player 2 wants v to be as *small* as possible, he seeks to maximize $1/v$. The formulation of the maximum problem is to find the maximum of

$$[1, 1, 1] \cdot \begin{bmatrix} \dfrac{q_1^*}{v} \\[2mm] \dfrac{q_2^*}{v} \\[2mm] \dfrac{q_3^*}{v} \end{bmatrix}$$

over the polyhedral convex set where $q_1^*/v \geq 0$, $q_2^*/v \geq 0$, $q_3^*/v \geq 0$, and

$$
\begin{bmatrix} 1 & 7 & 2 \\ 3 & 0 & 8 \\ 6 & 5 & 4 \end{bmatrix}
\begin{bmatrix} \dfrac{q_1^*}{v} \\ \dfrac{q_2^*}{v} \\ \dfrac{q_3^*}{v} \end{bmatrix}
\leq
\begin{bmatrix} 1 \\ 1 \\ 1 \end{bmatrix}
$$

Notice that the linear program for Player 2 is the dual of the linear program for Player 1. This is the first instance we have encountered in which the two dual problems have an obvious relation to each other. It is an important fact that the problems are dual since the simplex algorithm produces the solution to both problems simultaneously (see Exercise 4.8.3 for the minimax strategy and the value of this game).

Example 4.23   The game with payoff matrix

$$
\mathbf{M} = \begin{bmatrix} 3 & -1 & 4 \\ 6 & 7 & -2 \end{bmatrix}
$$

is solved by first adding 2 to each payoff resulting in

$$
\mathbf{M}^* = \begin{bmatrix} 5 & 1 & 6 \\ 8 & 9 & 0 \end{bmatrix}
$$

The initial simplex table for the maximum and dual minimum problems is

$\dfrac{q_1^*}{v^*}$	$\dfrac{q_2^*}{v^*}$	$\dfrac{q_3^*}{v^*}$	$\dfrac{p_1^*}{v^*}$	$\dfrac{p_2^*}{v^*}$	
5	1	6	1	0	1
8	9	0	0	1	1
−1	−1	−1	0	0	0

The final simplex table is

				$\dfrac{p_1^*}{v^*}$	$\dfrac{p_2^*}{v^*}$
$\dfrac{q_3^*}{v^*}$	$37/54$	0	1	1	$-1/54$	$4/27$
$\dfrac{q_2^*}{v^*}$	$8/9$	1	0	0	$1/9$	$1/9$
	$31/54$	0	0	$1/6$	$5/54$	$7/27$

Since the minimum and dual maximum value is $1/v^* = 7/27$, read from the lower right corner, we find $v^* = 27/7$. The maximum problem has the solution $q_1^*/v^* = 0$, $q_2^*/v^* = 1/9$, and $q_3^*/v^* = 4/27$ (these numbers appear in the last column of the table); while the minimum problem has the solution $p_1^*/v^* = 1/6$, $p_2^*/v^* = 5/54$ (these numbers appear along the bottom row of the table). Thus we find the minimax strategy

$$q_1^* = (27/7)(0) = 0$$

$$q_2^* = (27/7)(1/9) = 3/7$$

$$q_3^* = (27/7)(4/27) = 4/7$$

and the maximin strategy

$$p_1^* = (27/7)(1/6) = 9/14 \quad \text{and} \quad p_2^* = (27/7)(5/54) = 5/14$$

and the value

$$v = v^* - c = 27/7 - 2 = 13/7$$

Since the game of Example 4.23 is a 2×3 game, it could have been solved by the geometric method of Sec. 4.7. You should verify that the geometric method gives the same result as we obtained in Example 4.23.

Steps for Solving a Game

1. Is the game strictly determined? If it is, you are done.
2. Eliminate all dominated strategies.
3. If the lower value is positive, go on to step 4; otherwise add a positive number c to each payoff, obtaining a new payoff matrix M^* for which the lower value is positive. The number c is chosen to be any number which will make the lower value positive and leave the payoffs as convenient numbers to work with.
4. Solve the linear programing problem by the simplex method. The initial simplex table will have the form

		1
		1
M^*	I	$\vdots$
		1
$-1 -1 \cdots -1$	$0 \cdots 0$	0

5. The value v^* of the mixed extension of the game with payoff matrix M^* is the reciprocal of the maximum value of the linear programing problem. (This value is read from the lower right corner of the final simplex table.)

   If at step 3 an amount c was added to all payoffs, then the value of the mixed extension of the original game is $v^* - c$.

6. The minimax mixed strategy **q*** is obtained from the solution to the maximum problem (we read the values of $q_1^*/v, q_2^*/v, \ldots, q_n^*/v$ from the last column of the final simplex table). The maximin mixed strategy **p*** is pbtained from the solution to the dual minimum problem (the values of $p_1^*/v, p_2^*/v, \ldots, p_m^*/v$ are read from the bottom row of the final simplex table).

Exercise 4.8
1. Verify the solution to the game presented in Example 4.21.
2. Check the solution to the game presented in Example 4.23.
3. Solve the game of Examples 4.20 and 4.22. Check your maximin strategy and the value with that given in Example 4.20, and find the minimax solution for Example 4.22.
4. Use linear programing to solve the game of Exercise 4.3.1a.
5. Use linear programing to solve the game of Exercise 4.3.1b. This game is strictly determined. Compare the amount of work needed to solve the linear program to the simple job of checking whether a game is strictly determined.
6. Solve the game in Exercise 4.4.4 by linear programing. Compare the work of finding a graphical solution to the work of solving the linear program.

Follow the Steps for Solving a Game to solve the games of Exercises 4.8.7 to 4.8.10.

7. Solve the game in

a. Exercise 4.3.8   e. Exercise 4.6.1c
b. Exercise 4.5.5   f. Exercise 4.6.1d
c. Exercise 4.6.1a   g. Exercise 4.6.1e
d. Exercise 4.6.1b   h. Exercise 4.6.1f

8. Solve the game Morra in Exercise 4.2.5. You must also formulate the game.
9. Solve the dice game in Exercise 4.2.6. You must also formulate the game.
10. Solve the duopoly game in Exercise 4.2.7. You must also formulate the game.
11. A common childhood game is based on "stone dulls scissors, scissors cut paper, paper covers stone." Each player calls out one of "stone," "scissors," or "paper." Stone beats scissors, scissors beat paper, and paper beats stone. The loser pays the winner 1 cent; if both call the same word, no money changes hands. Formulate and solve this game.
12. Let **M*** be the payoff matrix obtained from **M** by adding c to each

payoff of **M**. If **M*** has the value v^*, show that **M** has the value $v = v^* - c$. (*Hint*: Write the value as the expected payoff when the minimax and maximin strategies are being used.)

13.   Let **M*** be the payoff matrix obtained from **M** by adding c to each payoff of **M**. If **M*** has minimax strategy q^* and maximin strategy p^*, show that q^* and p^* are also minimax and maximin, respectively, for **M**.

4.9   APPLICATIONS

It is fair to say that realistic applications of the theory of two-person zero sum games are less common than are applications of any of the other material in this book. Game theory is alleged to be extensively used in the study of military strategy, but most other practical applications seem to involve the extension of game theory which we will study in Chapter 5 under the name of decision theory. This dearth of applications is due in part to the assumption of strict conflict of interest made in zero sum games. Yet no mathematical development of recent years has had a greater qualitative impact on the social sciences. The theory of games has provided a way of thinking about competitive situations which is enlightening even when quantitative models cannot be given. We will give two applications in this section. The first is an anthropological example which illustrates both the difficulty of gathering data for a payoff matrix and the qualitative questions which game theory ideas can raise. The second illustrates how an economic problem can be solved by reducing it to a game, even though the original statement of the problem seems unrelated to game theory.

Application 4.1   A Fishing Village against Nature

One of the first applications of game theory outside economics is due to anthropologist William Davenport in "Jamaican fishing: A game theory analysis" (in *Papers in Caribbean Anthropology*, compiled by Sidney Mintz and reprinted by the Human Relations Area Files Press, New Haven, Conn., 1970). Davenport did extensive field work in a fishing village on the south shore of Jamaica, and later realized that his data allowed him to compute the payoff matrix of a game in which the village (Player 1) opposed "Nature" (Player 2).

The villagers fish by setting fish traps which are later drawn and reset. There are two fishing areas: outer banks and inner banks. The outer banks yield more and better fish, but are subject to unpredictable strong currents which often destroy the fish traps and damage canoes. These currents are not related to weather or other observable conditions, so that a captain who sets his traps on the outer banks cannot predict whether the current will appear before he draws them. The inner banks yield fewer fish and fish of lower quality, but are completely shielded from the current. Nature may be regarded as having two pure strategies, current (y_1) and no current (y_2). A fishing captain can set his traps inside (x_1) or outside (x_2). Davenport observed that many captains set about one-third of their traps on the inner banks and two-thirds on the outer banks, so he took this to be a third pure strategy (x_3). (Davenport does not discuss other possible allocations of pots between the inner and outer banks; these are apparently not used by the villagers.)

The payoff is the average net income per month of a fishing captain. The most important lesson of this example is that it is not at all easy to compute these payoffs. We will mention only a few of the factors discussed by Davenport. If there is no current, the traps set on the outer banks are very productive, and the large supply of high-quality fish drives down the market price of the fish caught on the inner banks. If there is a current, traps on the outer banks cannot be drawn, and the market price of fish from the inner banks rises. Captains who visit the outer banks (using either x_2 or x_3) must pay for new high-quality canoes regularly, while those who stay on the inner banks can purchase used canoes. But the new canoes offer a better chance to win cash prizes in the frequent canoe races in the area!

By gathering data on these and other factors (and by some judicious estimation), Davenport obtained the payoff matrix of Table 4.10 (the units are British pounds).

		2	
		y_1	y_2
1	x_1	17.3	11.5
	x_2	−4.4	20.6
	x_3	5.2	17.0

Table 4.10   Payoff matrix for Jamaican fishing

Let us solve this 3×2 game. The row minima are 11.5, −4.4, and 5.2, so that the lower value is 11.5. The column maxima are 17.3 and 20.6, so that the upper value is 17.3. The game is not strictly determined, and there are no dominated rows or columns. We must therefore solve the mixed extension.

Focusing on Nature (Player 2), who has only two pure strategies, we graph the expected payoffs of a mixed strategy $\mathbf{q} = [\beta, 1 - \beta]$ against x_1, x_2, and x_3. These expected payoffs as functions of β are

$$17.3\beta + 11.5(1 - \beta) = 11.5 + 5.8\beta$$
$$-4.4\beta + 20.6(1 - \beta) = 20.6 - 25\beta$$
$$5.2\beta + 17.0(1 - \beta) = 17.0 - 11.8\beta$$

These lines are graphed in Fig. 4.5.

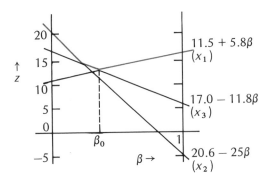

Figure 4.5 Expected payoffs in Jamaican fishing when Nature uses a mixed strategy and the village uses each of its pure strategies.

The blue line is a graph of

$$\max_{1 \le i \le 3} M^*(x_i, \mathbf{q})$$

as a function of β, and the value β_0 minimizes this maximum. Since the expected payoffs of $\mathbf{q}$ against x_1 and x_3 intersect at β_0, the solution of this game can be found from that of the 2×2 game

	y_1	y_2
x_1	17.3	11.5
x_3	5.2	17.0

We already see that the villagers should never use x_2. (*Translation*: Setting all your traps outside is too risky.) The formulas of Exercise 4.5.9 give as the solution of the 2×2 game

$$v = \frac{(17.3)(17.0) - (11.5)(5.2)}{17.3 + 17.0 - 11.5 - 5.2} = \frac{234.3}{17.6} = 13.3$$

$$\mathbf{p_0} = \left[\frac{17.0 - 5.2}{17.6}, \frac{17.3 - 11.5}{17.6} \right] = [0.67, 0.33]$$

$$\mathbf{q_0} = \left[\frac{17.0 - 11.5}{17.6}, \frac{17.3 - 5.2}{17.6} \right] = [0.31, 0.69]$$

Interpreting this result, Nature's minimax strategy is to send the current 31 percent of the time. Since the village has a number of captains with canoes, $\mathbf{p_0}$ says that the maximin strategy for the village is to have two-thirds of its canoes fish only the inside banks and one-third follow the "in-out" strategy x_3. If the villagers do this, and Nature happens to play minimax, the payoff will be 13.3 £/mo on the average.

It remains to compare the results of solving the game with the actual practice of the fishermen. Davenport observed that of 26 fishing canoes in the village, on the average, 18 fished inside and 8 fished both inside and outside. None placed fish traps only on the outer banks. So in fact the village uses the mixed strategy

$$\mathbf{p} = [^{18}/_{26}, 0, {}^{8}/_{26}]$$
$$= [0.69, 0, 0.31]$$

This is very close indeed to the maximin $\mathbf{p_0}$. It also happens that in fact the current runs 25 percent of the time, as against a minimax value of 31 percent.

The interpretation of these results raises some fascinating questions. It is reasonable to suspect that the village has learned over generations of experience to play this game against Nature well. This might explain why the actual practice is so close to the maximin ideal. But Nature is (we hope) not consciously minimizing the village's income, so it can only be accidental that Nature's actual practice is not very different from minimax. Are the villagers unconsciously adapting to nature, so that they are close to maximin only because Nature is close to minimax, or are they unconsciously playing maximin because it is the conservative strategy? In other words, if the actual frequency of current were far from the minimax 31 percent, would the villagers still play maximin or would they use a strategy which gave higher income against Nature's actual practice? (We will learn how to find such strategies in Chapter 5.) Only more field data (from situations in which Nature's practice is not close to minimax) will allow us to decide between these alternative suggestions.

Application 4.2   Optimal Assignment

A large organization faces the problem of assigning n people to n jobs. It has in hand numerical measures of the suitability of each person for each job, obtained by psychological testing or by other means. Denote the suitability score of the ith person for the jth job by s_{ij} and suppose that all $s_{ij} > 0$. The problem is to assign people to jobs so as to maximize the sum of the scores corresponding to the assignments made. Let us write the scores s_{ij} as entries in an $n \times n$ matrix $\mathbf{S}$ in which each row stands for a person and each column stands for a job. Each possible assignment of people to jobs is represented by a selection of n of the n^2 entries of $\mathbf{S}$. If s_{ij} is in the set of entries selected, this means that person i was assigned to job j. Since only one person can be assigned to each job, a set of n entries of $\mathbf{S}$ which represents a possible assignment must have exactly one entry in each row and one in each column. We want to make a selection of this kind which maximizes the sum of the entries selected. Such a selection represents an "optimal assignment" of people to jobs.

For example, Tom, Diane, and Harry are available for assignment; and jobs as draftsman, sales representative, and assembly supervisor are available. The matrix of suitability scores is given in Table 4.11.

	Draftsman	Sales	Assembly
Tom	1	③	2
Diane	③	4	1
Harry	½	1	②

Table 4.11   An example of optimal assignment

The optimal assignment consists of the circled entries—Tom goes to sales, Diane becomes a draftswoman, and Harry is assigned to assembly. The sum of scores for this assignment is 8. Notice that although Diane's suitability score for sales is the highest score in the matrix, she is not assigned to sales. If she were, you can see that the total score could be no greater than 7.

In this example, it is not hard to find the optimal assignment by trial and error. But in matching n people with n jobs, there are $n!$ possible assignments (Can you see why this is true?). The task of examining each of these soon becomes prohibitive, even for a computer. For $n!$ increases very rapidly as n increases—10! is already 3,628,800 and 50! is about 3.04×10^{64}. A clever way out was offered by the inventor of game theory, John von Neumann. He showed that solving the optimal assignment problem is equivalent to solving a game which we will call Hide and Seek. This game is $2n \times n^2$, but the computer time required to solve even a game of this size is very much less than that needed to examine all $n!$ possible assignments.

Here is the game of Hide and Seek. We are given the matrix **S** with entries s_{ij} as in the assignment problem. (This is *not* the payoff matrix for this game.) Think of the matrix **S** as a grid with n rows and n columns. There are n^2 cells in this grid, each located at the intersection of a row and a column. Player 2 "hides" in one of these cells. Player 1 "seeks" by looking along either one of the rows or one of the columns. If he sees Player 2, he wins the reciprocal of the entry s_{ij} for the cell in which Player 2 was hiding. If Player 1 chooses a row or column that misses the cell in which Player 2 is hiding, he collects nothing. Formally, Player 2 has n^2 pure strategies, corresponding to the n^2 positions $[i, j]$ in which he can hide. Player 1 has $2n$ pure strategies, since he can choose any of the n rows or any of the n columns. If Player 1 chooses row i or column j when Player 2 has chosen position $[i, j]$, the payoff is $1/s_{ij}$. Otherwise it is zero. For example, if **S** is the score matrix of Table 4.11, the payoff matrix for the associated game of Hide and Seek is that of Table 4.12.

The game of Hide and Seek is fun to play in practice (in the form of Table 4.11, not the payoff matrix form of Table 4.12), and the solution is not at all obvious. Von Neumann showed that finding a minimax strategy for Player 2 solves the assignment problem. We will discuss his reasoning briefly.

A mixed strategy for Player 2 is a probability distribution over all the

	2								
	[1, 1]	[1, 2]	[1, 3]	[2, 1]	[2, 2]	[2, 3]	[3, 1]	[3, 2]	[3, 3]
Row 1	1	$1/3$	$1/2$	0	0	0	0	0	0
Row 2	0	0	0	$1/3$	$1/4$	1	0	0	0
Row 3	0	0	0	0	0	0	2	1	$1/2$
Column 1	1	0	0	$1/3$	0	0	2	0	0
Column 2	0	$1/3$	0	0	$1/4$	0	0	1	0
Column 3	0	0	$1/2$	0	0	1	0	0	$1/2$

1

Table 4.12   The game of Hide and Seek associated with the assignment problem of Table 4.11

locations $[i, j]$ in which he can hide. Let $q(i, j)$ denote the probability that 2 will hide in location $[i, j]$ when using the mixed strategy **q**. If Player 1 guesses row k, he receives $1/s_{kj}$ if Player 2 hides in $[k, j]$ for any column $j = 1, 2, \ldots, n$. Player 2 chooses to hide in $[k, j]$ with probability $q(k, j)$, so the expected payoff when Player 1 guesses row k and Player 2 uses the mixed strategy **q** is

$$\sum_{j=1}^{n} \frac{q(k, j)}{s_{kj}}$$

Similarly, the expected payoff when Player 1 guesses column h and Player 2 uses **q** is

$$\sum_{i=1}^{n} \frac{q(i, h)}{s_{ih}}$$

These $2n$ numbers are the expected payoffs of **q** against each of Player 1's pure strategies: guess row k for any $k = 1, 2, \ldots, n$ or guess column h for any $h = 1, 2, \ldots, n$. By Theorem 4.1, a minimax **q** is one which minimizes the maximum of these $2n$ sums.

Von Neumann showed (it is not easy) that a **q** which does this can always be found among those probability distributions **q** which give positive probability to only n positions, one in each row and one in each column. This reduces Player 2's search for a minimax **q** to:

1. choosing a set B of n positions $[i, j]$, one in each row and column. All other positions receive probability 0
2. choosing probabilities $q(i, j)$ for each of the positions $[i, j]$ in the set B

He must make these choices in a way which minimizes his maximum possible payoff against Player 1's pure strategies. These payoffs are given by the sums we computed above. Since only one position in each row and each column is in B, only that position has a nonzero entry $q(i, j)$, and therefore each of the sums has only one positive term. So the maximum payoff is now simply

$$\max_{(i,j)\in B} \frac{q(i,j)}{s_{ij}}$$

Suppose that Player 2 has already chosen the set B of positions he will use. Then it is not hard to see that the maximum of $q(i,j)/s_{ij}$ is smallest when the $q(i,j)$ for positions $[i,j]$ in B are chosen to make all n of these ratios equal. For example, suppose that Player 2 in the game of Hide and Seek based on Table 4.11 chooses as B the set of circled positions,

$$B = \{[1, 2], [2, 1], [3, 3]\}$$

He must then choose $q(1, 2)$, $q(2, 1)$, and $q(3, 3)$ to be the nonnegative numbers with sum 1 (a probability distribution) which minimize

$$\max \left\{ \frac{q(1, 2)}{3}, \frac{q(2, 1)}{3}, \frac{q(3, 3)}{2} \right\}$$

If these three ratios are to be equal, he must use

$$q(1, 2) = 3/8 \qquad q(2, 1) = 3/8 \qquad q(3, 3) = 2/8$$

This makes all three ratios $q(i,j)/s_{ij}$ equal to $1/8$. Since the three $q(i,j)$ sum to 1 and the denominators are fixed, Player 2 can decrease one of the three only by increasing another—and this will increase the maximum. So the choice given is the best he can do. Notice that the resulting maximum, $1/8$, is the reciprocal of the sum of the entries s_{ij} for positions $[i,j]$ in B ($3 + 3 + 2 = 8$). (Exercise 4.9.4 shows that this will always happen.)

What about the choice of the set B of n positions? Since the maximum payoff for any B when the best $q(i,j)$ are used is the reciprocal of the sum of the entries s_{ij} for positions $[i,j]$ in B, this is minimized by choosing B to *maximize* the sum

$$\sum_{[i,j]\in B} s_{ij}$$

This shows that in finding a minimax strategy, Player 2 solves the assignment problem; for an optimal assignment is exactly a set B of entries, one in each row and column, having the largest possible sum.

In the course of this argument, we saw that the minimax strategy for Player 2 in the game of Table 4.12 is

hide in $[1, 2]$ with probability $3/8$

hide in $[2, 1]$ with probability $3/8$

hide in $[3, 3]$ with probability $2/8$

The value of the extended game is the reciprocal $1/8$ of the sum of the entries used. (This is the upper value, but we know by the fundamental theorem that the upper value is the value.) In this case we solved the game by first solving the optimal assignment problem of Table 4.11. When n is large, we would solve the game of Hide and Seek in order to solve the assignment problem.

It is easy to give examples in which there is more than one optimal assignment. In such cases Player 2 will have more than one minimax strategy for Hide and Seek.

The practical difficulty in application of these ideas lies not in solving large games, but in finding scores s_{ij} which accurately reflect suitability of candidates for jobs. Nonetheless, it is reported that the Army has used a version of this assignment technique. Another interpretation of the assignment problem is possible. Suppose that a company must assign n plants to n locations, with s_{ij} representing the profitability of plant i in location j. The goal is to maximize the total profitability, which is taken to be the sum of the s_{ij} for the assignment made. It turns out that the maximin strategy for Player 1 (whom we have ignored) has an economic interpretation in this plant-location version of the problem. The interpretation is somewhat exotic. Consider n individual entrepreneurs, each with one plant to locate and only his own profit in mind. The maximin strategy for Player 1 gives a set of rental values for plants and locations which would lead the n individuals to make the same assignment of plants to locations as would a single company controlling all n plants. Those with appetites for this and other economic tidbits can look at "Assignment problems and the location of economic activities" (by T. C. Koopmans and M. J. Beckman in *Econometrika* **25**, 1957, 53−76).

Exercise 4.9

1.   Compute the expected payoff when the Jamaican villagers and Nature use the mixed strategies Davenport actually observed them using. Compare your result with the value of the game.

2.   Suppose that a change in fishing conditions causes the payoff when a canoe fishes outside (x_2) and there is no current (y_2) to be 25 £/mo, without changing any other payoffs. Solve this revised version of the fishing game.

3.   An assignment problem has the following set of suitability scores

	Sales	Sales	Buyer	Storeroom
Peters	1	1	1	2
Rubin	3	3	1	1
Williams	2	2	4	1
Riley	1	1	2	1

Find an optimal assignment by trial and error. Write the payoff matrix of the game of Hide and Seek associated with this assignment problem. Using your knowledge of the optimal assignment, find the value of the mixed extension of this game and a minimax strategy for Player 2.

4.   In discussing Hide and Seek we observed that if we choose p_1, p_2, and p_3 to achieve

$$\frac{p_1}{s_1} = \frac{p_2}{s_2} = \frac{p_3}{s_3} = k$$

where the numbers s_i are given and p_i must satisfy $p_1 + p_2 + p_3 = 1$, then $k = 1/(s_1 + s_2 + s_3)$. Prove this. (*Hint*: $p_i/s_i = k$ means $s_i = p_i/k$.)

*4.10   SOME THEORY—THE FUNDAMENTAL THEOREM

The fundamental theorem of game theory, Theorem 4.2, says that the mixed extension of every finite game has a value. John von Neumann's original proof of this fact was very difficult indeed, but we will see that it is possible to give a proof using the geometric skills you acquired in Secs. 3.2 and 3.3. The proof is still not easy, but it will introduce you to a geometric interpretation of game theory which is important for more advanced study. On the way to the fundamental theorem, we will prove a number of other results. The first of these is Theorem 4.1. We will restate and prove only the first half of this theorem; the rest is handled in the same way.

THEOREM 4.1 If $y_1, \ldots, y_m$ are all Player 2's pure strategies and **p** is any mixed strategy for Player 1, then

$$\min_{\mathbf{q}} M^*(\mathbf{p}, \mathbf{q}) = \min_{1 \le j \le m} M^*(\mathbf{p}, y_j)$$

PROOF Suppose $\mathbf{q} = [q_1, \ldots, q_m]$ is a mixed strategy for Player 2. Let us define a random variable Z which has the following possible values with their probabilities of occurrence.

Value:	$M^*(\mathbf{p}, y_1)$	$M^*(\mathbf{p}, y_2)$	$\cdots$	$M^*(\mathbf{p}, y_m)$
Probability:	q_1	q_2	$\cdots$	q_m

Lemma 1.2 says that the expected value $E(Z)$ is at least as great as the smallest possible value of Z. The expected value of this random variable is just $M^*(\mathbf{p}, \mathbf{q})$ (see Exercise 4.10.3). So

$$M^*(\mathbf{p}, \mathbf{q}) \ge \min_{1 \le j \le m} M^*(\mathbf{p}, y_j)$$

Since this inequality holds for *any* probability vector **q**, it holds for the particular **q** which makes $M^*(\mathbf{p}, \mathbf{q})$ smallest. So

$$\min_{\mathbf{q}} M^*(\mathbf{p}, \mathbf{q}) \ge \min_{1 \le j \le m} M^*(\mathbf{p}, y_j)$$

Now each pure strategy y_j is the same as the mixed strategy **q** which puts probability 1 on y_j. So the minimum over all **q** can be no larger than the minimum over only the pure strategies (see Exercise 4.10.4), or

$$\min_{\mathbf{q}} M^*(\mathbf{p}, \mathbf{q}) \le \min_{1 \le j \le m} M^*(\mathbf{p}, y_j)$$

These two inequalities combine to give equality.

The second preliminary result we need is the basic fact that in any game the upper value is at least as great as the lower value. Our first lemma states this fact for mixed extensions of finite games, since these are the only games that concern us.

LEMMA 4.1   Let $\bar{v}$ be the upper value and $\underline{v}$ the lower value of the mixed extension of a finite game. Then $\bar{v} \geq \underline{v}$.

You can give a proof of Lemma 4.1 by filling in the outline provided in Exercise 4.10.5.
    The last preliminary result we need is a fact about convex sets. Recall that a convex set S contains the line segment joining any two points $\mathbf{s}_1$ and $\mathbf{s}_2$ of S. From Sec. 2.3 we know that this line segment consists exactly of points which can be expressed in the form

$$\beta \mathbf{s}_1 + (1 - \beta)\mathbf{s}_2$$

for some number β between 0 and 1. Lemma 4.2 which follows is a generalization of the fact that S contains all points of this form. As usual, we consider only polyhedral convex sets, though the Lemma is true for any convex set S.

LEMMA 4.2   Let S be a polyhedral convex set and suppose that $\mathbf{s}_1, \mathbf{s}_2, \ldots, \mathbf{s}_k$ are points in S and that $\beta_1, \beta_2, \ldots, \beta_k$ are numbers such that

1. $\beta_i \geq 0$ for $i = 1, 2, \ldots, k$
2. $\displaystyle\sum_{i=1}^{k} \beta_i = 1$

Then the point

$$\mathbf{s} = \sum_{i=1}^{k} \beta_i \mathbf{s}_i$$

also belongs to S.

PROOF   We first prove the lemma for the case when S is a half-space. Then the points $\mathbf{x}$ in S are exactly the points which satisfy an equation of the form

$$\mathbf{a} \cdot \mathbf{x} \geq b$$

Some simple algebra and the definition of dot product (Definition 2.5) will convince you (see Exercise 4.10.6) that if

$$\mathbf{s} = \sum_{i=1}^{k} \beta_i \mathbf{s}_i$$

then

$$\mathbf{a} \cdot \mathbf{s} = \sum_{i=1}^{k} \beta_i \mathbf{a} \cdot \mathbf{s}_i$$

Since all the s_i are in the half-space S, $\mathbf{a} \cdot \mathbf{s}_i \geq b$ for each i. The dot product $\mathbf{a} \cdot \mathbf{s}$ has the form of the expected value of a random variable Z which takes values $\mathbf{a} \cdot \mathbf{s}_i$ with probabilities β_i—conditions 1 and 2 assure that the β_i give a legitimate assignment of probabilities. Now Lemma 1.2 says that the expected value $E(Z) = \mathbf{a} \cdot \mathbf{s}$ is at least as great as the smallest of the values $\mathbf{a} \cdot \mathbf{s}_i$ of Z. Since all these satisfy $\mathbf{a} \cdot \mathbf{s}_i \geq b$, we have $\mathbf{a} \cdot \mathbf{s} \geq b$ also, so that $\mathbf{s}$ is in the half-space S. This means that the lemma is true when S is a half-space.

If S is any polyhedral convex set, it is the intersection of some number of half-spaces. If the points $\mathbf{s}_i$ are in S, they are all in each of these half-spaces. But we just proved that any half-space containing all the $\mathbf{s}_i$ contains $\mathbf{s}$ also. So $\mathbf{s}$ is in all these half-spaces and therefore in their intersection, which is S. This completes the proof of Lemma 4.2. Exercise 4.10.7 gives an example of the truth of this lemma.

With these tools in hand, we can turn to showing why the fundamental theorem is true. Rather than giving a formal proof, we will follow the reasoning for a particular game. Only a few fine points need be added to produce a complete proof. The game we will use is that of Example 4.17, which has the payoff matrix given in Table 4.13.

		2			
		y_1	y_2	y_3	y_4
1	x_1	1	12	4	0
	x_2	3	0	1	9

Table 4.13   The game of Example 4.17

The basic idea of the proof of the fundamental theorem is to represent the mixed extension of the game geometrically. We do this by associating with each mixed strategy $\mathbf{q}$ for Player 2 the point in the plane whose first component is the expected payoff of $\mathbf{q}$ against Player 1's first pure strategy x_1 and whose second component is the expected payoff of $\mathbf{q}$ against the second pure strategy x_2. We will call this point $\mathbf{s}(\mathbf{q})$; the *payoff point* of the mixed strategy $\mathbf{q}$. Since payoff points are points in the plane, we will write them as row vectors, so that

$$\mathbf{s}(\mathbf{q}) = [M^*(x_1, \mathbf{q}), M^*(x_2, \mathbf{q})]$$

If $\mathbf{q} = [\beta_1, \beta_2, \beta_3, \beta_4]$, then from Table 4.13 we can compute

$$M^*(x_1, \mathbf{q}) = 1 \cdot \beta_1 + 12 \cdot \beta_2 + 4 \cdot \beta_3 + 0 \cdot \beta_4$$
$$M^*(x_2, \mathbf{q}) = 3 \cdot \beta_1 + 0 \cdot \beta_2 + 1 \cdot \beta_3 + 9 \cdot \beta_4$$

If we write the payoff points as row vectors, these equations can be written

$$\mathbf{s}(\mathbf{q}) = \beta_1[1, 3] + \beta_2[12, 0] + \beta_3[4, 1] + \beta_4[0, 9]$$

The payoff points for Player 2's pure strategies y_1, y_2, y_3, and y_4 have as their components the payoffs of these pure strategies against Player 1's pure strategies. These payoffs are just the entries of the payoff matrix, so if we denote $s(y_i)$ by s_i Table 4.13 gives

$$s_1 = [1, 3] \qquad s_2 = [12, 0] \qquad s_3 = [4, 1] \qquad s_4 = [0, 9]$$

A glance at the vector equation we gave for $s(q)$ leads to an important conclusion: The *payoff point for any mixed strategy* q *for Player 2 is*

$$s(q) = \sum_{i=1}^{4} \beta_i s_i$$

Let S be the convex hull of the points s_1, s_2, s_3, and s_4. Figure 4.6 is a graph of this polyhedral convex set. Since q is a probability distribution, the β_i satisfy conditions 1 and 2 of Lemma 4.2. Lemma 4.2 then tells us that every payoff point $s(q)$ is a point in S.

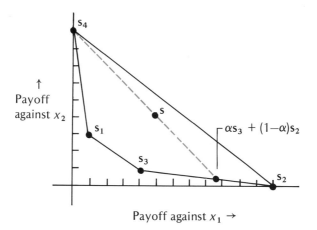

Figure 4.6   The set S of payoff points for the game of Table 4.13

It turns out that not only are all payoff points in S, but any point s in S is the payoff point for some q. To see why this is true, refer again to Fig. 4.6. Any point s in S lies on a line segment joining one of the s_i (s_4 in the case shown) with a point on the boundary of S. That boundary point in turn lies on the line segment joining two of the s_i (s_3 and s_2 in the case shown). The boundary point therefore has the form $\alpha s_3 + (1 - \alpha)s_2$ for some $0 \leq \alpha \leq 1$. So s in turn has the form

$$s = \lambda s_4 + (1 - \lambda)[\alpha s_3 + (1 - \alpha)s_2]$$

for some $0 \leq \lambda \leq 1$. Check that λ, $(1 - \lambda)\alpha$, and $(1 - \lambda)(1 - \alpha)$ add to 1, so that s is the payoff point for the mixed strategy

$$\mathbf{q} = [0, (1 - \lambda)(1 - \alpha), (1 - \lambda)\alpha, \lambda]$$

The conclusion is that *S is exactly the set of payoff points for mixed strategies* **q** *of Player 2.*

We have seen that each mixed strategy **q** corresponds to a point in *S*, and each point in *S* corresponds to at least one mixed strategy **q**. Two mixed strategies which have the same payoff point have the same payoffs whatever Player 1 does, so we will not bother to distinguish between them. So we can now act as though Player 2 chooses a point **s** in *S* rather than a mixed strategy **q**. Player 1 still chooses a probability vector **p** = [α, 1 − α]. The expected payoff is (by Exercise 4.10.3)

$$M^*(\mathbf{p}, \mathbf{q}) = \alpha M^*(x_1, \mathbf{q}) + (1 - \alpha)M^*(x_2, \mathbf{q})$$
$$= \mathbf{p} \cdot \mathbf{s}(\mathbf{q})$$

So in our new geometric representation of the mixed extension of the game, *Player 1 chooses a probability vector* **p**, *Player 2 chooses a point* **s** *in S, and the payoff is the dot product* **p·s**.

If you are properly suspicious, you should ask how this works for games larger than 2 × 4. Increasing the number of pure strategies for Player 2 causes no problem—it simply increases the number of points s_i of which *S* is the convex hull. If Player 1 has *n* pure strategies rather than 2, the payoff points are *n*-dimensional vectors. The set *S* of all payoff points is still a polyhedral convex set, but we cannot draw it. That is the only reason we have chosen a 2 × *m* game to illustrate the reasoning.

With this new version of the mixed extension in hand, we first find a minimax payoff point for Player 2 and the upper value $\bar{v}$ of the mixed extension. For any number *c*, let Q_c denote the "southwest quadrant" with "northeast corner" at the point [*c*, *c*]. More formally, let

$$Q_c = \{\mathbf{x} = [x_1, x_2]: x_1 \le c \text{ and } x_2 \le c\}$$

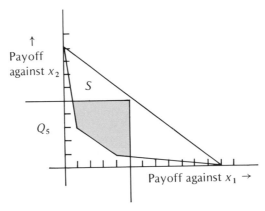

Figure 4.7 $Q_5 \cap S$, the set of all payoff points with largest component less than or equal to 5

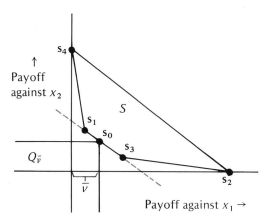

Figure 4.8   The minimax s_0 and upper value $\bar{v}$

Q_c can be described as the set of vectors whose largest component is no larger than c. If Q_c intersects the set S of payoff points, the points in $Q_c \cap S$ represent all the strategies for Player 2 whose maximum payoff against Player 1's pure strategies is no larger than c. Figure 4.7 illustrates this.

Player 2's goal is to find the point s_0 in S whose largest component is as small as possible. To accomplish this, he decreases c (which moves Q_c "southwest") until he reaches the smallest c for which Q_c intersects S: The geometry is illustrated in Fig. 4.8.

The point s_0 is minimax—any other point in S has larger payoff against one of Player 1's pure strategies. (Theorem 4.1 says we need only consider Player 1's pure strategies in searching for a minimax strategy for Player 2.) *The smallest value of c for which Q_c intersects S is the upper value $\bar{v}$.* For this value of c is the maximum payoff of the minimax strategy, and this is $\bar{v}$ by Definition 4.7.

In this particular case, s_0 is the point on the line segment joining $s_1 = [1, 3]$ and $s_3 = [4, 1]$ which has equal components. This line segment consists of all points of the form

$$\alpha[1, 3] + (1 - \alpha)[4, 1] = [4 - 3\alpha, 1 + 2\alpha]$$

for $0 \le \alpha \le 1$. The α for which the components are equal is found by solving

$$4 - 3\alpha = 1 + 2\alpha$$

and is $\alpha = \frac{3}{5}$. The point s_0 is therefore

$$[4 - 3(\frac{3}{5}), 1 + 2(\frac{3}{5})] = [\frac{11}{5}, \frac{11}{5}]$$

From Fig. 4.8 we see that s_0 is also the northeast corner of $Q_{\bar{v}}$, which means that $\bar{v} = \frac{11}{5}$.

Let us pause to review the general situation. It is not always the case that the minimax point $\mathbf{s}_0$ has equal components. Figure 4.9a illustrates a case in which this is false; Fig. 4.9b shows that there may be more than one minimax payoff point. But the construction we have given will always find the upper value $\bar{v}$ and all minimax payoff points.

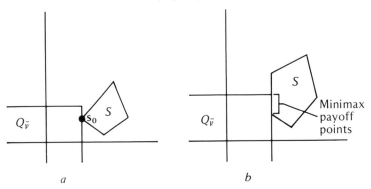

Figure 4.9   (a) The minimax $\mathbf{s}_0$ has unequal components. (b) More than one point $\mathbf{s}$ is minimax.

It remains to find the maximin $\mathbf{p}_0$ for Player 1 and with it the lower value $\underline{v}$ of the mixed extension, and to show that $\underline{v} = \bar{v}$. Returning to Fig. 4.8, notice that the line through $\mathbf{s}_1$ and $\mathbf{s}_3$ separates the convex sets S and $Q_{\bar{v}}$. This line can be written in vector form as $\mathbf{a}\cdot\mathbf{x} = b$, where

$$\mathbf{a}\cdot\mathbf{x} \le b \qquad \text{for all } \mathbf{x} \text{ in } Q_{\bar{v}}$$
$$\mathbf{a}\cdot\mathbf{s} \ge b \qquad \text{for all } \mathbf{s} \text{ in } S$$

The separating hyperplane theorem (Theorem 3.5) can be used to provide a hyperplane separating $Q_{\bar{v}}$ and S even in games in which Player 1 has $n > 2$ pure strategies and pictures fail us.

In our example the vector $\mathbf{a}$ can be any vector perpendicular to the vector

$$\mathbf{s}_3 - \mathbf{s}_1 = [4, 1] - [1, 3] = [3, -2]$$

One possible choice is $\mathbf{a} = [2, 3]$. Since the separating line $\mathbf{a}\cdot\mathbf{x} = b$ must pass through $\mathbf{s}_0 = [^{11}/_5, \, ^{11}/_5]$, we can find b by solving

$$b = \mathbf{a}\cdot\mathbf{s}_0 = [2, 3]\cdot[^{11}/_5, \, ^{11}/_5] = 11$$

Thus one expression for the separating line is

$$[2, 3]\cdot\mathbf{x} = 11$$

We could also use any positive multiple of $\mathbf{a}$ as the perpendicular vector to the separating line. (Only positive multiples can be used because we want $\mathbf{a}\cdot\mathbf{s} \ge b$ for any $\mathbf{s}$ in S.) Notice that *both components of $\mathbf{a}$ are positive*. This means that some positive multiple of $\mathbf{a}$ is a probability vector—just take $\mathbf{p}_0 = [p_1, p_2]$ with

$$p_1 = \frac{a_1}{a_1 + a_2} = \frac{2}{2+3} = \frac{2}{5}$$

$$p_2 = \frac{a_2}{a_1 + a_2} = \frac{3}{2+3} = \frac{3}{5}$$

If also we let

$$v = \frac{b}{a_1 + a_2} = \frac{11}{2+3} = \frac{11}{5}$$

then the separating line can also be described by the equation $\mathbf{p_0 \cdot x} = v$.

It turns out that if the hyperplane separating S and $Q_{\bar{v}}$ has the form $\mathbf{a \cdot x} = b$ with $\mathbf{a \cdot s} \geq b$ for $\mathbf{s}$ in S then all components of $\mathbf{a}$ are always nonnegative. So the separating hyperplane can always be written as $\mathbf{p_0 \cdot x} = v$ for some probability vector $\mathbf{p_0}$.

We claim that this $\mathbf{p_0}$ is a maximin strategy for Player 1 and that v is the value of the game. To begin with, $\mathbf{x} = [\bar{v}, \bar{v}]$ is always a point in $Q_{\bar{v}}$ (whether or not it is the minimax $\mathbf{s_0}$). Since $\mathbf{p_0 \cdot x} \leq v$ for any $\mathbf{x}$ in $Q_{\bar{v}}$ and $\mathbf{p} \cdot [\bar{v}, \bar{v}] = \bar{v}$ for any probability vector $\mathbf{p}$, it follows that $\bar{v} \leq v$. On the other hand, $\mathbf{p_0 \cdot s} \geq v$ holds for each $\mathbf{s}$ in S, and therefore

$$\min_s \mathbf{p_0 \cdot s} \geq v$$

Since the payoff of $\mathbf{p_0}$ against $\mathbf{s}$ is just the vector product $\mathbf{p_0 \cdot s}$, the lower value $\underline{v}$ of the mixed extension is the maximum over all mixed strategies $\mathbf{p}$ of

$$\min_s \mathbf{p \cdot s}$$

This maximum must be at least as large as the value for the particular strategy $\mathbf{p_0}$, so

$$\underline{v} \geq \min_s \mathbf{p_0 \cdot s}$$

The last two inequalities together say that $\underline{v} \geq v$. We have now shown that $\underline{v} \geq v \geq \bar{v}$. But Lemma 4.1 tells us that $\underline{v} \leq \bar{v}$ is always true. So it must be that $\underline{v} = v = \bar{v}$, which means that v is the value of the game. The last two displayed inequalities and the fact that $v = \underline{v}$ say that

$$\min_s \mathbf{p_0 \cdot s} = \underline{v}$$

which means that $\mathbf{p_0}$ is maximin.

In the example of Table 4.13, we have established that the maximin strategy is

$$\mathbf{p_0} = [{}^2\!/_5, \, {}^3\!/_5]$$

the minimax strategy is

$$\mathbf{q_0} = [{}^3\!/_5, \, 0, \, {}^2\!/_5, \, 0]$$

and the value of the mixed extension is

$$v = {}^{11}/_5$$

These are of course the results obtained in Example 4.17. The method of solution given there is usually quicker, but our intention here was to give a line of reasoning which shows that the mixed extension of any finite game has a solution. To review that reasoning, we first thought of Player 2 as choosing a point with coordinates equal to the payoffs of **q** against Player 1's pure strategies. The set of such points is a polyhedral convex set S. We next found the point or points in S whose maximum component is smallest. These are minimax payoff points, and the value of their maximum component is the value v of the mixed extension. Finally, we found the line separating S and the southwest quadrant Q_v. The probability vector perpendicular to this line is a maximin mixed strategy for Player 1.

Exercise 4.10   1.   Draw the payoff sets for the following games:

a.   Matching Fingers (Example 4.3)

b.

3	2	0	4	3
0	2	6	4	2

c.   Matching Fingers with clue (Exercise 4.7.5)

d.

3	4	4	5
1	0	2	2

2.   Solve each of the games a, b, c, and d in Exercise 4.10.1 by the geometric method used in demonstrating the fundamental theorem.

3.   Let Z be the random variable defined by the table given in the proof of Theorem 4.1. Write out the expected value $E(Z)$ from the definition of expected value. Then write each $M^*(\mathbf{p}, y_j)$ in terms of the quantities $M(x_i, y_j)$ and conclude that $E(Z) = M^*(\mathbf{p}, \mathbf{q})$.

4.   Suppose that A and B are sets of numbers and that $A \supset B$. Explain why

$$\min \{a \text{ in } A\} \leq \min \{b \text{ in } B\}$$

and

$$\max \{a \text{ in } A\} \geq \max \{b \text{ in } B\}$$

[In the proof of Theorem 4.1, the first part of this exercise is used. There A is the set of payoffs $M^*(\mathbf{p}, \mathbf{q})$ for a fixed $\mathbf{p}$, and all mixed strategies $\mathbf{q}$. B is the set of payoffs $M^*(\mathbf{p}, y_j)$ for all pure strategies y_j.]

*5.   Prove Lemma 4.1 by following this outline. Explain carefully why each step is true.

a.   For any fixed **q** and any fixed **p**′,

$$M^*(\mathbf{p}', \mathbf{q}) \leq \max_{\mathbf{p}} M^*(\mathbf{p}, \mathbf{q})$$

b.   Taking the minimum over all **q** on both sides in Exercise 4.10.5a preserves the inequality and gives

$$\min_{\mathbf{q}} M^*(\mathbf{p}', \mathbf{q}) \leq \bar{v}$$

for all fixed **p**′.

c.   Taking the maximum over all **p**′ in Exercise 4.10.5b gives $\underline{v} \leq \bar{v}$.

6.   Suppose that

$$\mathbf{s} = \tfrac{1}{2}[1, 1, 1] + \tfrac{1}{4}[-2, 1, -2] + \tfrac{1}{4}[4, -2, 2] + \tfrac{1}{2}[3, 4, -1]$$

Compute **s** by vector addition, and find the dot product **a·s** for **a** = [2, 4, −2]. Show that the result is the same as

$$\tfrac{1}{2}\mathbf{a}\cdot[1, 1, 1] + \tfrac{1}{4}\mathbf{a}\cdot[-2, 1, -2] + \tfrac{1}{4}\mathbf{a}\cdot[4, -2, 2] + \tfrac{1}{2}\mathbf{a}\cdot[3, 4, -1]$$

This should convince you that, in general, if

$$\mathbf{s} = \sum_{i=1}^{k} \beta_i \mathbf{s}_i$$

then

$$\mathbf{a}\cdot\mathbf{s} = \sum_{i=1}^{k} \beta_i \mathbf{a}\cdot\mathbf{s}_i$$

7.   Let S be the unit square in the plane,

$$S = \{[x, y]: 0 \leq x \leq 1 \text{ and } 0 \leq y \leq 1\}$$

This is a polyhedral convex set. Draw a graph of S and compute and plot the points [0, 0], [½, ½], [1, 0], [1, 1], and

a. $\tfrac{1}{2}[1, 1] + \tfrac{1}{2}[0, 0]$
b. $\tfrac{1}{3}[1, 1] + \tfrac{1}{3}[0, 0] + \tfrac{1}{3}[1, 0]$
c. $\tfrac{1}{4}[1, 1] + \tfrac{1}{2}[½, ½] + \tfrac{1}{4}[1, 0]$

Notice that points a, b, and c of this exercise all fall in S, as Lemma 4.2 says.

8.   In our proof of the fundamental theorem we did not need to find the minimax $\mathbf{q_0}$, although we did find the payoff point $\mathbf{s_0}$ of $\mathbf{q_0}$. In fact,

$$\mathbf{s_0} = \tfrac{3}{5}\mathbf{s_1} + \tfrac{2}{5}\mathbf{s_3}$$

Can you see from this what $\mathbf{q_0}$ is?

4.11   FOOD FOR THOUGHT—NON-ZERO SUM GAMES

If you have any taste for mathematical elegance, you should be impressed by the theory of two-person zero sum games. If you have any sense of reality, you should be concerned by the restrictiveness of this class of games. Yet the restrictions of strict competitiveness (zero sum) and two players are essential for the theory we have discussed. Games involving more than two players introduce several complications: The players can now form coalitions or alliances; and a "solution" to the game would have to describe what coalitions would form, how the total payoff would be shared among the coalitions, and how each coalition's share would be divided among the individual players. There is as yet no theory which offers "solutions" in this sense to games with more than two players, and it is doubtful that such a theory is possible. But there are several theories which concentrate on selected aspects of these games, and there are many applications in economics and political science. A very nice nontechnical description of both the theories and their applications is contained in Chapter 6 of *Game Theory* by Donald O. Morton (Basic Books, New York, 1970). We will focus our attention on the other restriction: that the game be strictly competitive.

Example 4.24   The Prisoner's Dilemma

Two suspects have been arrested and imprisoned (separately) for a robbery. As there is not enough evidence to convict them, the district attorney approaches each prisoner with a proposition: If he confesses and his partner does not, he will go free while his convicted partner will get 20 years. If neither confesses, both will get 1 year for possession of stolen property. If both confess, the judge will give each 5 years.

The players in this game are the two prisoners, and each has two strategies: to confess or to remain silent. Each prisoner knows the payoffs to himself and his "opponent," but the game is not zero-sum. Let us write (0, 20) to mean that Player 1 gets 0 yr (goes free) and Player 2 gets 20 yr. With this notation, the payoffs in years of prison can be given in the form of Table 4.14.

		2	
		Confess	Silence
1	Confess	(5, 5)	(0, 20)
	Silence	(20, 0)	(1, 1)

Table 4.14   Payoffs for the Prisoner's Dilemma

The interests of the players in the Prisoner's Dilemma are not strictly opposed, as in a zero sum game, for the "cooperative" approach of neither confessing is attractive to both. Notice that each player wishes to *minimize* his own payoff. The direct struggle of minimizer against maximizer is no longer present. Yet the essential element of a game is preserved, for the payoff either player receives depends on the actions of both.

Is there a rational strategy for a player in the Prisoner's Dilemma? Player 2, inspecting his payoffs, sees a game with the payoff matrix of Table 4.15.

		2	
		Confess	Silence
1	Confess	5	20
	Silence	0	1

Table 4.15   Payoffs for Player 2 in the Prisoner's Dilemma

If Player 2 considers only his own self-interest, then he is better off to confess, whatever Player 1 does. For if Player 1 confesses, Player 2 will get 20 years unless he also confesses, while if Player 1 clams up, Player 2 can go free by confessing. Player 1 can reason similarly, so it appears that two "rational" players will both confess—and each get 5 years. This is annoying, for it means that two students of game theory spend 5 years in prison while two dumb but stubborn crooks who refuse to confess get off with only 1 year. This suggests that following self-interest alone is not rational, as it always is in zero sum games. But there is no visible alternative description of "rational behavior" in this dilemma, even if the two prisoners can talk with each other before replying to the district attorney. For suppose they agree to refuse to confess. Each prisoner is tempted to double-cross the other and go free. And each is worried that the other will double-cross him, in which case he will have 20 years to meditate on the rewards of being trusting. So both have strong incentives to break their agreement and confess. Only if there is some way of enforcing agreements can we be sure that the prisoners will cooperate.

This discussion has not brought us closer to a "solution" of the Prisoner's Dilemma, but it has illuminated several ways in which non-zero sum games differ from zero sum games:

1. Rational behavior is not obviously identical with pursuit of self-interest.
2. The payoff matrix does not completely describe the game. Ground rules must be given which answer questions such as

   Can the players communicate?
   Can they make agreements?
   Can the agreements be enforced?
   Can the players split their payoffs by "under the table" payments?

The Prisoner's Dilemma is a famous game. It has incited a good deal of experimentation and a good deal of writing. The point of the Prisoner's Dilemma is that pursuit of individual self-interest may be injurious to all concerned. The game has been made the focus of discussions of everything from pollution control (self-interest says minimize costs by polluting, which harms all) to morality (the Golden Rule says to cooperate by not confessing). We will reluctantly leave it in order to illustrate by another example an unexpected possibility which may arise in non-zero sum games.

Example 4.25   **Organizational Infighting**

Player 2 is a junior employee under the supervision of Player 1. He is writing a report on a successful innovation in which he can give major credit either to himself (strategy y_1) or to his supervisor (strategy y_2). Player 1 is about to file a report on Player 2's personality which can help (strategy x_1) or endanger (strategy x_2) Player 2's career. The payoffs are given in Table 4.16.

		2	
		y_1	y_2
1	x_1	(1, 3)	(4, 1)
	x_2	(0, −50)	(2, −100)

Table 4.16   Payoffs for organizational infighting

The self-interest of the players seems to lead to an obvious conclusion. The employee does better to give himself credit (y_1) whatever his supervisor does, and the supervisor does better to write a favorable report (x_1) whatever the employee does. If no communication were possible, we would confidently predict the outcome (1, 3). But communication *is* possible, and the point of this example is that allowing communication greatly weakens Player 2's position. For the supervisor can now threaten, "If you don't give me credit (play y_2), I'll ruin your career (play x_2)." The employee knows that carrying out this threat is not in his supervisor's self-interest as given by the payoffs. But he has much more to lose than does his supervisor, and he cannot be sure the supervisor is only bluffing. He may well give in, resulting in the outcome (4, 1). [The thought that the supervisor might then double-cross him, with the result (2, −100), is too terrible to entertain.] Thus the possibility of communication may help both players by allowing cooperation, or it may reverse the relative power of the players, thus hurting one of them.

It is not surprising that there is as yet no acceptable theory of non-zero sum games. In fact, it is hard to imagine a theory covering so diverse a class of phenomena.

Exercise 4.11   1.   In the Prisoner's Dilemma, nothing was said about the true guilt or innocence of the prisoners. Does it make any difference?

2.   You may wish to have some of your friends play the Prisoner's Dilemma against each other. Be sure they are isolated from each other. [Almost certainly they will not cooperate, but this may be because they are more concerned with beating the other player than with their own self-interest. That might not be the case if the payoffs were real years in a real jail. The article on *Experimental Games* by Rapoport and Orwant (see Exercise 4.3.6) discusses a number of experiments with the Prisoner's Dilemma.]

3.   Make up a two-person, non-zero sum game in which one player would be tempted to offer the other ("under the table") a share of his payoff in return for playing a particular strategy. Can you invent a social situation for which your game might be a model?

4.   Suppose two players, isolated from each other, must each choose a point on or within a circle. If they choose the same point each wins $25, otherwise nothing. What point would you choose? (The idea here is that even if no communication is allowed two smart players may be able to cooperate with each other.)

5.   Can you invent several social situations in addition to those of Examples 4.24 and 4.25 and Exercise 4.11.3 which can be plausibly modeled by two-person, non-zero sum games?

COMPUTER PROJECTS

Project 1:   Simulation of Mixed Strategies

Mixed strategies for a game do not play the same pure strategy each time the game is repeated. Rather, a mixed strategy chooses a pure strategy according to the outcome of some random experiment. We have computed the expected payoff when both players use mixed strategies, and the law of large numbers tells us that the average of the actual payoffs in a long sequence of repetitions of the game will approach this expected payoff.

Your project is to verify this experimentally for the 2×4 game of Example 4.17. Your program should do the following.

1. Simulate a random experiment to choose Player 1's pure strategy according to the maximin mixed strategy $\mathbf{p}_0 = [\frac{2}{5}, \frac{3}{5}]$.
2. Simulate an independent random experiment to choose Player 2's pure strategy according to the minimax mixed strategy $\mathbf{q}_0 = [\frac{3}{5}, 0, \frac{2}{5}, 0]$.
3. Compute from the given payoff matrix the payoff which results.
4. Repeat the game n times (n is an input to the program). Printout the actual relative frequencies with which each pure strategy was chosen and the average of the payoffs made.

Compare your results with $\mathbf{p}_0$, $\mathbf{q}_0$, and the value $v = \frac{11}{5}$ of the mixed extension of this game.

Project 2:   A Simulation Game

"Games" in which players are exposed to a random environment simulated by a computer program are used in teaching and experimentation in areas such as human relations and military strategy. These are usually not two-person zero sum games, but your experience in game theory should help you grasp the ideas of simulation games. Here you are to program a simulation game which might be used to teach or study investment management; though of course, this version is too simple for that. Ideally, simulation games should be played interactively. Whether or not you can so program this one depends on the computer resources available to you. Here is a description of the "game."

1. *The states of nature:* The Xanadu Stock Exchange has three states: up, down, and sideways. These states succeed each other according to a Markov chain with the following transition matrix

$$\begin{array}{c} \\ U \\ D \\ S \end{array} \begin{array}{ccc} U & D & S \\ \begin{bmatrix} 0.3 & 0.5 & 0.2 \\ 0.2 & 0.3 & 0.5 \\ 0.5 & 0.2 & 0.3 \end{bmatrix} \end{array}$$

2. *The investor's strategies:* The investor begins the game with $10,000. (Yes, Xanadu values stocks in dollars.) He may invest any part of this in units of $100 in any or all of the three stocks listed on the exchange.

3. *The payoffs:* The payoff depends on both the investment and the state of the market. The value of a $100 investment at the end of a period for each stock and market state is as follows:

	U	D	S
Nanosecond Computers	600	0	50
Stodgysteel, Inc.	150	80	100
Slave Lake Gold Mines	80	200	90

Note that an entry of 100 means that no change has occurred.

The player can change his investment at the end of each period, if he has any money left. The Markov chain changes states at the end of each period. So the player knows the past state of the market from his investment results, but does not know the present state.

Your program should allow the player to play for several periods. If your computer installation lacks interactive capacity, this will require several job submissions. Your program must do the following:

1. Simulate the Markov chain, allowing any initial state as input.
2. Allow several players to invest simultaneously.
3. Printout the results of each player's investment decisions at the end of each period and keep account of each player's total wealth.

Project 3:   Tictactoe

This project should only be attempted if you have access to interactive computer facilities. It involves a program which will play tictactoe with a human player at the teletypewriter. We have explained that a strategy for a game is essentially a program which would enable a computer to play the game. Now we ask you to implement that idea. Your program will develop in the following order:

1. *Bookkeeping:*   The program should print

   YOUR MOVE
   MY MOVE IS _____
   YOU WIN
   I WIN
   WE DRAW
   ILLEGAL MOVE

   at the appropriate times. That means it must check the legality of the human player's moves and check for winning positions after each move. The human player is to move first. You must adopt a format to describe moves and a format in which the human player states his moves.

2. *A strategy:*   The program must always make legal moves. A list of squares remaining empty will help this and also the bookkeeping. The strategy can be very simple.

3. *Better bookkeeping (optional):*   Notice that any rotation of the tictactoe game ($1/4$, $1/2$, $3/4$, or 1 full turn) gives a position equivalent to the one being faced. Can you shorten your program by taking advantage of this symmetry? You should also consider ideas such as having the program print a tictactoe array with all moves shown after each move.

4. *Better strategy (optional):*   Once your program is running with a simple strategy, you will want to write a better strategy. You can write a strategy which will always win or draw. The symmetry explained in item 3 will help keep down the length of your strategy.

5   DECISION THEORY

5.1   INTRODUCTION

In the previous chapter we studied the theory of games played against a rational opponent. On occasion, however, we formulated games in which the opponent was Nature. In such games the payoff depends on the player's action and also on a "state of nature" not chosen by a rational opponent. For example, the profit of the manufacturer in Exercise 4.3.7 depends on his own product inspection strategy and also on whether the particular component is or is not defective. The profit of the snowmobile manufacturer in Exercise 4.3.8 depends on his production strategy and also on the severity of the coming winter. This chapter will be devoted to the study of games against Nature. Let us agree that *Nature will always be Player 2*, so that Player 1 can be called simply the "Player." This means that payoffs are always expressed as the Player's gains. The Player's losses are represented by negative payoffs. To introduce games against Nature, consider an example.

Example 5.1   A Personal Decision

Mr. Smith has a congenital hearing defect in one ear, caused by malformation of the bones of the inner ear. A surgeon states that an operation is available to correct this defect, but that the operation does not always succeed. In fact, the operation may correct Mr. Smith's hearing (y_1), have no effect (y_2), or destroy the partial hearing he now has (y_3). The surgeon cannot predict in advance which of these states of nature holds in Mr. Smith's case. Mr. Smith can of course agree to the operation (x_1) or refuse it (x_2).

Mr. Smith cannot easily assign dollar values to the possible outcomes, for he must consider the inconvenience of his present hearing loss and the greater inconvenience of a total hearing loss in one ear, as well as the cost of the operation. After careful consideration, however, he draws up a payoff matrix which reflects his personal feelings. It is given in Table 5.1.

		Nature		
		y_1	y_2	y_3
Mr. Smith	x_1	25	−15	−100
	x_2	−10	−10	−10

Table 5.1   Matrix of Mr. Smith's gains for Example 5.1

Example 5.1 illustrates several features which occur frequently in games against Nature. First, the payoffs are Mr. Smith's personal assessments of his gain in each situation. We mentioned in Sec. 4.2 that such personal numeric assessments of payoffs (called utilities) could be used. But in a game between two rational opponents, both players would have to agree in their assessment of utilities. This rarely happens, so payoffs in Chapter 4 were usually expressed in objective units such as dollars or number of casualties. We are of course happy to use such units in games against Nature whenever appropriate. But since Player is the only sentient partic- ipant in the game, we are also willing to use his utilities or personal assess- ments of gain or loss whenever he is able to express them.

Second, Nature is not a conscious opponent. In Example 5.1, we can see that x_2 (refuse the operation) is Mr. Smith's maximin strategy and that y_2 and y_3 are minimax strategies for Nature. The value of the game is −10, that being the utility associated with Mr. Smith's present hearing defect. So if Nature were an active opponent, no operation would ever succeed (nature would never choose y_1). Since Nature is not conscious, Mr. Smith cannot be sure that a minimax strategy will be used against him. This re- moves part of his incentive to choose his maximin strategy x_2. Of course, it remains true that no other strategy can guarantee as small a loss as can the maximin strategy. So maximin strategies remain valuable in games against Nature as conservative, "play it safe" courses of action.

We agree that Nature cannot be presumed to play minimax, as a con- scious opponent would. What is more, the Player may have some informa- tion about Nature's choice of strategy. In Example 5.1, the surgeon may in- form Mr. Smith that based on past experience 90 percent of operations of this type succeed, 5 percent have no effect, and 5 percent destroy the exist- ing hearing. This amounts to saying that Nature is using the mixed strategy $\mathbf{q} = [0.90, 0.05, 0.05]$. It is very helpful to know your opponent's mixed strategy—unless, of course, he is using a minimax strategy. Very often the Player in a game against Nature is faced with an opponent using a known mixed strategy which is not minimax. We will begin our study of games against Nature in the next section by seeing how Player can use this knowledge.

There is a third feature of games against Nature which is not brought out by Example 5.1. If Player wishes, he can often obtain additional information about Nature's strategy (the "true state of nature") by conducting experiments or otherwise taking observations. This opens a wide realm of applications, and is the feature which distinguishes decision theory from game theory. Section 5.3 discusses a mathematical model for decision theory, and the remainder of the chapter is devoted to the theory and application of decision making from the point of view of a game against Nature with information available from the outcome of a random experiment.

Exercise 5.1   1.   Table 5.1 gives utilities assigned by a Player with a moderate hearing loss. Give a payoff matrix for the same game which is a plausible assignment of utilities by a Player with a very severe hearing loss.

2.   Suppose that you must decide whether or not to enroll in ROTC before learning whether or not you will be drafted. If you enroll (x_1), you must serve 3 yr as a lieutenant. If you do not enroll (x_2) and are drafted (y_1), you must serve 2 yr as a private. If you do not enroll and are not drafted (y_2), you need not serve. Give a payoff matrix expressing your personal assignment of utilities to these outcomes.

3.   A hospital medical director realizes that use of a chemical cleanser reduces the incidence of staph infection of newborn infants, but causes a skin disorder if the infant has delicate skin. He assesses the utilities of x_1 (use cleanser) and x_2 (use ordinary soap) against

y_1:   staph exposure and sensitive skin

y_2:   staph exposure and tough skin

y_3:   no staph exposure and sensitive skin

y_4:   no staph exposure and tough skin

The result is the matrix of Table 5.2.

		Nature			
		y_1	y_2	y_3	y_4
Director	x_1	-6	-2	-5	-1
	x_2	-20	-20	0	0

Table 5.2   Matrix of utilities for the hospital director

Explain in layman's terms why solving this game need not be of much help to the director. Explain how the hospital director might know Nature's mixed strategy in this game. Explain in layman's language why knowing Nature's mixed strategy should help him make his decision.

4.   Suppose that in the situation of Exercise 5.1.2 a lottery will determine whether or not you are drafted. The 365 birthdates in the year are drawn at random from a drum, and you will be drafted if your birthdate is one of the first 50 drawn. What is Nature's mixed strategy in this game?

5.   Recall that an assignment of probabilities can be obtained by past frequency data (as in the case of Mr. Smith's hearing), by observing some balance or symmetry (as in Exercise 5.1.4), or by personal assessment of likelihood. Give an example of a game against Nature in which the Player does not have past frequency data or any symmetry from which to obtain Nature's mixed strategy, but would be willing to use personal probabilities to express Nature's mixed strategy.

6.   Figure 5.1 below is a graph of the utility of an amount of money on the vertical axis plotted against the amount itself on the horizontal axis. The dotted line is the graph of utility = amount of money. Can you explain why the utility of money for most people resembles the solid curve?

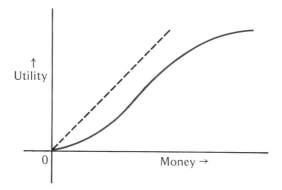

Figure 5.1   The utility of money

5.2   GAMES WITH OPPONENT'S STRATEGY KNOWN

We have seen that in a game against Nature, it is quite possible that the Player may know Nature's mixed strategy. This is certainly true in cases where long-term experience has established the frequency of the possible states of nature. An example is the surgeon's observation that 90 percent of operations of the type described in Example 5.1 succeed, 5 percent have no effect, and 5 percent destroy the existing hearing. Even if the operation were so new that long-term frequencies had not yet been observed, Player might be willing to accept the surgeon's informed estimate of the probability of success. This means using "subjective" or "personal" probabilities. Some of the decision problems in this chapter will involve probabilities interpreted as long-term frequencies. Others will involve probabilities representing personal assessment of likelihood. Since the mathematics of probability developed in Chapter 1 applies in both cases, we need not distinguish between them. You should be aware that there is considerable controversy concerning the extent to which personal probabilities should be used in practice. We will not enter this controversy, but will concentrate on the mathematics which must be learned whatever the philosophy.

Suppose, then, that Player knows Nature's mixed strategy. Since the payoff is the Player's gain, he wants to choose a strategy which will *maximize the expected payoff against Nature's known mixed strategy*. Let us return to the example of the last section.

EXAMPLE 5.1 Continued

Recall that Mr. Smith is faced with a game against Nature with payoffs assessed by him to be those in the matrix of Table 5.1. What is more, he now knows that Nature's strategy is

$$\mathbf{q} = [0.90, 0.05, 0.05]$$

If Mr. Smith decides to go ahead with the operation (chooses strategy x_1), his expected gain is

$$\begin{aligned} M^*(x_1, \mathbf{q}) &= (25)(0.90) + (-15)(0.05) + (-100)(0.05) \\ &= 22.50 - 0.75 - 5.00 \\ &= 16.75 \end{aligned}$$

If Mr. Smith refuses the operation (chooses strategy x_2), his expected gain is

$$\begin{aligned} M^*(x_2, \mathbf{q}) &= (-10)(0.90) + (-10)(0.05) + (-10)(0.05) \\ &= -10 \end{aligned}$$

Thus, if he knows the operation succeeds 90 percent of the time and his utilities are as stated, he should agree to the operation. For if he does, his expected gain is 16.75 units; and if he does not, he will lose 10 units. If the probability of a successful operation were smaller, Mr. Smith's decision might change (Exercise 5.1.1 illustrates this situation).

Notice that in Example 5.1 we only asked which of Mr. Smith's two pure strategies had the greater expected payoff against Nature's strategy $\mathbf{q}$. Could Mr. Smith have done better by using a mixed strategy? The answer is no. It is never to Player's advantage to use a mixed strategy when he knows Nature's strategy. The reason is that the expected payoff against $\mathbf{q}$ of any mixed strategy for Player is a weighted average of the expected payoffs against $\mathbf{q}$ of Player's pure strategies. It is therefore no greater than the greatest of the expected payoffs for the pure strategies, by Lemma 1.2. So Player might as well use the pure strategy with greatest expected payoff.

This result is a great relief. In many cases—such as Example 5.1—the Player intends to play the game only once and may feel uneasy about using a mixed strategy when his actual action depends on a random experiment such as a coin toss. To find a minimax strategy Player must often use mixed strategies. But to find a best possible strategy when his opponent's strategy is known, Player need never look beyond his pure strategies. So we have a simple rule.

How to Act When You Know Nature's Strategy

1. Determine the matrix of payoffs and Nature's mixed strategy $\mathbf{q}$.
2. Compute $M^*(x_i, \mathbf{q})$, the expected payoff when Nature uses $\mathbf{q}$ and Player uses the pure strategy x_i. Do this for all $i = 1, \ldots, n$ (for all Player's pure strategies).
3. Choose a pure strategy x_k for which the expected payoff is greatest. In symbols: choose x_k with

$$M^*(x_k, \mathbf{q}) = \max_i M^*(x_i, \mathbf{q})$$

We call x_k an *optimum strategy against* $\mathbf{q}$.

Example 5.2 An Economic Decision

A snowmobile manufacturer knows that the coming winter will be either severe (y_1), normal (y_2), or mild (y_3). What is more, meteorological records show that the probability of a normal winter is 50 percent and that mild and severe winters each occur 25 percent of the time. The manufacturer can determine the best production schedule to meet the demand of each type of winter, and he can compute the anticipated profit of each production schedule in each type of winter. Since the manufacturer is Player, we state the payoff in terms of his profits. The matrix of profits is given in Table 5.3, in units of $10,000.

		Nature		
		y_1	y_2	y_3
Manufacturer	x_1	120	50	-10
	x_2	100	100	50
	x_3	80	80	80

Table 5.3   Payoff matrix for the economic decision problem of Example 5.2

The expected profits against Nature's strategy $\mathbf{q} = [0.25, 0.50, 0.25]$ for the manufacturer's three pure strategies are:

$$M^*(x_1, \mathbf{q}) = (120)(0.25) + (50)(0.50) + (-10)(0.25)$$
$$= 30 + 25 - 2.5$$
$$= 52.5$$

$$M^*(x_2, \mathbf{q}) = (100)(0.25) + (100)(0.50) + (50)(0.25)$$
$$= 25 + 50 + 12.5$$
$$= 87.5$$
$$M^*(x_3, \mathbf{q}) = (80)(0.25) + (80)(0.50) + (80)(0.25)$$
$$= 80$$

The manufacturer expects to make a profit in any case. We have computed that his expected profit if he plans for a severe winter (x_1) is \$525,000; if he plans for a normal winter (x_2), it is \$875,000; and if he plans for a mild winter (x_3), it is \$800,000. So he should adopt production schedule x_2. Notice that his maximin strategy is x_3, the "play it safe" strategy of adjusting production to fit a mild winter. Following x_3 would cost the manufacturer an average of \$75,000 per season when compared to the optimum strategy x_2 against $\mathbf{q}$.

Exercise 5.2

1.   Suppose that the operation of Example 5.1 is so risky that it succeeds 50 percent of the time and destroys the existing hearing 50 percent of the time. Show that in this case the optimum strategy for a Player with the payoff matrix of Table 5.1 is to refuse the operation.

2.   The payoff matrix for Player in a game against Nature is

		Nature	
		y_1	y_2
	x_1	10	0
Player	x_2	5	5
	x_3	-5	10

Find Player's optimum strategy against each of the following mixed strategies for Nature:

a. (1, 0)   b. (0.5, 0.5)   c. (0.2, 0.8)

3.   The snowmobile manufacturer of Example 5.2 decides not to use the long-term frequencies of the three types of winter. Instead he accepts the assessment of a forecaster that the probability is 0.75 that the coming winter will be severe, 0.20 that it will be normal, and 0.05 that it will be mild. What is his optimum strategy in this case?

4.   A repairman is faced with a breakdown in a large radar receiver which he knows may be due to a defect in the microwave amplifier or in the power supply. The cost of examining the amplifier is 3 and that of examining the power supply is 1. He receives a fixed fee of 5 for the job. He must decide which section to examine first—if he guesses wrong, he must then examine the other. Write the matrix of the repairman's profits in this game against Nature.

Suppose next that from past experience the repairman knows that 80 percent of such breakdowns are due to a defect in the amplifier. Which section should he examine first?

5.   The owner of a Mandarin restaurant is considering preparing Peking ducks (which require a day's advance work) to satisfy customers who neglect to order in advance. His informed assessment of his situation results in the following distribution of requests per day for Peking duck without advance orders:

Number of requests	0	1	2	3	4
Probability	0.3	0.2	0.2	0.2	0.1

If a duck is sold, his profit is $2; a duck prepared but not sold costs him $1. The restaurant owner can prepare 0, 1, 2, 3, or 4 ducks in advance. Write down the 5×5 payoff matrix of the restaurant owner's profit or loss. How many ducks should he prepare in advance? Compare this decision with the maximin strategy for the restaurant owner.

6.   A plant pathologist observes a leaf wilt in his neighbor's pet tobacco plant which can be caused by either virus A or virus B. What is more, he knows that virus A cannot be treated, so that if the plant has virus A it will die, causing a loss of $15. There is a treatment (costing $5) which will save the plant if it has virus B. If not treated, the plant will die. The pathologist knows that the probability that the wilt is due to virus A is 0.20. Should he recommend that his neighbor treat the plant?

7.   In Exercise 5.2.6, how large must the probability that the plant has virus A be to make it uneconomical to treat the plant?

8.   You are starting out from your dwelling unit at 7:30 A.M. for a day of classes and cannot return until 5 P.M. The temperature is 40°, and the weather forecast predicts an 80 percent chance of rain. Nature can choose rain (y_1) or no rain (y_2). You can take both raincoat and umbrella (x_1), raincoat alone (x_2), or neither (x_3). The utilities depend on many factors not yet mentioned. Describe some of the factors that influence your utilities—for example, whether you walk to class or ride. Then make up a matrix of payoffs reflecting your utilities for a set of circumstances which you should briefly describe. Finally, find which strategy you should follow in those circumstances.

9.   Return to the game Jamaican fishermen against Nature discussed in Sec. 4.9. Find the fishermen's optimum strategy against the mixed strategy which Nature is using. How much would the village gain by using the optimum strategy rather than the strategy they now use?

5.3   FORMULATING DECISION THEORY PROBLEMS

The distinguishing feature of decision theory problems is that the decision maker can obtain information about Nature's pure strategy (the state of nature) from the outcome of a random experiment. More exactly, the decision maker has available a random experiment whose probability distribution depends on which state of nature happens to be the case. In this section we will construct a mathematical model for decision theory problems and gain experience in formulating such problems before learning how to solve them in the next section. We begin with an example.

Example 5.3   Drug Testing

A pharmaceutical manufacturer, Fizz Laboratories, has developed a new drug to treat high blood pressure. In order to market the drug Fizz must be able to claim that it works for 80 percent of all patients with high blood pressure. If the drug is marketed and does work at least 80 percent of the time, Fizz estimates a gain of $500,000. If it is marketed and is less effective than claimed, an estimated loss of $200,000 in goodwill and penalties will result. If the drug is not marketed, $20,000 in development costs will be lost. Fizz therefore decides to test the drug on 20 patients with high blood pressure before reaching a decision on marketing.

Example 5.3 has all the features we wish to incorporate in our model, except for the possibility that Fizz may know Nature's mixed strategy. We will ignore that possibility for now, but it will prove important in the next section. Let us ask: *What information is available to Fizz Laboratories in Example 5.3?* The answer to this question is the basis for our formulation of decision theory problems.

Consider first the experiment which Fizz Laboratories proposes to do. They will administer the drug to 20 patients and count the number of patients, say X, for whom it is effective in lowering blood pressure. This number X is of course a random variable. So our model for decision theory begins just as did our model for random experiments in Chapter 1—with the sample space of X and the probability distribution of X. The sample space is easy to find. It is

$$S = \{0, 1, 2, \ldots, 20\}$$

since the number of successes X can be any integer between 0 and 20 inclusive. What about the probability distribution of X? It is reasonable to assume that the trials on the 20 patients are independent and that each trial has the same probability p of succeeding. Here p is the proportion of patients for whom the drug is effective. So (from Sec. 1.8) the random variable X has a binomial distribution with number of trials $n = 20$ and probability of success p.

Notice that the number p is unknown—in fact, p is exactly what Fizz would like to know but does not. This is an important point: In probability theory we had a particular probability distribution which described a random experiment. In decision theory we have *many possible probability distributions,* and we do not know which is correct. What is more, the correct probability distribution for the random experiment depends on the true state of nature. The random experiment is informative exactly because the probability distribution describing it changes as the state of nature changes. In Example 5.3, the family of possible probability distributions is the family of binomial distributions with $n = 20$ and all values of p between 0 and 1. Knowing which of these is the correct distribution for X is the same as knowing the state of nature—the true proportion of patients for whom the drug is effective.

The next piece of information available in Example 5.3 is a list of all possible decisions which can be made. There are only two: Decide to market the drug, or decide not to market it.

Finally, we know what the decision maker will gain (or lose) for each possible value of p and each possible decision. From the point of view of game theory, Nature chooses a particular p (the state of nature); Fizz Laboratories makes one of its two possible decisions; and we know the payoffs for each pair of choices.

We can now generalize this discussion of Example 5.3 and state explicitly what we must do to formulate a problem in decision theory.

Elements of a Decision Theory Problem

> 1. The *sample space S* of the observed random variable X.
> 2. The *set W of possible states of nature.* Each state of nature w in W corresponds to a different probability distribution for the random variable X. W is Nature's set of pure strategies.
> 3. The set D of all possible decisions or actions available to the decision maker. D is called the *decision space.* It is Player's set of pure strategies in the game against Nature.
> 4. The *payoff function M(d, w),* which measures the decision maker's gain when he makes decision d in D and the true state of nature is w in W.

In practice the decision maker often begins knowing W, D, and M and must then *choose* a random experiment to provide more information. For simplicity, we will assume that this choice has been made so that S is known and each state of nature in W corresponds to a probability distribution on S.

Although we will not always be this formal, it is well to have these elements of a decision theory problem in mind when trying to solve the problem. Returning to Example 5.3, the formal statement of that problem is as follows:

1. $S = \{0, 1, 2, \ldots, 20\}$.
2. The state of nature is the true proportion p of all patients with high

blood pressure for whom the drug is effective. So $W = \{p: 0 \le p \le 1\}$. When a particular p is true, X has the binomial distribution with $n = 20$ and probability of success p.

3. $D = \{d_1, d_2\}$ where d_1 means "do not market the drug" and d_2 means "market the drug."

4. The payoff function is (in units of $10,000).

$$M(d_1, p) = -2 \quad \text{for any value of } p$$
$$M(d_2, p) = \begin{cases} -20 & \text{if } 0 \le p < 0.8 \\ 50 & \text{if } 0.8 \le p \le 1 \end{cases}$$

Notice that in this case Nature has infinitely many possible pure strategies (choices of p), so we cannot express the payoffs in matrix form.

Example 5.4   To Drill or Not to Drill

A wildcat oil driller has studied geological conditions surrounding producing wells and dry holes long and hard. He has observed two specific conditions which seem to be associated with oil. Call them conditions A and B, since he refuses to say what they are. More precisely, the wildcatter has observed that in past dry holes the following relative frequencies of various conditions occurred.

Condition	A only	B only	Both A and B	Neither
Probability	0.1	0.2	0.1	0.6

In past producing wells the corresponding probability distribution was as follows:

Condition	A only	B only	Both A and B	Neither
Probability	0.4	0.1	0.4	0.1

A producing well will bring the wildcatter a profit of $1 million and a dry hole will cost him $100,000. What should he do when he observes A? B? Both? Neither?

We cannot answer that (yet), but we can formulate the problem. When the wildcatter studies the geology of a potential drilling site, he observes one of four outcomes: A only, B only, both, or neither. We can think of the actually observed condition as the outcome of a random experiment. If you insist on a random variable, we can label the possible outcomes 1, 2, 3, 4. Then

$$S = \{1, 2, 3, 4\}$$

W contains only two states of nature, "dry hole" (strategy w_1 for Nature) and "oil" (strategy w_2 for Nature). Each of these states of nature corresponds to one of the probability distributions given above for the random experiment of checking for conditions A and B. The decision space D contains two decisions, d_1 (drill) and d_2 (don't drill). The payoff matrix (in units of $100,000) is given in Table 5.4. We take the payoffs in money, so that not drilling costs nothing. Most wildcatters would probably consider missing oil by not drilling a serious loss, and a payoff matrix expressed in terms of utilities would reflect this.

Table 5.4   Monetary payoffs for Example 5.4

Exercise 5.3   *Give S, W, D, and M for each of Exercises 5.3.1 to 5.3.4.*

1.   A cast-aluminum engine block may be perfect, flawed but repairable, or defective. As each block reaches a junction on the assemblyline, it must be sent ahead for use, shunted off for repair, or sent back to be melted down. Using a defective block costs $300, using a repairable block costs $50, and using a good block costs nothing. Shunting a good block to the repair shop costs $5, repairing a repairable block costs $20, and sending a defective block to the shop costs $5. Melting down a good block or a repairable block costs $100, while melting down a bad block costs nothing. (We are counting costs in excess of production costs already incurred.) An automatic electronic tester is used, but it is not foolproof. In fact, it gives the following probability distributions of results depending on the true condition of the engine block. If the block is in fact good,

Test result	Good	Repairable	Defective
Probability	0.9	0.1	0

If the block is in fact flawed but repairable,

Test result	Good	Repairable	Defective
Probability	0.6	0.3	0.1

If the block is in fact defective,

Test result	Good	Repairable	Defective
Probability	0.1	0.2	0.7

2.   An unscrupulous politician is facing an election in which he feels the voters will vote for or against him solely on the basis of his stand on allowing dogs in the local swimming pool. If his stand agrees with that of a

majority of the voters, he will be elected; otherwise he will be defeated. He decides that the utility of winning the election is 1 and that of losing is 0. To aid him in deciding whether to support or oppose dogs in the pool, he chooses 10 voters at random and observes how many support dogs.

3.   A surgeon has devised a rating system based on fitness, obesity, etc., which divides men into four classes related to susceptibility to heart attacks. He has observed the following proportions in the various classes among men who did (or did not) have heart attacks before age 60.

State of nature	Class 1	Class 2	Class 3	Class 4
Heart attack	0.1	0.1	0.4	0.4
No heart attack	0.4	0.2	0.2	0.2

The surgeon wants to choose men who will *not* have heart attacks before 60 for a study of another disease. His utilities are as follows: If he chooses a man who has a heart attack, he will lose 100. If he does not choose a man who does not have a heart attack, he loses nothing. Correct decisions have utility 5.

4.   The Big Brother Community Health Center examines large numbers of routine chest x-rays which must be classified as "normal," "lungs abnormal," "heart abnormal," or "both lungs and heart abnormal." The classification procedure uses a computer pattern recognition and diagnosis program. This program is not a perfect classifier, but past experience gives the following probability distributions of computer results for each of the four possible true states of the x-ray.

True State	Computer Result			
	Normal	Heart	Lungs	Both
Normal	0.9	0	0.1	0
Heart	0.2	0.4	0.2	0.2
Lungs	0.3	0.1	0.5	0.1
Both	0.1	0.2	0.3	0.4

Big Brother has assessed utilities taking into account risk to the patient and cost to the Health Center of errors in classification. Correct decisions cost nothing, and classifying "heart" or "lungs" as "both" or "lungs" as "heart" also cost nothing. Classifying "heart" as "lungs" costs 10, and classifying "heart" as normal costs 100. Classifying "lungs" as "normal" costs 50 and "both" as "normal" costs 110. Classifying a true "normal" as anything else costs 5.

5.   Ball bearings are inspected after manufacture and classified as grade 1 (which yields a profit of $3 on sale), grade 2 (yielding a profit of $1), or scrap (no profit). A machine tool company offers a new bearing production

process for sale, at a cost of $0.10 per bearing over the life of the machines. The proportions of bearings in each grade produced by present and new equipment are recorded in the table below. Should the new equipment be bought?

Equipment	Grade of Bearing		
	Grade 1	Grade 2	Scrap
Present	0.5	0.3	0.2
New	0.7	0.2	0.1

We claim that this is *not* a decision problem of the type formulated in this section. Discuss why it is not. Compute the expected profit per bearing under both present and new equipment, and tell the bearing manufacturer whether or not to purchase the new equipment.

6.   A bearing manufacturer uses an automatic inspection process which classifies bearings as grade 1, grade 2, or scrap. The inspection procedure is not always correct, but makes the following proportions of various decisions for bearings which in fact fall in each class.

True State of Nature	Inspection Decision		
	Grade 1	Grade 2	Scrap
Grade 1	0.80	0.15	0.05
Grade 2	0.40	0.50	0.10
Scrap	0.05	0.20	0.75

Correct classification involves no loss. Classifying grade 1 as grade 2 costs 2 and as scrap costs 3. Classifying grade 2 as grade 1 costs 3 and as scrap costs 1. Classifying scrap as grade 1 costs 5 and as grade 2 costs 3. The manufacturer must decide how to classify each bearing after the inspection decision is made.

Is this a decision problem of the type discussed in this section? If so, give S, W, D, and M. If not, can you solve it by other means?

7.   An investor must decide whether or not to sell a stock in which he has a $50,000 long-term capital gain. He believes that if he waits a year, his gain will be $100,000 with probability 0.6, $50,000 with probability 0.3, and $20,000 with probability 0.1. But there is probability 0.1 that Congress will repeal the tax advantage of capital gains during this year. This event is independent of the amount of his gain, but if it occurs, his net gain will be only one-half of the gross gain because of taxes. Faced with these six possible states of nature, the investor must decide whether to sell now or hold for a year.

Is this a decision theory problem? If so, formulate it. Can you solve the investor's problem?

5.4 BAYES' DECISIONS

In Sec. 5.2 we described games against Nature in which the Player knew Nature's mixed strategy. In Sec. 5.3 we formulated decision theory problems as games against Nature (with the decision maker as Player) in which the decision maker can obtain information about the state of nature from the outcome of a random experiment. Let us now suppose that in a decision theory problem the decision maker knows Nature's mixed strategy—often because he is willing to use subjective or personal probabilities. So we add to the decision theory problem a probability distribution over the states of nature. We call this the *prior distribution* to emphasize that it represents knowledge available *before* any experiments are conducted.

When a prior distribution is available, it is possible to solve decision theory problems. The first step is to ask how the information gained from the outcome of a random experiment should be combined with the prior distribution. The answer is given by Bayes' formula (Sec. 1.11), as the following example illustrates.

Example 5.5   To Drill or Not to Drill

The wildcat oil driller of Example 5.4 is considering drilling at a particular point. Before conducting any field observations, he knows that the probability of a producing well at a given point in this region is 0.05. This is a prior probability. He now studies the site and observes condition A but not condition B, an event we label M. He knows from past experience (see page 338) that event M has probability 0.1 if the site is a dry hole (state of nature w_1) and probability 0.4 if a producing well (state of nature w_2). Since we are now thinking of the state of nature as chosen by Nature according to some mixed strategy, w_1 and w_2 are events; and we will think of 0.1 as the *conditional probability* of event M given that event w_1 occurs. So we know

$$P(w_1) = 0.95 \qquad P(w_2) = 0.05$$
$$P(M|w_1) = 0.1 \qquad P(M|w_2) = 0.4$$

Observing the event M changes the wildcatter's estimate of the probability of oil from $P(w_2)$ to $P(w_2|M)$. By Bayes' formula (Theorem 1.9) this is

$$P(w_2|M) = \frac{P(M|w_2)P(w_2)}{P(M|w_1)P(w_1) + P(M|w_2)P(w_2)}$$
$$= \frac{(0.4)(0.05)}{(0.1)(0.95) + (0.4)(0.05)}$$
$$= \frac{0.02}{0.02 + 0.095} = 0.17$$

Thus observing event M has increased the probability of oil from 0.05 to 0.17. This is called the *posterior probability* of oil, to indicate that it is computed *after* the outcome of the random experiment is available. The posterior probability of a dry hole is of course

$$P(w_1|M) = 1 - P(w_2|M) = 1 - 0.17 = 0.83$$

The role of Bayes' theorem in a decision theory problem is illustrated by Fig. 5.2. It combines the information available from the prior distribution with the result of experimentation or data collection to produce a posterior distribution over the possible states of nature.

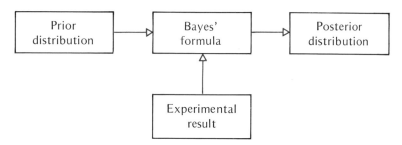

Figure 5.2   Bayes' formula in action in a decision theory problem

Should the wildcatter of Example 5.4 drill? More generally, how should the decision maker use posterior probabilities once he has them? The answer is refreshingly simple: *Exactly as prior probabilities were used in Sec. 5.2.* Once the decision maker knows the probabilities of Nature's pure strategies, he should make the decision that maximizes his expected payoff. This is true whether he has only his prior knowledge of Nature's mixed strategy or prior knowledge plus observed data. A decision which maximizes the expected payoff against the posterior distribution of the states of nature is called a *Bayes' decision* or *Bayes' strategy* for the decision maker.

EXAMPLE 5.5 Continued
The payoff matrix for the wildcatter was given in Table 5.4 and is repeated as Table 5.5. The unit is $100,000.

		Nature	
		w_1	w_2
Wildcatter	d_1	−1	10
	d_2	0	0

Table 5.5   Payoffs for the wildcatter in Example 5.5

Using the posterior probabilities for the states of nature, the expected payoff if the wildcatter drills (decision d_1) is

$$-1 \cdot P(w_1|M) + 10 \cdot P(w_2|M) = (-1)(0.83) + (10)(0.17)$$
$$= -0.83 + 1.7 = 0.87$$

Thus the expected payoff for drilling is $87,000. Not drilling always has a payoff of 0, so based on the available information the wildcatter should drill.

Example 5.6 **Medical Diagnosis**

A physician is faced with a patient exhibiting symptoms which may be the result of (1) cancer of the prostate, (2) primary cancer of the liver, or (3) minor ailments not requiring surgery. He knows that 8 percent of such patients have cancer of the prostate, 6 percent have cancer of the liver, and the remaining 86 percent have minor ailments. The physician assesses the cost of various diagnoses (and the treatment that follows) in terms of risk to the patient, and comes up with the matrix of utilities given in Table 5.6. ($S1$ is cancer of the prostate, and so on.)

	Nature		
	$S1$	$S2$	$S3$
d_1	−1	−6	−2
d_2	−5	−2	−2
d_3	−15	−20	0

Diagnosis d_1, d_2, d_3

Table 5.6   Utilities for the patient in Example 5.6

Before making his decision, the physician conducts some specialized blood chemistry tests. He groups the outcomes of these tests into three disjoint categories, A, B, and C. Past experience has shown the following incidence of these outcomes in cases later diagnosed by surgery. (These are conditional probabilities of the various test outcomes given the state of nature.)

State of Nature	Test Outcome		
	A	B	C
$S1$	0.57	0.22	0.21
$S2$	0.31	0.42	0.27
$S3$	0.11	0.15	0.74

The blood test for his present patient has outcome A. What diagnosis should the physician make? The prior probability distribution is

$$P(S1) = 0.08 \qquad P(S2) = 0.06 \qquad P(S3) = 0.86$$

The posterior probabilities of the states of nature are therefore

$$P(S1|A) = \frac{P(A|S1)P(S1)}{P(A|S1)P(S1) + P(A|S2)P(S2) + P(A|S3)P(S3)}$$

$$= \frac{(0.57)(0.08)}{(0.57)(0.08) + (0.31)(0.06) + (0.11)(0.86)}$$

$$= \frac{0.046}{0.159} = 0.29$$

$$P(S2|A) = \frac{P(A|S2)P(S2)}{P(A|S1)P(S1) + P(A|S2)P(S2) + P(A|S3)P(S3)}$$

$$= \frac{(0.31)(0.06)}{0.159} = 0.12$$

$$P(S3|A) = 1 - P(S1|A) - P(S2|A) = 1 - 0.29 - 0.12 = 0.59$$

The posterior expected payoffs for the three possible decisions are

for d_1 $(-1)(0.29) + (-6)(0.12) + (-2)(0.59) = -2.19$
for d_2 $(-5)(0.29) + (-2)(0.12) + (-2)(0.59) = -2.87$
for d_3 $(-15)(0.29) + (-20)(0.12) + (0)(0.59) = -6.75$

Decision d_1 has the maximum utility here—all expected payoffs are negative, and the payoff for d_1 is least negative. So the surgeon should diagnose cancer of the prostate and take the appropriate action.

In these examples we have found the Bayes' strategy given specific outcomes of the random experiment—condition A alone in Example 5.5 and test outcome A in Example 5.6. To give a complete *Bayes' decision procedure* we must find the Bayes' strategy for each possible outcome of the random experiment. In Example 5.6, that means executing our computations for test outcomes B and C, as well as A.

Computing Bayes' Decisions

1. Determine the prior probabilities of all states of nature and the probability distribution of the random experiment under each state of nature.
2. Compute the posterior probability of each state of nature given the observed outcome of the random experiment. Use Bayes' formula.
3. Determine the payoff for each possible decision in the presence of each state of nature.
4. Compute the expected payoff of each decision, using the posterior probabilities of the states of nature.
5. The decision with the greatest expected payoff is the Bayes' decision.

Several comments on solving decision theory problems are in order. First, we have solved decision theory problems only when a prior distribution over the states of nature is available. Decision makers who are not willing to use subjective probability will often find that a prior distribution is not available. In such a case there is no generally accepted "solution" to the decision problem, though a conservative decision maker can use a maximin strategy.

Second, you should be aware that using a Bayes' strategy in a decision theory problem does not guarantee making the right decision. The choice of decision is based in part on the outcome of a random experiment which gives information about the state of nature. This outcome *may* be misleading. For example, the wildcatter of Example 5.5 knows that condition A alone occurs in 40 percent of all producing wells and in only 10 percent of all dry holes, so he is quite optimistic when he observes it. But condition A alone *may* occur in a dry hole, and that would lead to the wrong decision. This is not a drawback of Bayes' strategies in particular—no scheme for decision making in the presence of uncertainty can guarantee to make the right decision.

Third, there is an elaborate mathematical theory of decision problems which makes it clear that Bayes' decisions are the best possible decisions in terms of expected payoff, when a prior distribution is known. Section 5.6 (which is optional) outlines this mathematical theory.

Exercise 5.4   1.   Compute the Bayes' decision in the following situations:

a.   Example 5.5, when condition B alone is observed
b.   Example 5.5, when conditions A and B are both observed
c.   Example 5.5, when neither A nor B is observed

2.   Complete the computation of the Bayes' decision procedure for Example 5.6 by finding the Bayes' decision when

a.   the blood test results fall in category B
b.   the blood test results fall in category C

3.   Suppose that because of equipment failure the physician in Example 5.6 cannot make a blood test. He is now faced with a game against Nature with no experimental data to aid him. What should his decision be? (Note that this is exactly the situation analyzed in Sec. 5.2.)

4.   The Bayes' decision makes use of prior knowledge as well as experimental data. Show in Example 5.5 that if the prior probability of oil is only 0.01 then the wildcatter should not drill if he observes event M (condition A alone).

5.   A biochemist at Fizz Laboratories (Example 5.3) has considerable experience with the new high blood pressure remedy. His informed assess-

ment of the likelihood that the unknown p will take various values is given below.

p value	0.6	0.7	0.8	0.9
Prior probability	0.10	0.20	0.30	0.40

When the drug is tried on 20 patients, it is effective in 14 cases. Find the Bayes' decision given this information. (Use Table A of binomial probabilities.)

6.   Exercise 5.3.1 describes an assemblyline decision problem. You are asked to give instructions on which action to take with an engine block given each of the three possible test results. Past experience has shown that 80 percent of blocks are perfect, 15 percent are flawed but repairable, and 5 percent are defective. Find the Bayes' decision procedure.

7.   The surgeon of Exercise 5.3.3 knows that in his locality 10 percent of all men have heart attacks before age 60. Using that prior probability find the Bayes' decision procedure for the surgeon. That is, from which fitness classes should he choose men for his study?

*8.   Return to Example 5.5. Show that the posterior probability of oil (whatever the prior) is the same when both conditions A and B are observed as when condition A only is observed. (*Hint*: The tables on page 338 should make this clear.) The posterior probability of oil is smaller when B alone or neither A nor B is observed. The wildcatter wants to know the smallest prior probability of oil for which he would decide to drill if he observed the most favorable conditions (either A alone or A and B). Find that probability—it is the prior probability which makes the posterior expected loss (given A alone is observed) equal for both possible decisions.

9.   The bearing manufacturer of Exercise 5.3.6 knows that in the past 50 percent of his bearing production has been grade 1, 40 percent has been grade 2 and 10 percent scrap. Combine this information with the result of Exercise 5.3.6 to find how a bearing should be classified given each possible result of the inspection procedure.

10.   The Big Brother Community Health Center (Exercise 5.3.4) knows that about 90 percent of routine chest x-rays are normal, 4 percent show heart abnormalities, 5 percent show lung abnormalities, and 1 percent show both. How should Big Brother classify an x-ray

a.   when the computer classifies it as normal?
b.   when the computer classifies it as having heart abnormal?

5.5   APPLICATIONS TO MANAGEMENT DECISION MAKING

The application of the concepts of probability theory, game theory, and decision theory to making business decisions is now becoming widespread.

The basic reason for adopting these new methods is that factors involved in making business decisions are often subject to uncertainty. A company considering a large investment in a new plant, for example, will need estimates of such factors as useful life of capital equipment, share of the market a new product will capture, construction costs, etc. These quantities are all subject to uncertainties. The conventional approach to decision making would demand a "best estimate" of each of these factors, and would make an investment decision on the basis of these estimates. The new approach begins by asking instead for a range of possible values of each factor with associated probabilities. This has the great advantage of making the uncertainty about (say) market share explicit. Under the conventional approach the sales manager when asked his opinion might say, "I think our new toothpaste will capture about 15 percent of the market." He can express his uncertainty much more clearly by giving a probability distribution, such as the following:

Market share (percentage)	30	25	20	15	10	5	0
Probability	0.05	0.1	0.2	0.4	0.2	0.05	0

The use of probabilistic estimates rather than conventional "best estimates" does not require elaborate mathematics to understand. We gave an example of the use of such probabilities in Sec. 1.17, in which only basic probability theory was applied. The additional information provided by a probability distribution over a range of values is so obvious that such distributions are widely used by sophisticated companies. General Electric, for example, requires that all investment requests exceeding $500,000 be supported by probabilistic assessments of factors such as rate of return (*Harvard Business Review*, May–June 1970, p. 83).

These probabilistic assessments of key factors are usually subjective probability distributions. That is, they represent the informed judgment of responsible executives rather than long-term relative frequencies. There is a very good reason for this, namely that long-term relative frequency information is rarely available for random variables such as the market share that a new product will capture. It is essential that the subjective probabilities used result from careful thought by the officers best qualified to judge the issue. There are no shortcuts. It is tempting to say, "I don't have any information about the likelihood of the various possible outcomes, so I'll just make them equally likely." That's cheating, for the equally likely distribution is a claim that you *do* know the likelihood of the possible outcomes. At a more advanced level, it is tempting to choose subjective probability distributions which are mathematically convenient. This is legitimate in textbooks, but is cheating in practice.

A subjective probability distribution for a key factor, or more commonly a joint probability distribution for several key factors, is a prior distribution in our present terminology. The next step in applying decision theory is to identify the possible actions or decisions and their consequences under each state of nature. This is not as simple in practice as textbook problems suggest. In fact, decision theory enthusiasts in industry report that the discipline of trying to explicitly list all decision alternatives, list all important uncertainties, and quantify the consequences of various actions often clarifies the problem to such an extent that a full analysis is not required. Thus the disadvantage of having a complex mathematical model carries with it the advantage of forcing the decision makers to be explicit about all aspects of their problem.

When prior probabilities for the states of nature, a list of available decisions, and numerical values for the consequence of each decision in the presence of each state of nature are available, we have a "no data" decision problem. This is a game against Nature as discussed in Sec. 5.2. If data (such as a market survey) are available to supplement the prior distribution, we have a full decision problem as discussed in Secs. 5.3 and 5.4. We do *not* want to condition you to force all decision problems into these molds. The basic ideas of decision theory are applicable to many problems in which the full mathematical models we have discussed are not present. The sales strategy example in Sec. 1.17 is such a case. It is the concepts of

1. probabilistic estimates of key factors
2. identification of all possible alternative actions
3. numeric evaluation of consequences of actions

which will inevitably spread through business decision making, not necessarily a particular model. With that caveat, let us turn to an example. This is a variation of a case study presented by John F. Magee (*Harvard Business Review*, July–August 1964, pp. 126–138) as an example of the type of problem—capital investment decisions—to which decision theory is currently most often applied.

Application 5.1 A Plant Investment Decision

Stygian Chemicals, Inc., has developed a new commercial cleaning solvent with 10 years expected market life. The demand for this product is unpredictable, but can be described in a given year as either "low" or "high." Management must decide whether to build a large plant (in anticipation of high demand) or a small plant to produce the solvent. What is more, management fears that an initially high demand may give way to low demand after a year or two. The company president therefore leans toward building a small plant, then deciding after a year's experience whether or not to expand it. Let us do a decision-theoretic evaluation of the two proposed actions:

d_1: Build a large plant initially.

d_2: Build a small plant initially; if first year demand is high, expand the plant during the second year so that large production
capacity is available in the third and later years.

The possible states of nature are taken to be:

$S1$: first year demand high, subsequent demand high

$S2$: first year demand high, subsequent demand low

$S3$: first year demand low, subsequent demand low

$S4$: first year demand low, subsequent demand high

(*Comment*: In this example management is certain that demand will not
change after the second year. A realistic problem would involve more states
of nature and more possible actions. In practice, a corporation which uti-
lizes decision theory has available computer routines to handle the tedious
calculations which prevent us from enlarging this example.)

A meeting of development and marketing officers results in agreement on
an assessment of the likelihood of these four states of nature. This is a prior
probability distribution, and in this case is as follows:

State of nature	$S1$	$S2$	$S3$	$S4$
Probability	0.6	0.1	0.3	0.0

(*Comment*: Very often the development manager will offer a more optimistic assessment of demand than the sales manager. With computerized
calculation it is very easy to analyze the problem using both prior distributions to see if the decision changes. Another advantage of the decision
theory approach is that it facilitates such a sensitivity analysis—an investiga-
tion of how sensitive the decision is to changes in the input.)

The annual profit after operating expenses from a large plant operating
under high demand is estimated to be $1 million. If the demand is low, the
profit from a large plant will be only $250,000 as the high fixed overhead
cuts into profits. The small plant will be fully utilized at low demand, with
resulting annual profit of $400,000. A small plant with high demand will
yield $450,000 in each of the first two years but only $350,000 in succeeding years as competitors draw off the excess demand. If the small plant is
expanded, the resulting facility will be less efficient than a large plant and
will yield $700,000 per year with high demand and only $50,000 per year
with low demand. The fixed costs of opening a large plant are $3 million.
It will cost $1,300,000 to open a small plant and $2,200,000 to enlarge it
if that becomes necessary.

These data allow us to compute the payoffs, which we will take to be
total net profit over a 10-yr period. Management of course wants to maximize the expected payoff. Let us compute the total expected profit if a large

plant is built (decision d_1) and demand is high each year (state of nature $S1$). In that case a net profit after operating expenses of $1 million per year will be realized, for a total of $10 million. From this we must subtract the opening costs of $3 million, leaving $7 million total net profit.

As a second example, suppose a small plant is built initially (decision d_2) and that demand is consistently high ($S1$) again. We will then realize $450,000 in each of the first 2 yr. At the beginning of the third year the expanded plant will be in operation and will yield $700,000 in each of the remaining 8 yr. The total profit after operating costs is

$$2(\$450,000) + 8(\$700,000) = \$6,500,000$$

From this must be subtracted the capital costs of opening and expanding the plant,

$$\$1,300,000 + \$2,200,000 = \$3,500,000$$

The net profit is therefore $3 million. The results of these and similar calculations are given in the payoff matrix of Table 5.7. Since state of nature $S4$ has 0 prior probability, we have saved ourselves unnecessary work by omitting it from the payoff matrix.

		State of nature		
		$S1$	$S2$	$S3$
Decision	d_1	70	2.5	−5
	d_2	30	−22.5	27

Table 5.7 Net profits of Stygian Chemicals in units of $100,000

It is now clear that either decision runs the risk of a substantial loss in some circumstances. Stygian's management therefore decides to investigate a third possible action, d_3: build a small plant initially, wait 2 yr, and expand the plant during the third year if demand was high in each of the first 2 yr. By the time the expanded plant is available (the fourth year) competitors have appeared, so that the expanded plant will realize a profit of only $500,000 in subsequent years of high demand. If $S1$ holds, d_3 will yield $450,000 in each of the first 2 yr, $350,000 in the third year, and $500,000 in each of the fourth through tenth years. From this total of $4,750,000 we must subtract the $3,500,000 cost of opening and expanding the plant, so that total net profit is $1,250,000. If $S2$ holds, the strategy d_3 of waiting 2 yr to decide on expansion means that no expansion will occur. Thus d_3 yields $450,000 the first year when demand is high, then $400,000 in each of the following 9 yr of low demand. Subtracting the opening costs gives a net profit of $2,750,000. In the case where $S3$ holds, d_2 and d_3 give the same result. The expanded payoff table is given in Table 5.8.

		State of nature		
		S1	S2	S3
	d_1	70	2.5	−5
Decision	d_2	30	−22.5	27
	d_3	12.5	27.5	27

Table 5.8   Enlarged payoff matrix of profits for Stygian Chemicals in units of $100,000

(*Comment*: The calculation of net profit here is greatly oversimplified. In practice a company would discount future profits to put dollar values in different years on the same basis and would take tax differences into account before subtracting capital plant costs from net profits. Any such calculation of payoffs would be done by the financial department. What is more, the "best estimates" of profits and costs should be replaced by probabilistic estimates, just as probabilistic estimates of future demand were used. We have not done this because it would greatly complicate our computations.)

We now have a "no data" decision problem, and can evaluate the alternatives by computing expected net profits using the prior distribution. They are (in units of $100,000):

for action d_1: $(0.6)(70) + (0.1)(2.5) + (0.3)(−5) = 40.75$
for action d_2: $(0.6)(30) + (0.1)(−22.5) + (0.3)(27) = 23.85$
for action d_3: $(0.6)(12.5) + (0.1)(27.5) + (0.3)(27) = 18.35$

Thus the expected profit is maximized (at $4,075,000) by building the large plant to start with. [*Comment*: This does *not* necessarily mean that Stygian Chemicals will build the large plant. This depends on the attitude of the company toward risk. Building a large plant will incur a $500,000 loss in the case of low demand. A small company which cannot afford such a loss might prefer the safe strategy d_3, which guarantees a profit in all circumstances; that is, net profits may not represent the *utility* of an outcome. The utilities—including attitude toward risk in this case—should be determined with great care, since management may be more concerned about the effect of a loss (or a large profit) on their career than on the company.]

Stygian's management is unsatisfied with their available information, so they commission a market survey. The survey report will project either a strong or a weak market. Call these possible outcomes S and W. Management estimates that if demand will in fact be consistantly high over a 10-yr period, the probability of a strong market report is 0.7. If the demand will be high only in the first year, the market survey will still report a strong market with probability 0.5. If demand is consistently low, the probability of a (erroneous) strong market report is 0.05. This information gives the probability distribution of the outcomes of a random experiment—the market survey—under each state of nature. In formal terms,

$$P(S|S1) = 0.7 \qquad P(W|S1) = 0.3$$
$$P(S|S2) = 0.5 \qquad P(W|S2) = 0.5$$
$$P(S|S3) = 0.05 \qquad P(W|S3) = 0.95$$

(*Comment*: Notice that the probability distribution of the market survey outcome is expressed in terms of subjective probabilities.)

Bayes' formula now allows us to combine our prior knowledge with the result of the market survey to obtain the posterior distribution of the state of nature. We will compute the complete Bayes' decision procedure so that we know what to do for either survey result. Some arithmetic with Bayes' formula shows that if the survey reports strong demand the posterior distribution is

$$P(S1|S) = 0.87 \qquad P(S2|S) = 0.10 \qquad P(S3|S) = 0.03$$

If the survey reports weak demand, the posterior distribution is

$$P(S1|W) = 0.35 \qquad P(S2|W) = 0.10 \qquad P(S3|W) = 0.55$$

Returning to Table 5.8, the posterior expected profits if the survey reports strong demand are:

for action d_1: $(70)(0.87) + (2.5)(0.10) + (-5)(0.03) = 61.0$
for action d_2: $(30)(0.87) + (-22.5)(0.10) + (27)(0.03) = 24.66$
for action d_3: $(12.5)(0.87) + (27.5)(0.10) + (27)(0.03) = 14.435$

Thus, if the market survey reports strong demand, the best decision is to build the large plant. The case for this decision has been strengthened, for with the additional information the expected net profit is now $6,100,000.

If the market survey reports weak demand, the posterior expected payoffs are:

for action d_1: $(70)(0.35) + (2.5)(0.10) + (-5)(0.55) = 22$
for action d_2: $(30)(0.35) + (-22.5)(0.10) + (27)(0.55) = 23.1$
for action d_3: $(12.5)(0.35) + (27.5)(0.10) + (27)(0.55) = 21.975$

All three actions yield nearly equal expected profits with this information. Action d_2 is slightly preferable, but notice that even a small aversion to risk would lead to choosing d_3, which guarantees a profit in all circumstances and yields only $112,500 less expected profit over a 10-yr period.

(*Comment*: The Bayes' decision is therefore to build a large plant if the market report is positive, and to build a small plant with a decision on expansion to be made after 1 yr if the market report is negative. The "no data" optimal procedure is to build the large plant. It is helpful to compute the Bayes' procedure in advance to help decide the worth of the market survey. If the Bayes' procedure always chooses to build the large plant, the result of the survey would never change the "no data" decision. In that case the survey would be a waste of money, unless the more accurate estimates of future profit which the survey can give are needed.)

The current situation regarding business applications of decision theory is summed up in "Do managers find decision theory useful?" by Rex V. Brown (*Harvard Buisiness Review*, May–June 1970). Here is Brown's conclusion (DTA is his shorthand for a version of decision theory equivalent to ours but presented somewhat differently).

> As might be expected when a radically new approach is used, business executives have often found DTA methods frustrating and unrewarding. Nevertheless, there is a steadily growing conviction in the management community that DTA should and will occupy a very important place in the executive's arsenal of problem-solving techniques. Only time can tell if, as some enthusiasts claim, decision theory will be to the executive of tomorrow what the slide rule is to the engineer of today. But clearly, in my opinion, the *potential* impact of DTA is great.

Exercise 5.5   1.   A corporation has just switched to probabilistic estimates of rate of return on investment. Give an example of distributions of rate of return for investment 1 and investment 2 such that both

a.   the "most probable" value of the return on investment 1 exceeds the "most probable" value for investment 2, and also,

b.   investment 2 shows a higher chance of a high rate of return and a lower chance of a loss

In such a case the conventional "best estimate" approach would choose investment 1, while the probability distributions might well lead management to prefer investment 2.

2.   Complete Table 5.8 by computing the net profit of each action under state of nature $S4$ (low demand in the first year followed by 9 yr of high demand.) Does adding this column to Table 5.8 change the expected net profits for the various actions under the prior distribution? What is the posterior probability of $S4$ if the market survey reports strong demand? Weak demand? (You should now understand why a state of nature with prior probability 0 can be ignored.)

3.   Using Bayes' formula, compute the posterior distributions of the states of nature given both strong and weak market reports to check that the figures given on page 353 are correct.

4.   A pessimistic executive disagrees with the prior distribution given on page 350. His assessment of likelihoods is as follows:

State of nature	$S1$	$S2$	$S3$	$S4$
Probability	0.3	0.3	0.4	0

Does this prior distribution result in a different decision in the "no market survey" case?

5.   Read the article by Rex V. Brown cited on page 354, and list the obstacles and difficulties to practical use of decision theory which he mentions.

6.   Enlarge Table 5.8 by computing payoffs for action d_4, build a small plant and make no changes in it. Is this action ever chosen in the "no data" problem? When a market survey is taken?

*5.6   SOME THEORY—A MATHEMATICAL MODEL FOR DECISION THEORY

In discussing decision theory, we began with a game against Nature and added to it the feature that the decision maker could obtain information about Nature's pure strategy from the outcome of a random experiment. This was a major change of emphasis, and seemed to separate decision theory from game theory. In this section we want to introduce a game theory model which is adequate to describe decision theory. In this model Nature's pure strategies (the possible states of nature) are as before, and a mixed strategy for Nature is a prior distribution. The decision maker's pure strategies in the new game are more complicated, in order to take account of the information he obtains from the random experiment. Specifically, a pure strategy for the decision maker is a rule which tells him what decision to make for every possible outcome of the random experiment. We considered such "decision procedures" briefly in Sec. 5.4, and will introduce them formally here. The payoff function remains the decision maker's expected gain. The most important result we will prove is that in this new game a Bayes' decision procedure is an optimum strategy against Nature's known mixed strategy. This greatly strengthens the justification for using Bayes' procedures when a prior distribution is known. In addition, our new model allows us to discuss decision procedures other than Bayes' procedures.

We begin with a decision theory problem as formulated in Sec. 5.3. That is, we know the sample space S of the observed random variable X, the set W of states of nature, and the probabilities $P(X = x|w)$ for all possible outcomes x in S when any state of nature w in W is true. We also know the set D of possible decisions and the payoff $M(d, w)$ to the decision maker when he makes decision d in D and w in W is the true state of nature. Here is a formal definition of a decision procedure, which we denote by a lowercase Greek "delta."

DEFINITION 5.1   A decision procedure δ is a rule specifying which decision in D is to be made for any outcome in S. Thus δ can be expressed as a function assigning a d in D to every x in S.

Example 5.7   In Example 5.6 and Exercise 5.4.2 we computed the Bayes' decision procedure for a medical diagnosis problem when a certain prior distribution was given. It was:

If blood test outcome is A, diagnose cancer of the prostate (d_1).

If blood test outcome is B, diagnose cancer of the liver (d_2).

If blood test outcome is C, diagnose a minor ailment (d_3).

In the form of a function, this decision procedure is

$$\delta(A) = d_1 \qquad \delta(B) = d_2 \qquad \delta(C) = d_3$$

Example 5.8   Estimating Proportions

Ball bearings are produced by an automated process, and some unknown proportion w fail to meet specifications. An inspector examines 10 bearings chosen from each hour's production and counts the number which are defective. He must estimate w. That means that the inspector's decision is a number d which he guesses to be the true value of w. The inspector (or his employer) suffers a loss for inaccurate estimates which increases with the distance of the estimate d from the true proportion w.

Here the state of nature is the true proportion of defective bearings, so that $W = \{w: 0 \leq w \leq 1\}$. The inspector can estimate any proportion to be defective, so his set of possible decisions is $D = \{d: 0 \leq d \leq 1\}$. One possible choice of payoff function is

$$M(d, w) = -(d - w)^2$$

This means that the inspector loses nothing if his estimate is correct, and otherwise loses an amount which depends only on the distance of d from w and increases rapidly as the estimate d gets farther from the true proportion w. The observed random variable X is the number of defective bearings out of 10 inspected, so $S = \{0, 1, 2, \ldots, 10\}$. It is reasonable to assume that choosing and inspecting 10 bearings represents 10 independent trails, so when w is true, our model is that X has the binomial distribution with $n = 10$ and $p = w$. Formally,

$$P(X = x|w) = \binom{10}{x} w^x (1 - w)^{10-x}$$

for x in S. The decision theory problem is now formulated.

The inspector may estimate the proportion of defectives w in an hour's production to be the same as the proportion $X/10$ among the 10 bearings he inspects. This is a decision procedure which makes decision $x/10$ when $X = x$ is observed. Expressed as a function, this procedure is

$$\delta_1(x) = \frac{x}{10}$$

Another possible procedure (used by an inspector who spent the hour in the coffee shop) is to simply guess that $1/10$ are defective without looking at X. This procedure can be expressed as

$$\delta_2(x) = 1/10$$

for all x in S.

As we stated earlier, our model for decision theory is a game against Nature in which Nature's pure strategies are the various w in W, and the decision maker's pure strategies are decision procedures δ. We will denote the set of all possible decision procedures δ by Δ, a capital Greek "delta."

So the strategy spaces for the game are W and Δ. The payoff is the decision maker's expected gain. We must see how to calculate that quantity. When the decision maker uses procedure δ, the decision he actually makes is $\delta(X)$, which depends on the observed random variable X. So his gain when the state of nature is w is $M(\delta(X), w)$, and this is a random variable. The payoff for our new game is the expected value of this gain—which of course depends on w as well as on δ.

DEFINITION 5.2 The payoff when decision procedure δ is used and the true state of nature is w is

$$R(\delta, w) = E[M(\delta(X), w)]$$
$$= \sum_x M(\delta(x), w)P(X = x|w)$$

[The sum is over all x for which $P(X = x|w)$ is positive.]

EXAMPLE 5.7 Continued
Referring back to the matrix of payoffs for the medical diagnosis problem (Table 5.6), we see that the payoff for the procedure δ described in Example 5.7 when the state of nature is $S1$ (prostate cancer) is

$$R(\delta, S1) = M(\delta(A), S1)P(A|S1) + M(\delta(B), S1)P(B|S1)$$
$$+ M(\delta(C), S1)P(C|S1)$$
$$= M(d_1, S1)P(A|S1) + M(d_2, S1)P(B|S1) + M(d_3, S1)P(C|S1)$$
$$= (-1)(0.57) + (-5)(0.22) + (-15)(0.21)$$
$$= -4.82$$

The probabilities $P(A|S1)$, etc., used in this calculation were given in the unnumbered table on page 344.

EXAMPLE 5.8 Continued
The payoff when the inspector uses δ_1 and w is the true proportion of defective bearings is

$$R(\delta_1, w) = -E\left[\left(\frac{X}{10} - w\right)^2\right]$$

This can be evaluated with little work by the use of some facts from Chapter 1. Since the binomial random variable X has expected value $10w$ (page 83), the random variable $X/10$ has expected value w. So $M(\delta_1, w)$ is just the negative of the variance of $X/10$, by the definition of variance. Since X has variance $10w(1 - w)$ by Exercise 1.16.6, $X/10$ has variance $10w(1 - w)/10^2$ by Theorem 1.11. This means that

$$R(\delta_1,\ w) = -\frac{w(1-w)}{10}$$

For the lazy man's procedure δ_2, the payoff against w is

$$R(\delta_2,\ w) = -E[(^1/_{10} - w)^2] = -(^1/_{10} - w)^2$$

Notice that since the loss here did not depend on X, finding its expected value was trivial.

Competing procedures for the same decision theory problem are evaluated by studying their payoffs against all states of nature w. For example, in Fig. 5.3 we graph the payoffs for procedures δ_1 and δ_2 in Example 5.8.

Calculation shows that the payoff using δ_2 is greater (less negative) than that using δ_1 if the true w falls between approximately 0.04 and 0.23. So δ_2 is preferable for these values of w. Elsewhere δ_1 is preferable, and δ_1 does very much better than δ_2 for w greater than $^1/_2$. What is more, if more than 10 bearings (say n) were inspected, the payoff for $\delta_1 = X/n$ would be less negative for all w, while the payoff for δ_2 would not change. So δ_1 would probably be used unless we knew in advance that w was near $^1/_{10}$. This example illustrates the difficulty of "solving" decision theory problems— which of two competing procedures is better depends on the unknown state of nature w.

For now we do not need to consider mixed strategies for the decision maker in our new game with strategy spaces Δ and W and payoff function $R(\delta, w)$. Mixed strategies for Nature are prior distributions. The payoff when the decision maker uses δ and Nature uses a prior distribution $\mathbf{q}$ which puts probabilities $q_1, \ldots, q_k$ on states of nature $w_1, \ldots, w_k$ is, of course, the expected payoff

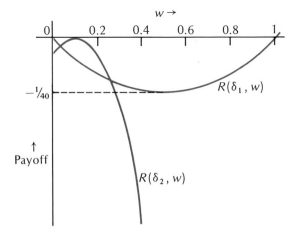

Figure 5.3 Payoffs for procedures δ_1 and δ_2 graphed as functions of the state of nature w

$$R^*(\delta, \mathbf{q}) = \sum_{i=1}^{k} R(\delta, w_i)q_i$$

This is not a new definition—it is forced on us by our previous work with mixed extensions of games. The set W of states of nature may be infinite, but for simplicity we require that any prior distribution $\mathbf{q}$ give positive probability to only finitely many states of nature, which we have labeled $w_1, w_2, \ldots, w_k$ above.

EXAMPLE 5.8 Continued
Suppose the inspector, on the basis of past experience with his job, believes that the following prior distribution on w is true:

w_i	0.0	0.1	0.2	0.3
q_i	0.2	0.5	0.2	0.1

The expected payoff of δ_1 against this prior distribution is

$$
\begin{aligned}
R^*(\delta_1, \mathbf{q}) &= -\sum_{i=1}^{4} \frac{w_i(1 - w_i)}{10} q_i \\
&= -\frac{(0)(1)}{10}(0.2) - \frac{(0.1)(0.9)}{10}(0.5) - \frac{(0.2)(0.8)}{10}(0.2) \\
&\qquad\qquad\qquad\qquad - \frac{(0.3)(0.7)}{10}(0.1) \\
&= -0.0298
\end{aligned}
$$

The expected payoff of procedure δ_2 is

$$
\begin{aligned}
R^*(\delta_2, \mathbf{q}) &= -\sum_{i=1}^{4} (w_i - \tfrac{1}{10})^2 q_i \\
&= -(\tfrac{1}{10})^2(0.2) - (0)^2(0.5) - (\tfrac{1}{10})^2(0.2) - (\tfrac{2}{10})^2(0.1) \\
&= -0.008
\end{aligned}
$$

So procedure δ_2 is better than δ_1 against this prior distribution. This particular prior distribution tells the inspector in advance that w is quite near $\frac{1}{10}$, so that simply guessing $\frac{1}{10}$ turns out to be better than using the reasonable procedure δ_1. Of course, it is easy to find prior distributions for which δ_1 is better than δ_2.

Now in practice we would use neither δ_1 nor δ_2 in this example. We would use the optimum strategy against Nature's known mixed strategy $\mathbf{q}$, as discussed in Sec. 5.2. We are now ready to prove that in the game with strategy spaces Δ and W and payoff function $R(\delta, w)$ the optimum strategies against a given $\mathbf{q}$ are exactly the Bayes' decision procedures we learned to compute in Sec. 5.4. That is the content of Theorem 5.1.

THEOREM 5.1 Suppose $\mathbf{q}$ is a prior distribution giving positive probability to only finitely many states of nature. A decision procedure δ^* is an optimum strategy against $\mathbf{q}$ if, and only if, $\delta^*(x)$ is a Bayes' decision against $\mathbf{q}$ for every x in S.

PROOF Suppose that $\mathbf{q}$ gives prior probability q_i to state of nature w_i for $i = 1, 2, \ldots, k$. Then when $X = x$ is observed, the posterior probability $p(w|x)$ is 0 for all states of nature except the w_i (see Exercise 5.5.2). So if $\delta(x)$ is a Bayes' decision against $\mathbf{q}$ when $X = x$ is observed, then $\delta(x)$ maximizes the posterior expected payoff

$$\sum_{i=1}^{k} M(d, w_i)p(w_i|x)$$

over all procedures d in D. We must show that optimum decision procedures δ^* against $\mathbf{q}$ are exactly these Bayes' decision procedures.

A procedure δ^* is optimum against $\mathbf{q}$ if it maximizes the expected payoff

$$R^*(\delta, \mathbf{q}) = \sum_{i=1}^{k} R(\delta, w_i)q_i$$

over all procedures δ in Δ. We know further that

$$R(\delta, w_i) = \sum_{x} M(\delta(x), w_i)P(X = x|w_i)$$

Substituting this into the expression for R^* gives the expected payoff as a repeated sum

$$R^*(\delta, \mathbf{q}) = \sum_{i=1}^{k} \sum_{x} M(\delta(x), w_i)P(X = x|w_i)q_i$$

which we rewrite by summing first over i and then over x as

$$R^*(\delta, \mathbf{q}) = \sum_{x} \left[\sum_{i=1}^{k} M(\delta(x), w_i)P(X = x|w_i)q_i \right]$$

To maximize R^*, the optimum δ^* must make a decision $\delta^*(x)$ for each x in S which maximizes the expression in brackets above. It must do this for each individual x, otherwise the sum over x will not be as large as possible. So we conclude: δ^* *is optimum against* $\mathbf{q}$ *if and only if for each x in S, $\delta^*(x)$ is a decision which maximizes*

$$\sum_{i=1}^{k} M(d, w_i)P(X = x|w_i)q_i \tag{5.1}$$

over all decisions d in D.

Bayes' formula expresses the posterior distribution when $X = x$ is observed as

$$p(w_i|x) = \frac{P(X = x|w_i)q_i}{\sum_j P(X = x|w_j)q_j} = \frac{P(X = x|w_i)q_i}{P(X = x)}$$

The denominator $P(X = x)$ is the same for all i when x is fixed. So we see that *choosing a decision d to maximize* Eq. (5.1) *is exactly the same as choosing a d to maximize the posterior expected gain*

$$\sum_{i=1}^{k} M(d, w_i)p(w_i|x)$$

That completes the proof of Theorem 5.1.

You should recall from Sec. 5.2 the remark that when **q** is known, no mixed strategy for the Player has greater expected payoff against **q** than an optimum pure strategy. In the present context this means that the decision maker need only consider his pure strategies (decision procedures δ in Δ) when a prior distribution is known. If no prior distribution is known, the decision maker may decide to use a maximin strategy, and this may be a mixed strategy. The strategy space Δ may be infinite, but we require that any mixed strategy **p** for the decision maker give positive probability p_i to only finitely many decision procedures δ_i in Δ. In that case the payoff function for the mixed extension of the game given by Δ, W, and $R(\delta, w)$ is most easily expressed as

$$R^*(\mathbf{p}, \mathbf{q}) = \sum_{i=1}^{m} R^*(\delta_i, \mathbf{q})p_i$$

where δ_i are the procedures in Δ to which **p** gives positive probability.

Since Δ may be large and complicated, finding a maximin mixed strategy is often very difficult. An alternative is to eliminate mixed strategies which are dominated by other mixed strategies—do no better whatever the state of nature may be. In the language of decision theory such mixed strategies are called *inadmissible*.

DEFINITION 5.3 A strategy **p** for the decision maker is *inadmissible* if there is another strategy **p*** such that

$$R^*(\mathbf{p}^*, w) \geq R^*(\mathbf{p}, w)$$

for all states of nature w in W, and strict inequality holds for at least one w. A strategy which is not inadmissible is called *admissible*.

There are usually many admissible strategies in a decision theory problem, just as there were often many nondominated strategies in the games of Chapter 4. One can begin the study of a decision theory problem when no prior distribution is known by trying to find all the admissible mixed strategies for the decision maker. It turns out that there is a close connection between the set of all admissible strategies and the set of all Bayes' strategies for all possible prior distributions. This means that Bayes' strategies are very useful even to decision makers who are not willing to use subjective probabilities and so often have no prior distribution.

We will prove only a few simple results. For the rest of this section we will assume that $D = \{d_1, \ldots, d_m\}$ and $W = \{w_1, \ldots, w_k\}$ are both finite. We know that for any prior distribution q we can always find a *pure* strategy (a decision procedure δ) which maximizes $R^*(p, q)$ over all mixed strategies p. Theorem 5.1 tells us that a Bayes' procedure against q will do this. There may, however, be mixed strategies p which do just as well (not better) against q. *Let us agree to call all strategies—pure or mixed—which maximize $R^*(p, q)$ Bayes' strategies against q.*

Are all Bayes' strategies admissible? Not quite, but the following theorem and Exercise 5.6.9 show that most are.

THEOREM 5.2   If p^* is a Bayes' strategy against a prior distribution q which gives positive probability to every w in W, then p^* is admissible.

PROOF   Suppose that p^* is *inadmissible*. We will show that it cannot be a Bayes' strategy against q, and this contradiction means that p^* must be admissible.

If p^* is inadmissible, there is a strategy p such that

$$R^*(p, w_i) \geq R^*(p^*, w_i)$$

for each $i = 1, 2, \ldots, k$ and for at least one j

$$R^*(p, w_j) > R^*(p^*, w_j)$$

From these two inequalities and the fact that $q_i > 0$ for all i, it follows that the expected payoffs against q satisfy

$$R^*(p, q) = \sum_{i=1}^{k} R^*(p, w_i)q_i$$
$$> \sum_{i=1}^{k} R^*(p^*, w_i)q_i = R^*(p^*, q)$$

This shows that p^* is not a Bayes' strategy against q, for p has larger expected payoff.

There is a converse to Theorem 5.2 which is the most important theorem in decision theory. This theorem states that when D and W are finite, *all admissible strategies are Bayes' strategies*—not, of course, all against the same prior distribution. This result can be proved by a geometric argument related to that used in Sec. 4.10 to prove the fundamental theorem of game theory.

Are maximin strategies admissible? Usually, but not always. If you studied Sec. 4.10, you can see that Fig. 4.9*b* illustrates a situation in which there are many minimax strategies, one of which dominates all the others. The same situation can occur for maximin strategies. Here is a simple result on admissibility of maximin strategies.

THEOREM 5.3   If $\mathbf{p}^*$ is the only maximin strategy for the decision maker, then $\mathbf{p}^*$ is admissible.

PROOF   We will again give a proof by contradiction by showing that if $\mathbf{p}^*$ is inadmissible, it cannot be the only maximin strategy. If $\mathbf{p}^*$ is inadmissible, then there is a $\mathbf{p}$ with

$$R^*(\mathbf{p}, w) \geq R^*(\mathbf{p}^*, w)$$

for all w in W. This means that if $\mathbf{q}$ is any mixed strategy for Nature

$$R^*(\mathbf{p}, \mathbf{q}) \geq R^*(\mathbf{p}^*, \mathbf{q})$$

But then taking the minimum over all $\mathbf{q}$ gives

$$\min_{\mathbf{q}} R^*(\mathbf{p}, \mathbf{q}) \geq \min_{\mathbf{q}} R^*(\mathbf{p}^*, \mathbf{q})$$

This means that $\mathbf{p}^*$ cannot be the only maximin strategy, for the minimum payoff using $\mathbf{p}$ is at least as large.

A COMMENT ON TERMINOLOGY   We have taken the decision maker to be Player 1 (the maximizer), as is usual in business decision making. It is usual for mathematicians working on decision theory to take the decision maker to be Player 2, so that the payoff is his loss. This is a matter of taste. If Nature is Player 1 and the decision maker is Player 2, the expected payoff $R(w, \delta)$ of a decision procedure δ is called the *risk*. The expected payoff $R^*(\mathbf{p}, \delta)$ against a prior distribution $\mathbf{p}$ is called the *Bayes' risk* of the procedure δ against $\mathbf{p}$. You will meet this terminology if you study mathematical treatments of decision theory.

Exercise 5.6   1.   Here is a very simple decision theory problem. The Player knows that a coin is loaded so that either heads will come up with probability $\frac{2}{3}$ (state of nature w_1) or tails will come up with probability $\frac{2}{3}$ (state of nature w_2). Player must guess that w_1 is true (decision d_1) or that w_2 is true (decision d_2). Correct decisions gain 1 and incorrect decisions lose 1, so that $M(d, w)$ is given by

	w_1	w_2
d_1	1	−1
d_2	−1	1

The Player tosses the coin once and observes either heads ($X = 1$) or tails ($X = 0$), where X is the number of heads observed.

Show that there are only four possible decision procedures in this case, say δ_1, δ_2, δ_3, and δ_4. Compute the payoff $R(\delta, w)$ for each δ and each w, and express the result as a payoff matrix. You have now completely described the game which is a model for this decision problem.

2.　Return to Example 5.7. What is the expected payoff $R^*(\delta, \mathbf{q})$ of the procedure δ against the prior distribution $\mathbf{q} = [0.08, 0.06, 0.86]$?

3.　Suppose that under normal conditions one-tenth of the ball bearings produced by a manufacturing process are defective. Once in a while a machine breakdown causes one-half of the bearings to be defective. An inspector is to decide which of these possibilities is true on the basis of the number X of defective bearings in a sample of size 10. Here $W = \{1/10, 1/2\}$ since these are assumed to be the only possible proportions of defectives. The decision space is $D = \{d_1, d_2\}$ where

d_1:　means decide $1/10$ is true

d_2:　means decide $1/2$ is true

Take the loss to be 0 for a correct decision and 1 for an incorrect decision. More formally

$$M(d_1, w) = \begin{cases} 0 & \text{if } w = 1/10 \\ -1 & \text{if } w = 1/2 \end{cases}$$

$$M(d_2, w) = \begin{cases} -1 & \text{if } w = 1/10 \\ 0 & \text{if } w = 1/2 \end{cases}$$

a.　Show that in this problem the payoff $R(\delta, w)$ of a procedure δ when w is true is just the negative of the probability of an incorrect decision when w is true.

b.　Using Table A of binomial probabilities, find the payoffs of the following procedures against both values of w.

$$\delta_1(X) = \begin{cases} d_1 & \text{if } X = 0 \text{ or } 1 \\ d_2 & \text{if } X \geq 2 \end{cases}$$

$$\delta_2(X) = \begin{cases} d_2 & \text{if } X = 0 \text{ or } 1 \\ d_1 & \text{if } X \geq 2 \end{cases}$$

$$\delta_3(X) = d_1 \quad \text{for all values of } X$$

c.　Do your calculations show any of these procedures to be inadmissible? If so, can you explain intuitively why the inadmissible procedure is not reasonable?

4.　Find a prior for the proportion of defective bearings w in Example 5.8 such that the expected payoff $R^*(\delta_1, \mathbf{q})$ is less negative than the expected payoff $R^*(\delta_2, \mathbf{q})$.

5. In Example 5.7 we found the Bayes' decisions under all three possible outcomes of the blood test for Example 5.6. Go through the steps of the proof of Theorem 5.1 with the actual numbers from this example to convince yourself of the validity of that proof.

6. Explain carefully where the proof of Theorem 5.2 breaks down if some $q_i = 0$.

7. Here is an example of an inadmissible procedure. Suppose that a new inspector in the setting of Example 5.8 is so unskilled that he can inspect only six bearings per hour. He counts the number Y of defectives among these, a binomial random variable with $n = 6$ and $p = w$. He then estimates the overall proportion of defectives to be $\delta_3(Y) = Y/6$. Imitate Example 5.8 to find the expected payoff $R^*(\delta_3, w)$. Plot $R^*(\delta_3, w)$ and $R^*(\delta_1, w)$ on the same graph. Conclude that δ_3 is inadmissible.

8. Suppose that δ^* is an admissible procedure and that δ^* has constant risk; that is, $R^*(\delta^*, w) = a$ for some constant a and all w in W. Prove that δ^* is maximin.

9. Prove that if δ^* is the *only* Bayes' procedure against a prior $\mathbf{q}^*$, then δ^* is admissible. (*Hint*: Give a proof by contradiction similar to the proof of Theorem 5.2.)

5.7 FOOD FOR THOUGHT—DECISION THEORY AND STATISTICS

Statistics is often defined as "a body of methods for making wise decisions in the face of uncertainty," to quote the first sentence of an outstanding expository book on the subject, *Statistics—A New Approach* (W. Allen Wallis and Harry V. Roberts, Free Press, 1956). That sounds very much like what we have studied in this chapter. And in fact, the mathematical model of the last section has been widely used as a vehicle for the theory of statistics. There is, however, much more to statistics than can easily be grasped using our background in decision theory. In this section we first discuss some contrasts between statistics and decision theory and then return to Example 5.8 to treat that problem from the statistical point of view.

The first thing to be said is that the decision theory model does not cover all the activities of statisticians. One important statistical activity is the interpretation of data, trying to make sense of data when a precise model is not available. Decision theory has little to offer here, and many applied statisticians are somewhat hostile to people who mention payoff functions. The decision theorist's reply is that interpreting data without a model is an art at best.

Second, and more important, decision theory has not been widely successful in suggesting good procedures. Competing procedures usually have expected payoffs similar to those in Fig. 5.3, so that neither procedure dominates the other. In such cases it is not at all clear how to decide which procedure to use on decision-theoretic grounds. In the earlier sections of this chapter we have simply used Bayes' procedures. Everyone agrees that a Bayes' procedure should be used if the state of nature can be conceived to be a random variable and the distribution of that random variable is known. An influential group of statisticians (called Bayesians) feels that these conditions are almost always satisfied. As should be clear from the examples in this chapter, this means willingness to use subjective or personal probabilities. Other statisticians are quite unwilling to use probabilities which do not have a clear-cut relative frequency interpretation. We are not going to enter this controversy. But it is fair to say that subjective probability is widely accepted in management decision making and other situations in which the partial information and skilled judgment of the decision maker play a role. By contrast, the use of personal probability is often frowned upon in scientific and engineering problems. The quality control example of the last section displays a situation in which the Bayes' approach is usually *not* used in practice. After all, the purpose of the inspection is to detect divergence from the expected proportion of defective bearings, and most engineers feel that personal likelihoods and even past frequencies should play no role in the decision.

Nonetheless, the decision-theoretic formulation of statistical problems has had great influence on the theory of statistics even when decision theory cannot pinpoint a "best" procedure. For example, here is a standard problem of statistics: You have observations on n ball bearings chosen independently from a large lot of bearings. Based on these observations you must estimate the overall proportion of defective bearings in the lot. This problem is exactly the decision theory problem of Example 5.8 (where $n = 10$) with one significant omission. The omission is that no payoff function is given, and this is typical of "practical" problems of statistics. One of the services decision theory performs is to stress the importance of assessing the losses due to wrong decisions. Thus the payoff function in Example 5.8,

$$M(d, w) = -(w - d)^2$$

is probably not correct for this problem, since it says that errors of the same magnitude in either direction cost the same amount. It seems likely that underestimating the proportion of defectives will cost the manufacturer more than overestimating, and if this is so, the payoff function should reflect it.

The "standard" procedure in this statistical problem is to estimate w, the proportion of defectives in the lot, by the proportion of defectives in the sample of n bearings which are inspected. Let us survey how statisticians defend this procedure.

Let X stand for the number of defective bearings in the sample of size n. Then X is a binomial random variable with n known and $p = w$ unknown. The standard estimator for w is X/n. First notice that since X has expected value nw, the estimator has expected value w, whatever the true value w is. Statisticians say that X/n is an *unbiased* estimator of w—on the average it guesses right. Next, by referring to Sec. 1.18 we can see that the law of large numbers tells us that as n increases, X/n must approach the true value of w. Statisticians say that X/n is a *consistent* estimator of w—it will be very likely to give an estimate close to the true value if you can afford to take enough observations. Now X/n need not be the only unbiased and consistent estimator of w, so we have only begun to defend its use. Statistical theory goes on to show that X/n for any fixed sample size n has the smallest variance among all unbiased estimators of w. So X/n is called the *best unbiased* estimator of w. That says that among all estimators that have expected value w (are unbiased), X/n is most concentrated about the true w (has smallest variance).

The last remark has a decision-theoretic interpretation. We saw in the last section that if the payoff function is the squared error $-(w - d)^2$, then the expected loss of X/n is just its variance. It is easy to see that this is true of any unbiased estimator, so we can say that X/n has the smallest expected loss against any value of w among all unbiased estimators. This introduction of expected payoffs forces us to think about several issues we might otherwise evade. First, if the payoff is *not* squared error (and in this problem it may not be), it is no longer obvious that the "best" unbiased estimator is the one having smallest variance. We might even prefer a *biased* estimator with *larger* variance if it avoided costly underestimation of w. This can only be settled by giving a more appropriate payoff function. Second, even if the payoff is squared error, we cannot compare all estimators by comparing their variances. After all, variance is a measure of spread about the mean, and if the estimator does not have mean w, we may get small spread about a wrong guess. To compare estimators in general, we must compare expected payoffs. Only for unbiased estimators does the expected loss become the variance.

The final item in the statistician's defense of the estimator X/n is more specialized. Many statisticians are attracted by a particular method of finding estimators called the *maximum-likelihood* method. This is as follows: Estimate w to be that value which would make the actually observed value of X most probable. It turns out that in this problem X/n is the resulting estimate of w. So a statistician, when asked why he uses X/n to estimate the proportion w of defectives, will reply, "Because it is best unbiased, consistent, and is the maximum-likelihood estimator."

We want to stress that none of the items in this defense of the estimator X/n has any essential connection with decision theory, and decision theorists do not think that the statistician's defense is strong. Only comparison of procedures in terms of expected payoffs is decision theoretic. Even our remarks about variances and expected losses are not in the spirit of standard statistics, although the cautions they led to are. Can we defend X/n from the standpoint of decision theory as well? To a certain extent, yes. First, X/n is admissible, so that there is no estimator which is better for all w in terms of risks. The second point is more technical. For squared-error payoff

$$M(d, w) = -(w - d)^2$$

X/n is not maximin, but the maximin estimator turns out to be

$$\frac{X + \frac{1}{2}\sqrt{n}}{n + \sqrt{n}}$$

When the sample size n is large, $\sqrt{n}$ is small relative to n, so that X/n is very close to being maximin when the sample size is large. If we change the payoff function to be

$$M(d, w) = -\frac{(w - d)^2}{w(1 - w)}$$

then X/n *is* maximin.

This discussion is typical of the relation between decision theory and the theory of statistics. The decision-theoretic viewpoint helps formulate problems and can make helpful comments about suggested procedures. But unless a prior distribution is known, decision theory does not tell the statistician which procedure to use. Statistical theory suggests general methods of finding procedures (such as the maximum-likelihood method) and studies properties of procedures which may not be directly related to payoff functions (such as unbiasedness and consistency of estimators). But statistical theory also does not usually point to a single "best" procedure. Statistics remains full of knotty theoretical problems in addition to the fascinating practical problems of data interpretation.

Exercise 5.7

1.   Exercise 5.6.8 says that an admissible procedure with constant expected payoff is maximin. Show that X/n has constant expected payoff if the payoff function is

$$M(d, w) = -\frac{(w - d)^2}{w(1 - w)}$$

2.   Show that

$$\frac{X + \frac{1}{2}\sqrt{n}}{n + \sqrt{n}}$$

has constant expected payoff if the payoff function is $M(d, w) = -(d - w)^2$. (*Comment*: These estimators are admissible by Theorem 5.2 but we cannot prove this because they are Bayes' procedures with respect to certain *continuous* prior distributions for w. Section 1.19 briefly discusses continuous probability distributions.)

3.   Show that $(X/n)^2$ is *not* an unbiased estimator for w^2. Use your knowledge of $E(X)$ and $E(X^2)$ to find an unbiased estimator for w^2.

4.   Here is an example of a maximum-likelihood estimator. Suppose you observe two bearings independently and count the number X of defectives. If w is the true proportion of defective bearings in the lot, the distribution of X is

x	$P(X = x)$
0	$(1 - w)^2$
1	$2w(1 - w)$
2	w^2

What value of w makes $P(X = x)$ largest when $x = 0$? When $x = 1$? When $x = 2$? The resulting procedure is the maximum-likelihood estimator of w.

COMPUTER PROJECTS

Project 1:   Solution of a Game by Fictitious Play

A two-person zero sum game can be solved by a method called *fictitious play*. A description of this method will be given as an algorithm for repeated play of the game. The algorithm is as follows:

1.   Player 1 and Player 2 each arbitrarily select a strategy for the first play of the game. Both players know which strategy was used.
2.   On each subsequent play of the game, each player acts as though the relative frequency of occurrence for his opponent's strategies on all previous plays of the game were his opponent's mixed strategy. He therefore plays the strategy which is optimum against this mixed strategy, as described in Sec. 5.2.

It turns out that if both players proceed in this manner, the relative frequencies with which the various pure strategies are played eventually approach the maximin and minimax strategies, and the average payoff actually realized approaches the value of the game.

   You are to implement this method of solution by a computer program. The input should consist of:

1.   the number of rows and columns in the payoff matrix
2.   the number of times the game is to be played (The upper limits on these numbers will be determined by the computer facilities available to you.)
3.   the entries of the payoff matrix
4.   the plays made by both players on the first repetition

The output should include:

1.   for each repetition of the game, a table of the relative frequencies with

which each pure strategy has been played on all repetitions up to and including the present one

2. similarly, for each repetition the average of the actual payoffs on all repetitions up to and including the present one

Apply your program to the game of Table 4.5. You should be able to see the sequence of results in items 1 and 2 approaching their limiting values as more and more repetitions are made.

Project 2:   A Routine for Computing Bayes' Decisions

When a decision theory problem is formulated and a prior distribution is known, we have seen that Bayes' decisions are appropriate. Your project is to write a program which will accept as input the elements of a decision theory problem and a prior distribution, and compute Bayes' decisions for a given value (or, on request, for all possible values) of the observed random variable.

Your program should

1. accept S, W, and D of any size up to some maximum determined by the resources available to you
2. contain a subroutine for implementing Bayes' formula

SUPPLEMENTARY READINGS

Chapter 1

1. Blakeslee, David, and William Chinn, *Introductory Statistics and Probability*, Houghton Mifflin, 1971. (Despite the title, the book is devoted almost entirely to probability, done very clearly at a lower level than the present text. A good recommendation for students who have trouble with our Chapter 1.)

2. Mosteller, Frederick, Robert Rourke, and George Thomas, *Probability with Statistical Applications*, Addison-Wesley, 2d ed., 1970. (An excellent text on a level roughly the same as this book. Contains material on combinatorial probability, normal distribution, and classical statistics, as well as the subjects of our Chapter 1.)

Chapter 2

1. Kemeny, John G., and Laurie J. Snell, *Finite Markof Chains*, Van Nostrand, 1960. (For the advanced student who wishes to pursue additional theory, this is an excellent source. The main thrust of the exposition is toward the use of algebraic arguments to derive the desired results.)

Chapter 3

1. Cooper, William W., and Abraham Charnes, "Linear programming," *Mathematics in the Modern World; Readings from Scientific American*, Freeman, 1968. (This nontechnical exposition is written by leaders in the applications of linear programming to economics. A detailed explanation of the linear programing method is given for a manufacturing example.)

2. Hadley, G., *Linear Programming*, Addison-Wesley, 1963. (A good source for the theory which is lacking in this chapter. The level is distinctly more advanced than the present text.)

3. Lipschutz, Seymour, *Finite Mathematics*, Schaum's Outline Series, McGraw-Hill, 1966. (The exposition, at about the same level as this text, provides more routine problems and emphasizes the use of the simplex algorithm.)

Chapter 4   1. Davis, Morton D., *Game Theory, A Nontechnical Introduction,* Basic Books, 1970. (True to its title, this book contains no mathematics at all. An excellent source of concepts, examples, and applications, especially for non-zero sum and *n*-person games.)

2. Luce, R. Duncan, and Howard Raiffa, *Games and Decisions: Introduction and Critical Survey,* John Wiley & Sons, 1957. (At once more advanced and less mathematical than Chapter 4. Recommended for comments on philosophy and for a good treatment of utility, which is omitted in this book for lack of space.)

Chapter 5   1. Chernoff, Herman, and Lincoln E. Moses, *Elementary Decision Theory,* John Wiley & Sons, 1959. (Probability, utility, and decision theory from the mathematician's point of view at a level comparable to that of the present text.)

2. Raiffa, Howard, *Decision Analysis,* Addison-Wesley, 1968. (This book is informal, Bayesian, and oriented toward management science rather than mathematics. Covers rather more than our Chapter 5, using quite different notation.)

APPENDIX

Table A Binomial probabilities

								p						
	k	0.01	0.1	0.2	0.3	0.33	0.4	0.5	0.6	0.67	0.7	0.8	0.9	0.99
$N=1$	0	0.990	0.900	0.800	0.700	0.670	0.600	0.500	0.400	0.330	0.300	0.200	0.100	0.010
	1	0.010	0.100	0.200	0.300	0.330	0.400	0.500	0.600	0.670	0.700	0.800	0.900	0.990
$N=2$	0	0.980	0.810	0.640	0.490	0.445	0.360	0.250	0.160	0.111	0.090	0.040	0.010	
	1	0.020	0.180	0.320	0.420	0.444	0.480	0.500	0.480	0.444	0.420	0.320	0.180	0.020
	2		0.010	0.040	0.090	0.111	0.160	0.250	0.360	0.445	0.490	0.640	0.810	0.980
$N=3$	0	0.970	0.729	0.512	0.343	0.301	0.216	0.125	0.064	0.036	0.027	0.008	0.001	
	1	0.030	0.243	0.384	0.441	0.444	0.432	0.375	0.288	0.219	0.189	0.096	0.027	
	2		0.027	0.096	0.189	0.219	0.288	0.375	0.432	0.444	0.441	0.384	0.243	0.030
	3		0.001	0.008	0.027	0.036	0.064	0.125	0.216	0.301	0.343	0.512	0.729	0.970
$N=4$	0	0.961	0.656	0.410	0.240	0.202	0.130	0.063	0.026	0.012	0.008	0.002		
	1	0.039	0.292	0.410	0.412	0.397	0.346	0.250	0.154	0.096	0.076	0.026	0.004	
	2		0.048	0.154	0.265	0.293	0.346	0.375	0.346	0.293	0.265	0.154	0.048	
	3		0.004	0.026	0.076	0.096	0.154	0.250	0.346	0.397	0.412	0.410	0.292	0.039
	4			0.002	0.008	0.012	0.026	0.063	0.130	0.202	0.240	0.410	0.656	0.961
$N=5$	0	0.951	0.590	0.328	0.168	0.135	0.078	0.031	0.010	0.004	0.002			
	1	0.048	0.328	0.410	0.360	0.332	0.259	0.156	0.077	0.040	0.028	0.006		
	2	0.001	0.073	0.205	0.309	0.328	0.346	0.313	0.230	0.161	0.132	0.051	0.008	
	3		0.008	0.051	0.132	0.161	0.230	0.313	0.346	0.328	0.309	0.205	0.073	0.001
	4			0.006	0.028	0.040	0.077	0.156	0.259	0.332	0.360	0.410	0.328	0.048
	5				0.002	0.004	0.010	0.031	0.078	0.135	0.168	0.328	0.590	0.951

Table A *(continued)*

								p						
	k	0.01	0.1	0.2	0.3	0.33	0.4	0.5	0.6	0.67	0.7	0.8	0.9	0.99
N = 6	0	0.941	0.531	0.262	0.118	0.090	0.047	0.016	0.004	0.001	0.001			
	1	0.057	0.354	0.393	0.303	0.267	0.187	0.094	0.037	0.016	0.010	0.002		
	2	0.002	0.098	0.246	0.324	0.330	0.311	0.234	0.138	0.080	0.060	0.015	0.001	
	3		0.015	0.082	0.185	0.216	0.276	0.313	0.276	0.216	0.185	0.082	0.015	
	4		0.001	0.015	0.060	0.080	0.138	0.234	0.311	0.330	0.324	0.246	0.098	0.002
	5			0.002	0.010	0.016	0.037	0.094	0.187	0.267	0.303	0.393	0.354	0.057
	6				0.001	0.001	0.004	0.016	0.047	0.090	0.118	0.262	0.531	0.941
N = 7	0	0.932	0.478	0.210	0.083	0.061	0.028	0.008	0.002					
	1	0.066	0.372	0.367	0.247	0.209	0.131	0.055	0.017	0.006	0.004			
	2	0.002	0.124	0.275	0.318	0.309	0.261	0.164	0.077	0.037	0.025	0.004		
	3		0.023	0.115	0.227	0.253	0.290	0.273	0.194	0.125	0.097	0.029	0.003	
	4		0.003	0.029	0.097	0.125	0.194	0.273	0.290	0.253	0.227	0.115	0.023	
	5			0.004	0.025	0.037	0.077	0.164	0.261	0.309	0.318	0.275	0.124	0.002
	6				0.004	0.006	0.017	0.055	0.131	0.209	0.247	0.367	0.372	0.066
	7						0.002	0.008	0.028	0.061	0.083	0.210	0.478	0.932
N = 8	0	0.923	0.430	0.168	0.058	0.041	0.017	0.004	0.001					
	1	0.075	0.383	0.336	0.198	0.160	0.090	0.031	0.008	0.002	0.001			
	2	0.002	0.149	0.294	0.296	0.276	0.209	0.109	0.041	0.016	0.010	0.001		
	3		0.033	0.147	0.254	0.272	0.279	0.219	0.124	0.066	0.047	0.009		
	4		0.005	0.046	0.136	0.167	0.232	0.273	0.232	0.167	0.136	0.046	0.005	
	5			0.009	0.047	0.066	0.124	0.219	0.279	0.272	0.254	0.147	0.033	
	6			0.001	0.010	0.016	0.041	0.109	0.209	0.276	0.296	0.294	0.149	0.002
	7				0.001	0.002	0.008	0.031	0.090	0.160	0.198	0.336	0.383	0.075
	8						0.001	0.004	0.017	0.041	0.058	0.168	0.430	0.923
N = 9	0	0.914	0.387	0.134	0.040	0.027	0.010	0.002						
	1	0.083	0.387	0.302	0.156	0.121	0.060	0.018	0.004	0.001				
	2	0.003	0.172	0.302	0.267	0.238	0.161	0.070	0.021	0.007	0.004			
	3		0.045	0.176	0.267	0.273	0.251	0.164	0.074	0.032	0.021	0.003		
	4		0.007	0.066	0.172	0.202	0.251	0.246	0.167	0.099	0.074	0.017	0.001	
	5		0.001	0.017	0.074	0.099	0.167	0.246	0.251	0.202	0.172	0.066	0.007	
	6			0.003	0.021	0.032	0.074	0.164	0.251	0.273	0.267	0.176	0.045	
	7				0.004	0.007	0.021	0.070	0.161	0.238	0.267	0.302	0.172	0.003
	8					0.001	0.004	0.018	0.060	0.121	0.156	0.302	0.387	0.083
	9							0.002	0.010	0.027	0.040	0.134	0.387	0.914
N = 10	0	0.904	0.349	0.107	0.028	0.018	0.006	0.001						
	1	0.092	0.387	0.268	0.121	0.090	0.040	0.010	0.002					
	2	0.004	0.194	0.302	0.233	0.199	0.121	0.044	0.011	0.003	0.001			
	3		0.057	0.201	0.267	0.262	0.215	0.117	0.042	0.015	0.009	0.001		
	4		0.011	0.088	0.200	0.225	0.251	0.205	0.111	0.055	0.037	0.006		
	5		0.001	0.026	0.103	0.133	0.201	0.246	0.201	0.133	0.103	0.026	0.001	
	6			0.006	0.037	0.055	0.111	0.205	0.251	0.225	0.200	0.088	0.011	

Table A *(continued)*

	k	0.01	0.1	0.2	0.3	0.33	0.4	0.5	0.6	0.67	0.7	0.8	0.9	0.99
	7			0.001	0.009	0.015	0.042	0.117	0.215	0.262	0.267	0.201	0.057	
	8				0.001	0.003	0.011	0.044	0.121	0.199	0.233	0.302	0.194	0.004
	9						0.002	0.010	0.040	0.090	0.121	0.268	0.387	0.092
	10							0.001	0.006	0.018	0.028	0.107	0.349	0.904
$N = 15$	0	0.860	0.206	0.035	0.005	0.002								
	1	0.130	0.343	0.132	0.031	0.018	0.005							
	2	0.009	0.267	0.231	0.092	0.063	0.022	0.003						
	3	0.001	0.129	0.250	0.170	0.134	0.063	0.014	0.002					
	4		0.043	0.188	0.219	0.198	0.127	0.042	0.007	0.001	0.001			
	5		0.010	0.103	0.206	0.215	0.186	0.092	0.024	0.006	0.003			
	6		0.002	0.043	0.147	0.176	0.207	0.153	0.061	0.021	0.012	0.001		
	7			0.014	0.081	0.111	0.177	0.196	0.118	0.055	0.035	0.003		
	8			0.003	0.035	0.055	0.118	0.196	0.177	0.111	0.081	0.014		
	9			0.001	0.012	0.021	0.061	0.153	0.207	0.176	0.147	0.043	0.002	
	10				0.003	0.006	0.024	0.092	0.186	0.215	0.206	0.103	0.010	
	11				0.001	0.001	0.007	0.042	0.127	0.198	0.219	0.188	0.043	
	12						0.002	0.014	0.063	0.134	0.170	0.250	0.129	0.001
	13							0.003	0.022	0.063	0.092	0.231	0.267	0.009
	14								0.005	0.018	0.031	0.132	0.343	0.130
	15									0.002	0.005	0.035	0.206	0.860
$N = 20$	0	0.818	0.122	0.012	0.001									
	1	0.165	0.270	0.058	0.007	0.003								
	2	0.016	0.285	0.137	0.028	0.015	0.003							
	3	0.001	0.190	0.205	0.072	0.045	0.012	0.001						
	4		0.090	0.218	0.130	0.095	0.035	0.005						
	5		0.032	0.175	0.179	0.149	0.075	0.015	0.001					
	6		0.009	0.109	0.192	0.184	0.124	0.037	0.005	0.001				
	7		0.002	0.055	0.164	0.181	0.166	0.074	0.015	0.003	0.001			
	8			0.022	0.114	0.145	0.180	0.120	0.036	0.009	0.004			
	9			0.007	0.065	0.095	0.160	0.160	0.071	0.023	0.012			
	10			0.002	0.031	0.052	0.117	0.176	0.117	0.052	0.031	0.002		
	11				0.012	0.023	0.071	0.160	0.160	0.095	0.065	0.007		
	12				0.004	0.009	0.036	0.120	0.180	0.145	0.114	0.022		
	13				0.001	0.003	0.015	0.074	0.166	0.181	0.164	0.055	0.002	
	14					0.001	0.005	0.037	0.124	0.184	0.192	0.109	0.009	
	15						0.001	0.015	0.075	0.149	0.179	0.175	0.032	
	16							0.005	0.035	0.095	0.130	0.218	0.090	
	17							0.001	0.012	0.045	0.072	0.205	0.190	0.001
	18								0.003	0.015	0.028	0.137	0.285	0.016
	19									0.003	0.007	0.058	0.270	0.165
	20										0.001	0.012	0.122	0.818

ANSWERS TO
ODD-NUMBERED EXERCISES

CHAPTER 1

Section 1.3

1. a. $\{100, 101, 102, \ldots\}$ b. $\{0, 1, 2, \ldots, 100\}$ c. $\{\ldots, 148, 149, 150\}$
   d. $\{\ldots, -3, -2, -1, 101, 102, \ldots\}$ e. $\{151, 152, \ldots\}$

5. a. True. If B occurs whenever A does ($A \subset B$), then whenever B does not occur, A does not occur ($A^c \supset B^c$).   b. True. It is not possible for A to both occur and not occur.   c. False. It can happen that $A \cap B$ is impossible but that $A^c \cap B^c$ is a possible event. You give an example.

7. a. $A^c \cap B \cap C$ b. $A \cap B \cap C^c$ c. $(A \cap B^c \cap C^c) \cup (A^c \cap B \cap C^c) \cup (A^c \cap B^c \cap C)$ d. $(A^c \cap B \cap C) \cup (A \cap B^c \cap C) \cup (A \cap B \cap C^c)$

9. If A and B cannot occur together, and A occurs whenever C does, then C and B cannot occur together.

Section 1.4

1. He has assigned a total probability of 1.7, but the set of all outcomes must always have a probability of 1.

3. $P(A) = 1/6$, $P(B) = 1/2$, $P(C) = 1/2$, $P(D) = 1/3$, $P(A \cup D) = 1/2$, $P(B \cup C) = 1$, $P(A \cap B) = 1/6$, $P(A^c) = 5/6$

5. a. $S = \{$white white collar, white blue collar, nonwhite white collar, nonwhite blue collar$\}$ b. $P(\text{nonwhite}) = 0.11$, $P(\text{white collar}) = 0.48$

7. a. 0.45   b. $A_0 \cup A_1 \cup A_2$; 0.85   c. $A_3 \cup A_4$; 0.15   d. Because the two events are disjoint and their union is the entire sample space the probability is 1.

9. a. 40 percent   b. 70 percent   c. Surprise. This cannot be answered from the information given.

Section 1.5

1. 0.248
3. Each column contains positive entries with sum 1; P(white collar) = 0.507 for whites; P(white collar) = 0.279 for nonwhites
5. *a.* P(Bob is hired) = $1/4$, P(Andrew is hired) = $1/4$, P(Carol is hired) = $1/2$ *b.* P(Bob is hired) = 0.1, P(Andrew is hired) = 0.5, P(Carol is hired) = 0.4
7. *a.* $P(1) = 0.112$, $P(2) = P(3) = \cdots = P(9) = 0.111$ *b.* The digits 0, 1, 2, . . . , 9 have a probability of 0.1 each.
9. P(at least one) = 0.8, P(neither) = 0.2
11. Take $A = \{1,2,3\}$ and $B = \{3,4,5\}$.

Section 1.6

1. $P(C|B) = 2/3$
3. *a.* 0.290   *b.* 0.326   *c.* 0.010
5. 0.41
7. *a.* 33 percent   *b.* 23 percent   *c.* Note that $A \cap B$ and $A \cap B^c$ are disjoint and that their union is A.   *d.* 56 percent
9. $1/2$
13. $2/9$

Section 1.7

1. No
3. The large battle
5. Local leaders get 0.19; school board gets 0.73.
7. *a.* $25/32$ *b.* $7/32$

Section 1.8

1. *a.* Yes   *b.* No   *c.* Debatable—probably yes   *d.* Yes   *e.* Yes   *f.* No   *g.* No   *h.* Yes
5. 0.677
7. *a.* 0.172   *b.* 0.382
9. *a.* 0.317   *b.* 0.365   *c.* 0.213   *d.* 0.245

Section 1.9

1. *a.* 0.512   *b.* 0.128
3. *a.* $1/32$ *b.* $5/32$ *c.* $1/32$ *d.* $1/2$
5. $2/3$

Section 1.10

1. 47 percent
3. 56 percent
5. 0.44

Section 1.11

1. P(Republican) = 0.936, P(Democrat) = 0.032, P(Independent) = 0.032
3. 0.348
5. $P(A|D) = 0.519$, $P(B|D) = 0.296$, $P(C|D) = 0.185$

Section 1.12

3. The number of defectives in a sample can be no larger than the size of the sample (n) and also can be no larger than the total number (D) of defectives in the lot.

5. The probability of acceptance is 0 for $p > 0.02$ and 1 for $p \leq 0.02$.
7. a. We can decide he does not have ESP when in fact he does; or we can decide he does have ESP when in fact he does not.   b. 0.0026   c. 0.4119
9. 0.105
11. $1/32$
13. $\alpha^2 p$

Section 1.13

1. a. No   b. Yes   c. No   d. Yes
3. a. Single $= 0$, Married, husband present $= 1$, Married, husband not present $= 1$, Widowed or divorced $= 2$.   b. $P(X = 0) = 0.22$, $P(X = 1) = 0.64$, $P(X = 2) = 0.14$
5. a. P(white head) $= 0.77$, P(female head) $= 0.53$ b. No   c. Probably not

Section 1.14

1. 2.32 days
5. Expected weight is 1.2 oz in both. Variety 2 has more variability in weight than variety 1.

Section 1.15

1. a. 65 min   b. 70 min
3. 39.9 sec
5. 0.5
7. a. 0.2182   b. No, as long as $p = 0.8$ is constant
9. $E(X) = 1/p$; the expected number of trials is 90.9.

Section 1.16

1. \$1,650,000 (standard deviation is \$1284.)
3. X has variance 70; Y has variance 25.
5. a. $P(Y = 0) = 1/4$, $P(Y = 1) = 3/4$ b. Both have mean $3/4$ and variance $3/16$ c. $P(Z = 0) = 1/28$, $P(Z = 1) = 12/28$, $P(Z = 2) = 15/28$ d. $E(Z) = 3/2$; Var $Z = 9/28$. Neither variances nor standard deviations are additive.
7. Mean 40, variance 36, standard deviation 6
9. Mean 100, standard deviation $16/5$
11. $E(X - m)$ is always 0.

Section 1.17

3. All three strategies have the same minimum sales, \$500,000.
5. a. $S = \{0,1,3,4\}$, $w_1 = (1/2, 0, 1/2, 0)$, $w_2 = (0, 3/4, 0, 1/4)$ b. $E_1 = 3/2$, $E_2 = 7/4$ c. $w_3 = (1/4, 3/8, 1/4, 1/8)$, $E_3 = 13/8$ which is the average of $E_1 = 12/8$ and $E_2 = 14/8$.   d. $P(w_1) = 12/39$, $P(w_2) = 14/39$, $P(w_3) = 13/39 = 1/3$
7. Check Lemma 1.2. $E_1 + E_2 + E_3$ must be positive, which is the case if any wager has a nonzero probability of winning an amount greater than 0.

Section 1.18

1. The Chebyshev bound is 2.6 (useless), while the true probability is 0.11.
3. $96/n$

Section 1.19

1. a. $1/2$ b. $1/2$ c. 0.1
3. $P(X < 0) = 1/2$, $P(X < 1/2) = 7/8$, $P(X > 1/4) = 9/32$
5. $1/4$

CHAPTER 2

Section 2.1   3.   *a.* $\frac{1}{3}$ *b.* $\frac{1}{3}$ *c.* $\frac{1}{3}$
5.   *a.* An inefficient week is followed by a satisfactory week.   *b.* A satisfactory week is followed by an inefficient week.   *c.* An inefficient week is followed by an inefficient week.

Section 2.2   1.   *a.* $p_{i,e}$ *b.* $p_{i,e}p_{e,s}$ *c.* $p_{i,i}p_{i,s} + p_{i,s}p_{s,s} + p_{i,e}p_{e,s}$ *d.* $p_{i,e} + p_{i,i}p_{i,s} + p_{i,s}p_{s,s}$
3.   0.024
5.   0.68
7.   The figure required has 27 outcomes. To extend this to 4 wk requires 81 outcomes.

Section 2.3   1.   *a.* $[4, 5, 2, -2]$ *b.* $[2+b, 6+a, 4]$ *c.* $[0, 7, -3]$ *d.* $[4, 4, 5, 11]$ *e.* $[1, 0, 2, 2]$ *f.* $[0, 0, 0]$
3. $[-1, 4]$
7.   *a.* $[3, -2] \cdot \begin{bmatrix} x \\ y \end{bmatrix} = 4$ *b.* $[3, 0, -2] \cdot \begin{bmatrix} x_1 \\ x_2 \\ x_3 \end{bmatrix} = 4$ *c.* $[c_1, c_2, c_3] \cdot \begin{bmatrix} p_1 \\ p_2 \\ p_3 \end{bmatrix} = 5$

d. $[p_{1,1}, p_{1,2}, p_{1,3}] \cdot \begin{bmatrix} p_{1,3} \\ p_{2,3} \\ p_{3,3} \end{bmatrix} = p$

Section 2.4   1.   *a.* $\begin{bmatrix} -9 \\ 2 \end{bmatrix}$ *b.* $[-4, -2, 3]$ *c.* $\begin{bmatrix} -2 \\ 0 \end{bmatrix}$ *d.* $[1, -2, 6]$
3. $6x + 3y = 0, 2x - y = 7, 4x + 7y = 3$
5.   The product, equal to $\begin{bmatrix} 0.33 \\ 0.605 \end{bmatrix}$, gives the cost of two inventories.

Section 2.5   1.   *a.* $\mathbf{p}^{(1)} = [\frac{1}{2}, \frac{3}{8}, \frac{1}{8}]$ and $\mathbf{p}^{(2)} = [\frac{37}{96}, \frac{11}{24}, \frac{5}{32}]$ *b.* $\mathbf{p}^{(1)} = [\frac{1}{4}, \frac{1}{2}, \frac{1}{4}]$ and $\mathbf{p}^{(2)} = [\frac{1}{3}, \frac{49}{96}, \frac{5}{32}]$
3.   *a.* $\frac{3}{64}$ *b.* $\frac{1}{16}$ *c.* $\frac{11}{32}$ *d.* $\frac{7}{32}$ *e.* $\frac{1}{2}$
5.   *a.* $\mathbf{p}^{(0)} = [0, 1, 0, 0]$, $\mathbf{p}^{(1)} = [\frac{1}{2}, 0, \frac{1}{2}, 0]$, and $\mathbf{p}^{(2)} = [\frac{1}{2}, \frac{1}{4}, 0, \frac{1}{4}]$.   *b.* If $\mathbf{p}^{(0)} = [0, 0, 1, 0]$, you start with two pennies, then $\mathbf{p}^{(1)} = [0, \frac{1}{2}, 0, \frac{1}{2}]$ and $\mathbf{p}^{(2)} = [\frac{1}{4}, 0, \frac{1}{4}, \frac{1}{2}]$. No, the vector $[\frac{1}{4}, 0, \frac{1}{4}, \frac{1}{2}]$ gives the chances for my opponent who has already won with a probability of $\frac{1}{2}$ while I have won with a probability of $\frac{1}{4}$ as shown in item *a*, and with a probability of $\frac{1}{4}$, the game continues with the original (time 0) position.

Section 2.6   1.   Both computations give $\mathbf{p}^{(3)} = [\frac{77}{192}, \frac{115}{192}]$
5.   *a.* $\mathbf{p}^{(1)} = [\frac{2}{7}, \frac{3}{7}, \frac{2}{7}]$ *b.* $\mathbf{p}^{(2)} = [\frac{2}{7}, \frac{3}{7}, \frac{2}{7}]$ *c.* Yes, since the same product results each time we have $\mathbf{p}^{(n)} = [\frac{2}{7}, \frac{3}{7}, \frac{2}{7}]$

9. $AB = \begin{bmatrix} 5 & 11 \\ 11 & 25 \end{bmatrix}$ and $BA = \begin{bmatrix} 10 & 14 \\ 14 & 20 \end{bmatrix}$

Section 2.7

3. a. 0.032   b. 0.115   c. 0.096
5. This may be interpreted as $P(X_1 = 1, X_2 = 1, X_3 = 1 | X_0 = 1) = 0.343$.
7. 0.746
9. 0.00094
11. The expected number of desirable responses is 5.3872, the expected number of uninformative responses is 1.9696, and the expected number of undesirable responses is 0.6432.

Section 2.8

1. a. $\mathbf{p}^{(1)} = [\,^1/_2,\ ^1/_2\,]$, $\mathbf{p}^{(2)} = [\,^1/_2,\ ^1/_2\,]$ b. $\mathbf{p}^{(n)} = [\,^1/_2,\ ^1/_2\,]$ c. $[\,^1/_2,\ ^1/_2\,]$
3. $[\,^9/_{32},\ ^8/_{32},\ ^{15}/_{32}\,]$
5. $\mathbf{P}^2 = \begin{bmatrix} ^1/_2 & ^1/_2 & 0 \\ ^1/_2 & ^1/_2 & 0 \\ 0 & 0 & 1 \end{bmatrix}$ and $\mathbf{P}^3 = \begin{bmatrix} 0 & 0 & 1 \\ 0 & 0 & 1 \\ ^1/_2 & ^1/_2 & 0 \end{bmatrix} = \mathbf{P}$
7. $^2/_5$
9. a. $\mathbf{P}^2 = \begin{bmatrix} ^3/_4 & ^1/_4 \\ ^1/_2 & ^1/_2 \end{bmatrix}$ b. $\mathbf{P}^4 = \begin{bmatrix} ^{11}/_{16} & ^5/_{16} \\ ^{10}/_{16} & ^6/_{16} \end{bmatrix}$ c. $\mathbf{P}^8 = \begin{bmatrix} ^{171}/_{256} & ^{85}/_{256} \\ ^{170}/_{256} & ^{86}/_{256} \end{bmatrix}$ d. Yes   e. $[\,^2/_3,\ ^1/_3\,]$
11. $[\,^{837}/_{1243},\ ^{306}/_{1243},\ ^{100}/_{1243}\,] = [0.67337, 0.24619, 0.08044]$

Section 2.9

1. a. $b_0 = ^{13}/_{133}$ b. $^{264}/_{133}$
3. $b_0 = ^1/_{111}$, the expected inventory size is $^{245}/_{111}$.
7. $b = [\,^{19}/_{177},\ ^{30}/_{59},\ ^{68}/_{177}\,] = [0.107, 0.508, 0.385]$, $\mathbf{p}^{(1)} = [0.25, 0.5, 0.25]$, $\mathbf{p}^{(2)} = [0.133, 0.525, 0.342]$, $\mathbf{p}^{(3)} = [0.111, 0.517, 0.372]$, $\mathbf{p}^{(4)} = [0.107, 0.512, 0.381]$.

Section 2.10

1. $\mathbf{P}$ is a matrix for which $\mathbf{P}^2 = \mathbf{P}$
3. a. $s^{(1)} = ^8/_3, L^{(1)} = 1^1/_2$ b. For $m = 2$ we have $a = ^7/_{32}$ c. $s^{(3)} = 4.0, L^{(3)} = 4.4$ d. From Eq. (2.1) we have $s^{(3)} \geq 3.28$, $L^{(3)} \leq 4.86$.

Section 2.11

1. $b_{1,0} = ^3/_4$, $b_{2,0} = ^1/_2$, $b_{3,0} = ^1/_4$
3. $b_{3,5} = 0.153$

Section 2.12

1. $n_{1,1} = ^3/_2$, $n_{1,2} = 1$, $n_{1,3} = ^1/_2$
3. $n_{3,3} = 2.76$

Section 2.13

1. a. 20   b. 23.3
3. $\alpha p / [1 - p(1 - \alpha)]$
5. $n_{0,0} + n_{0,0'}$ equals 669.43 when $b = 0.01$, 34.45 when $b = 0.05$, and 11.37 when $b = 0.1$.
7. 578,804 when $b = 0.01$, 4651 when $b = 0.05$, and 671 when $b = 0.1$.

Section 2.14   3. a. $\mathbf{p}^2 = \begin{bmatrix} 1/3 & 1/3 & 1/3 & 0 & 0 & 0 \\ 1/3 & 1/3 & 1/3 & 0 & 0 & 0 \\ 5/9 & 1/3 & 1/9 & 0 & 0 & 0 \\ 0 & 0 & 0 & 7/18 & 2/9 & 7/18 \\ 0 & 0 & 0 & 1/2 & 2/9 & 5/18 \\ 0 & 0 & 0 & 11/18 & 2/9 & 1/6 \end{bmatrix}$

b. $[\,13/33,\ 1/3,\ 3/11\,]$ and $[\,95/198,\ 2/9,\ 59/198\,]$

c. $\begin{bmatrix} 13/33 & 1/3 & 3/11 & 0 & 0 & 0 \\ 13/33 & 1/3 & 3/11 & 0 & 0 & 0 \\ 13/33 & 1/3 & 3/11 & 0 & 0 & 0 \\ 0 & 0 & 0 & 95/198 & 2/9 & 59/198 \\ 0 & 0 & 0 & 95/198 & 2/9 & 59/198 \\ 0 & 0 & 0 & 95/198 & 2/9 & 59/198 \end{bmatrix}$

d. $\begin{bmatrix} 0 & 0 & 0 & 95/198 & 2/9 & 59/198 \\ 0 & 0 & 0 & 95/198 & 2/9 & 59/198 \\ 0 & 0 & 0 & 95/98 & 2/9 & 59/198 \\ 13/33 & 1/3 & 3/11 & 0 & 0 & 0 \\ 13/33 & 1/3 & 3/11 & 0 & 0 & 0 \\ 13/33 & 1/3 & 3/11 & 0 & 0 & 0 \end{bmatrix}$

CHAPTER 3

Section 3.1   1. $\begin{aligned} 20x + 20y &= 1 \\ 15x + 30y &= 1 \end{aligned} \rightarrow \begin{aligned} 30x + 30y &= 3/2 \\ 15x + 30y &= 1 \end{aligned} \rightarrow 15x = 1/2$

Thus $x = 1/30$ so that $30x + 30y = 3/2 \rightarrow 30y = 1/2$, or $y = 1/60$.

3. $13/12$

5. $1/15{,}000$

7. $x = 1/30$, $y = 1/60$

9. $10x + 15y \geq 1$ results from vitamin A requirement. $8x + 20y \geq 1$ results from vitamin B_1 requirement. $20x + 8y \geq 1$ results from vitamin B_{12} requirement.

Section 3.2   1. a. $[3, -9]$ b. $[-4, 1]$ c. $[-2, 1]$ d. $[-3, -4]$ e. $[1, 5, 2]$
f. $[3, 2, 0]$

3. $[2, 4] - [1, 3] = [1, 1]$ and $[1, 1] \cdot [-1, 1] = 0$

5. a. $[3, 0] - [2, -1] = [1, 1]$ b. $[5, 2] - [2, -1] = 3[1, 1]$ c. $[1, 1] - [1/2, 1/2] = 1/2\,[1, 1]$ d. $[c, c] - [1/2, 1/2] = (c - 1/2)[1, 1]$.

Section 3.3   1.

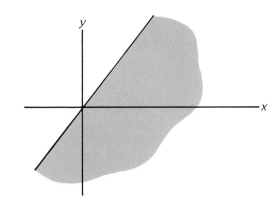

3.

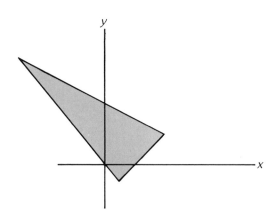

5.   At the point C the hyperplanes $x = 0$, $y = 0$, and $x + y + z = 1$ intersect.

Section 3.4   1. $f(0, 1) = -1$, $f(2, 1) = 1$ a. $f(\frac{1}{4}[0, 1] + \frac{3}{4}[2, 1]) = \frac{1}{2}$ b. $f(2[0, 1] - [2, 1]) = -3$ c. $f(\frac{1}{2}[0, 1] + \frac{1}{2}[2, 1]) = 0$ d. $f([0, 1] + \frac{1}{3}\{[2, 1] - [0, 1]\}) = -\frac{1}{3}$

3.   a. The maximum is 6 at $[0, 3]$, and the minimum is -5 at $[5, 0]$.   b. The maximum is 4 at $[2, 3]$, and the minimum is 0 at the origin $[0, 0]$.   c. The maximum is 6 at $[0, 3]$ while the minimum is -1 at $[1, 0]$.   d. The maximum is -1 at $[1, 0]$, and the minimum is -7 at two extreme points, $[9, 1]$ and $[7, 0]$, and consequently along the line segment joining those points.

Section 3.5   1. $[-2, 1] \cdot \mathbf{v} = -2$. At the point b we have many choices of hyperplanes. One is $[1, 3] \cdot \mathbf{v} = 8$.

3.   a. $[2, 0] \cdot \mathbf{v} = 12$ (there are many choices available).   b. $x = 5$ (there are many choices).

Section 3.6

1.
$$\begin{bmatrix} 3 & 20 \\ 12 & 2 \\ 4 & 10 \\ -1 & 0 \\ 0 & -1 \end{bmatrix} \begin{bmatrix} x \\ y \end{bmatrix} \leq \begin{bmatrix} 1200 \\ 1000 \\ 800 \\ 0 \\ 0 \end{bmatrix}$$

determines the convex set of feasible solutions, and we wish to maximize $f(x, y) = 80x + 50y$.

3. Find the maximum of $f(x, y) = 80x + 50y$ where

$$\begin{bmatrix} 2 & 15 \\ 9 & 3 \\ 5 & 9 \\ -1 & 0 \\ 0 & -1 \end{bmatrix} \begin{bmatrix} x \\ y \end{bmatrix} \leq \begin{bmatrix} 1600 \\ 1000 \\ 1200 \\ 0 \\ 0 \end{bmatrix}$$

5. Find the minimum of $30s + 80m + 15v + 300d$ where

$$\begin{bmatrix} 0.5 & 0.6 & 0.15 & 0.05 \\ 0.1 & 0.3 & 0.35 & 0.15 \\ 0.4 & 0.1 & 0.5 & 0.8 \\ 1 & 0 & 0 & 0 \\ 0 & 1 & 0 & 0 \\ 0 & 0 & 1 & 0 \\ 0 & 0 & 0 & 1 \end{bmatrix} \begin{bmatrix} s \\ m \\ v \\ d \end{bmatrix} \geq \begin{bmatrix} 4 \\ 2 \\ 1 \\ 0 \\ 0 \\ 0 \\ 0 \end{bmatrix}$$

7. Find the minimum of $\frac{2}{3}s + L$ where

$$\begin{bmatrix} 30 & 50 \\ 170 & 150 \\ 1 & 0 \\ 0 & 1 \end{bmatrix} \begin{bmatrix} s \\ L \end{bmatrix} \geq \begin{bmatrix} 350 \\ 650 \\ 0 \\ 0 \end{bmatrix}$$

9. Maximize $100c + 135s$ where

$$\begin{bmatrix} 15 & 20 \\ 20 & 20 \\ 5 & 7 \\ 1 & 0 \\ 0 & 1 \\ -1 & 0 \\ 0 & -1 \end{bmatrix} \begin{bmatrix} c \\ s \end{bmatrix} \leq \begin{bmatrix} (13,000)60 \\ (17,000)60 \\ (5200)60 \\ 50,000 \\ 2500 \\ 0 \\ 0 \end{bmatrix}$$

11. For this situation the convex set is unchanged; only the function to be made a maximum will change. This function is $3b + 12g + 2A + 7B + 1m$.

13. We will seek to find a maximum of the savings of manufacturing over purchase costs. Thus we focus on the maximum of the function $4c + 3m$ where

$$\begin{bmatrix} 3 & 4 \\ 2 & 1 \\ 1 & 0 \\ 0 & 1 \\ -1 & 0 \\ 0 & -1 \end{bmatrix} \begin{bmatrix} c \\ m \end{bmatrix} \le \begin{bmatrix} 2500 \\ 1000 \\ 300 \\ 500 \\ 0 \\ 0 \end{bmatrix}$$

Section 3.7

1. $[\,{}^{21}\!/_{13},\ {}^{40}\!/_{13}\,]$, $[\,{}^{175}\!/_{31},\ {}^{24}\!/_{31}\,]$, $[0, 2]$, $[5, 0]$, $[0, 0]$
3. The minimum value is ${}^{37}\!/_{27}$ at the point $[1,\ {}^{5}\!/_{9}\,]$.
5. The maximum is 7,450,000,000 at the point $[70{,}000,\ 30{,}000]$.
7. At the extreme point $[\,{}^{13}\!/_{663},\ {}^{500}\!/_{663},\ {}^{150}\!/_{663}\,]$, the maximum is ${}^{661}\!/_{663}$.

Section 3.8

1. The maximum value is 10 when $x = 2$, $x = 0$, $z = 0$.
3. The maximum value of $3x + 5y - 2z$ is 11 at $x = 2$, $y = 1$, $z = 0$. The minimum value of $3r + 4s$ is 11 at $r = \frac{1}{5}$, $s = \frac{13}{5}$.
5. The maximum is 37 when $x = 0$, $y = 5$, and $z = 12$.

Section 3.9

1. *b.* In the simplex table we have a negative number in the bottom row, in the column labeled y, but there is no row on which we can pivot since all entries are negative. The instructions for carrying out the simplex algorithm do not tell us how to proceed.

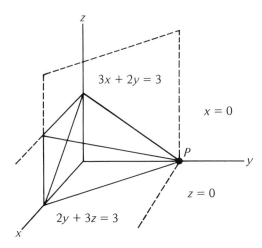

3. P is the point of intersection of the four planes $x = 0$, $z = 0$, $3x + 2y = 3$, and $2y + 3z = 3$.

Section 3.10

1. The same solution results. The first three conditions are not used to constrain the solution.

3. Again 25 hr of fishing and $^{95}/_3$ hr of farming produce the requirements.
5. The solution is the same as the solution given in the text.
7. If a goal is unattainable, the set of feasible solutions must be empty (there will be no points in the convex set).

Section 3.11   3. A sends 10 to e, B sends 7 to e, C sends 3 to e, and C sends 15 to d. The cost of this solution is 297.

CHAPTER 4

Section 4.1   1.

X	O	O
	O	
X	X	X

3. There are many possible strategies, we will write a very simple one.
   (1) Place an O in 5 unless it is filled. If 5 is filled place an O in 7.
   (2) If 5 has an O in it place an O in 7. If 5 does not have an O in it place an O in 4. Otherwise place an O in the lowest-numbered empty space.
   (3) Complete the game by placing an O in the lowest-numbered empty space.

   The result of the two strategies is

X	O	Ø
	Ø	X
Ø	X	X

Section 4.2   1. The matrix of Player 2's sales volumes is

	y_1	y_2	y_3
x_1	9	3	8
x_2	7	10	2
x_3	6	5	4

   The second part of this exercise now follows by subtracting the entries of this table from 10 to obtain 2's losses.
3. The payoff matrix is the same since losses to each side are equal and cancel each other.
5. Each player has six strategies: (1) show one finger and call 2; (2) show one finger and call 3; (3) show one finger and call 4; (4) show two fingers and call 2; (5) show two fingers and call 3; (6) show two fingers and call 4.
7. The strategies are labeled according to the settlement chosen.

2

	y_1	y_2	y_3
x_1	3	4	3
x_2	2	3	3
x_3	3	3	3

1

The payoff is the total trade for Player 1; a loss for Player 2 is then the difference between 6 and his total trade (see Exercise 4.2.1).

9. a. Yes. Losses might be the number of ICBMs which score a hit for example. b. No. Presumably an agreement is mutually beneficial here. c. Yes. A duopoly results if there are only two possible dealers. d. Yes. e. No. Here there are probably more than two players. f. Yes.

Section 4.3

1. a. y_3 is minimax and x_3 is maximin. b. y_3 is minimax and x_2 is maximin— this game is strictly determined.

3. The matrix below gives payoffs for a game which has two maximin strategies

	y_1	y_2	y_3
x_1	1	2	4
x_2	1	3	1

The matrix below gives payoffs for a game which has two maximin and two minimax strategies.

	y_1	y_2	y_3
x_1	1	1	1
x_2	1	1	2
x_3	0	−1	0

5. a. If the game is strictly determined, the plot is impossible since no strategy can improve on the "best" one. b. If the game is not strictly determined, the plot is possible as the discussion following Table 4.5 indicates; the worse strategy against maximin can be the minimax.

7.

	y_1	y_2
x_1	−50	50
x_2	0	0
x_3	−10	40
x_4	−60	−10
x_5	−60	40
x_6	−10	−10

Player 1 has the maximin strategy x_2, while Player 2 has the minimax strategy y_1. If we eliminate the absurd strategies x_2, x_4, x_5, and x_6, the remaining game has the matrix

	y_1	y_2
x_1	-50	50
x_3	-10	40

The solution to the game is x_3, y_1, and the value is -10.

9. The payoffs for x_0 against y_0, y_1, y_2, y_3 are

	y_0	y_1	y_2	y_3
x_0	1	2	1	0

and so x_0 is also maximin as well as x_4. Now the minimax strategy for Rommel is y_0, and the upper value is 1; the lower value is 0, and the game is not strictly determined.

11. In the game, x_1 is maximin, y_1 is minimax, and the value is 0; yet the strategies x_2 and y_2 have the payoff 0 as well. Neither player can profit by playing the strategy x_2 or y_2, however.

Section 4.4

1. $M^*(\mathbf{p}, \mathbf{q}) = -\frac{1}{4}$. Using the formula found in Example 4.11, $(2\alpha - 1)(2\beta - 1) = -\frac{1}{4}$.

3. The "mixed strategy" $\mathbf{p} = [1, 0]$ has the intuitive interpretation—always play strategy x_1—thus it should be the same as strategy x_1.

5. The expected payoff is $2\alpha(1 - 2\beta - \gamma) + 2\beta + \gamma$.

7. $\mathbf{Mq}$ is a column vector with ith component

$$\sum_{j=1}^{m} M(x_i, y_j)q_j$$

and the dot product of $\mathbf{p}$ with $\mathbf{Mq}$ is the result

$$\sum_{i=1}^{n} p_i \sum_{j=1}^{m} M(x_i, y_j)q_j$$

so that $\sum_{i=1}^{u} \sum_{j=1}^{m} p_i M(x_i, y_j)q_j$ is the expected payoff.

9. In Exercise 4.4.4 we had $\mathbf{p} = [\frac{1}{2}, \frac{1}{2}]$ and $\mathbf{q} = [\frac{1}{6}, \frac{1}{6}, \frac{2}{3}]$, thus

$$M^*(\mathbf{p}, \mathbf{q}) = [\frac{1}{2}, \frac{1}{2}] \begin{bmatrix} 0 & 1 & 2 \\ 2 & 1 & 0 \end{bmatrix} \begin{bmatrix} \frac{1}{6} \\ \frac{1}{6} \\ \frac{2}{3} \end{bmatrix} = 1$$

Section 4.5   1.

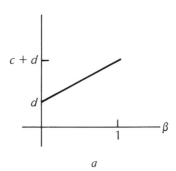
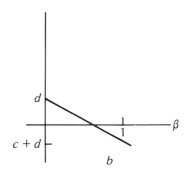

3. We know that the minimax strategy $\mathbf{q}_0 = [\beta, 1 - \beta]$ is one which minimizes
$$\max_{i=1,2} M^*(x_i, \mathbf{q}_0) = \max\{M^*(x_1, \mathbf{q}_0), M^*(x_2, \mathbf{q}_0)\}.$$

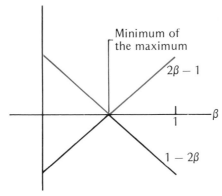

The maximum of the two functions is the part of the two lines above the β axis. The minimum occurs at the point of intersection, when $\beta = \frac{1}{2}$.

5. Since $M^*(\mathbf{p}, \mathbf{q}) = (3\alpha - 1)(2\beta - 1) + 2$, 1 can guarantee no more than 2, since $(3\alpha - 1)(2\beta - 1)$ may be negative, and he can guarantee 2 by making that product equal to 0, so he takes $\alpha = \frac{1}{3}$. Similarly, 2 makes the product zero by taking $\beta = \frac{1}{2}$ and guaranteeing at most 2. Thus $\mathbf{p}_0 = [\frac{1}{3}, \frac{2}{3}]$ and $\mathbf{q}_0 = [\frac{1}{2}, \frac{1}{2}]$ give the solution, and the value is 2.

7. a. Definition 4.7 tells us that the minimum for all $\mathbf{q}$ when $\mathbf{p}_0$ is used will be v, and so we have $M^*(\mathbf{p}_0, y_j) \geq v$ for all y_j.   b. Definition 4.7 tells us that the maximum for all $\mathbf{p}$ when $\mathbf{q}_0$ is used will be v and so $M^*(x_i, \mathbf{q}_0) \leq v$.

9. Consider $\mathbf{p}_0 = \left[\dfrac{d - c}{a + d - b - c}, \dfrac{a - b}{a + d - b - c}\right]$, and the two strategies y_1, y_2.
$M^*(\mathbf{p}_0, y_1) = a(d - c)/(a + d - b - c) + c(a - b)/(a + d - b - c) = (ad - bc)/(a + d - b - c) = v$, and $M^*(\mathbf{p}_0, y_2) = b(d - c)/(a + d - b - c) + d(a - b)/(a + d - b - c) = (ad - bc)/(a + d - b - c) = v$. Thus we have $M^*(\mathbf{p}_0, y_j) \geq v$ for every y_j. Next you can verify part a of Exercise 4.5.8 to finish the claim.

Section 4.6   1. a. Strategy x_3 dominates x_2. The resulting game is not strictly determined, and none of 2's strategies are dominated.   b. None of 1's strategies are dominated. y_2 is dominated by y_1; y_3 is dominated by y_1. In the reduced game x_1 is dominated by x_2, x_3 is dominated by x_2, and x_4 is dominated by x_2. In this further reduction of the game, y_4 dominates y_1, and the solution is x_2, y_4 with a value of 1. This game is of course strictly determined.   c. There are no dominated strategies.   d. x_2 dominates x_3. In the reduced game y_2 dominates both y_1 and y_3. Then x_2 dominates x_1, and the solution of this strictly determined game is x_2, y_2 and the value is 48.   e. x_2 dominates x_1. In the reduced game y_1 dominates y_3. The game thus reduces to a 2×2 game with matrix

$$\begin{bmatrix} 1 & 0 \\ -1 & 2 \end{bmatrix}$$

The solution to this game uses the mixed strategies $\mathbf{p} = [3/4, \ 1/4]$, $\mathbf{q} = [1/2, \ 1/2]$, and the value is $1/2$.   f. y_3 dominates y_2 and y_4. In the reduced game x_2 dominates both x_1 and x_3, and then y_3 dominates y_1 so that the game is determined. The solution is x_2, y_3 and the value is 0.

3. x_4 dominates x_1, x_2, and x_3; and in the reduced game, y_0 dominates y_1, y_2, and y_3. The game is strictly determined, and the value is 0.

5. $M^*(\mathbf{p}, \mathbf{q}^*) = 1\,2/3$. $M^*(\mathbf{p}, \mathbf{q}) = 2\,5/9$. Thus $M^*(\mathbf{p}, \mathbf{q}^*) < M^*(\mathbf{p}, \mathbf{q})$.

Section 4.7   1. The three expected payoffs are 4α, $2 - \alpha$, and $3 - 4\alpha$. α_0 is the value of α when $4\alpha = 3 - 4\alpha$ or $\alpha = 3/8$.

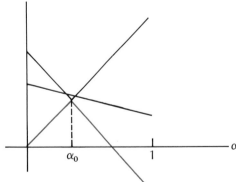

3. a. $\mathbf{p} = [2/3, \ 1/3]$ and $\mathbf{q} = [2/3, \ 1/3]$, $v = 2$.   b. Thus $\mathbf{p} = [5/7, \ 2/7]$ and $\mathbf{q} = [0, \ 6/7, \ 1/7]$ and $v = 12/7$.   c. $\mathbf{p} = [1/3, \ 2/3]$ and $\mathbf{q} = [0, 1, 0]$ so that $v = 1$ or $\mathbf{p} = [5/6, \ 1/6]$ and $\mathbf{q} = [0, 1, 0]$ and again $v = 1$.   d. $\mathbf{p} = [5/7, \ 2/7]$, $\mathbf{q} = [0, \ 6/7, \ 1/7, \ 0]$, and $v = 12/7$.   e. y_4 dominates y_1, y_2, and y_3 and then x_2 dominates x_1, so the game is strictly determined and the value is 0.   f. $\mathbf{p} = [2/3, \ 1/3]$, $\mathbf{q} = [2/3, \ 0, \ 1/3]$, and $v = 2$.

5. a. Note that y_2 dominates y_3. The solution is $p = [\frac{1}{4}, \frac{3}{4}]$, $q = [0, 1, 0, 0]$, and $v = -\frac{1}{2}$.

Section 4.8 *The authors elected to give answers to even-numbered exercises in Sec. 4.8.*

4. $\mathbf{p} = [\frac{9}{16}, \frac{7}{16}, 0]$, $\mathbf{q} = [0, \frac{1}{3}, \frac{1}{2}]$, and $v = \frac{5}{2}$.

6. The simplex algorithm causes some trouble with this exercise; we find that the solution is $\mathbf{p} = [\frac{1}{2}, \frac{1}{2}]$, $\mathbf{q} = [0, 1, 0]$, and $v = 1$.

8. The payoff matrix is

	y_1	y_2	y_3	y_4
x_1	0	2	−3	0
x_2	−2	0	0	3
x_3	3	0	0	−4
x_4	0	−3	4	0

where the strategies are

x_1, y_1: Show one finger and call 2
x_2, y_2: Show one finger and call 3
x_3, y_3: Show two fingers and call 3
x_4, y_4: Show two fingers and call 4

$\mathbf{p} = [0, \frac{3}{5}, \frac{2}{5}, 0]$, $\mathbf{q} = [0, \frac{3}{5}, \frac{2}{5}, 0]$, and $v = 0$.

10. If the strategies x_1 and y_1, x_2 and y_2, x_3 and y_3 stand for "builds in settlement" 1, 2, 3 respectively, then the payoff matrix is

	y_1	y_2	y_3
x_1	3	4	3
x_2	2	3	3
x_3	3	3	3

Notice that x_1 dominates x_2 and x_3 and y_1 dominates y_2 and y_3. Thus the game is strictly determined and the value is 3.

Section 4.9

1. The expected payoff with the strategies $\mathbf{p} = [\frac{18}{26}, 0, \frac{8}{26}]$ and $\mathbf{q} = [0.25, 0.75]$ is 13.28.

3. By trial and error we find the solution to be Peters to storeroom, Rubin to sales, Williams to buyer, and Riley to sales. This assignment is easy to justify as a solution since the maximum possible score is 11; but that is not allowed since it involves assigning both Williams and Riley to buyer. Thus we can do no better than 10, and our assignment gives 10. The payoff matrix with zeros omitted is

	(1,1)	(1,2)	(1,3)	(1,4)	(2,1)	(2,2)	(2,3)	(2,4)	(3,1)	(3,2)	(3,3)	(3,4)	(4,1)	(4,2)	(4,3)	(4,4)
Row 1	1	1	1	$\frac{1}{2}$												
Row 2					$\frac{1}{3}$	$\frac{1}{3}$	1	1								
Row 3									$\frac{1}{2}$	$\frac{1}{2}$	$\frac{1}{4}$	1				
Row 4													1	1	$\frac{1}{2}$	1
Column 1	1				$\frac{1}{3}$				$\frac{1}{2}$				1			
Column 2		1				$\frac{1}{3}$				$\frac{1}{2}$				1		
Column 3			1				1				$\frac{1}{4}$				$\frac{1}{2}$	
Column 4				$\frac{1}{2}$				1				1				1

Section 4.10   1.

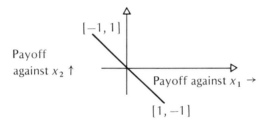

a

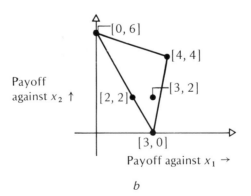

b

CHAPTER 5

Section 5.1

3. The game is strictly determined and the strategies are x_1, y_1, but we know that nature will not use its minimax strategy. The hospital director can know from past experience the relative frequency of y_1, y_2, y_3, y_4 among newborn infants.

5. Many examples can be thought of which involve the use of weather prediction or accident frequency.

Section 5.2

1. We compute the expected payoffs for the strategy $\mathbf{q} = [0.5, 0, 0.5]$:

$$M^*(x_1, \mathbf{q}) = -37.5$$
$$M^*(x_2, \mathbf{q}) = -10$$

and so $M^*(x_1, \mathbf{q}) < M^*(x_2, \mathbf{q})$.

3. $M^*(x_1, \mathbf{q}) = 99.5$, $M^*(x_2, \mathbf{q}) = 97.5$, $M^*(x_3, \mathbf{q}) = 80$, and x_1 is optimum.

5.

	y_0	y_1	y_2	y_3	y_4
x_0	0	0	0	0	0
x_1	-1	2	2	2	2
x_2	-2	1	4	4	4
x_3	-3	0	3	6	6
x_4	-4	-1	2	5	8

$M^*(x_0, \mathbf{q}) = 0$, $M^*(x_1, \mathbf{q}) = 1.1$, $M^*(x_2, \mathbf{q}) = 1.6$, $M^*(x_3, \mathbf{q}) = 1.5$, $M^*(x_4, \mathbf{q}) = 0.8$. The best strategy is x_3 which gives the largest expected payoff.

7. Consider the strategy $(\alpha, 1 - \alpha)$ for nature. $M^*(x_1, \mathbf{q}) = -15\alpha - 5$. This will equal the payoff for strategy x_2 if $\alpha = 2/3$. Thus, if the probability for virus A, y_1, is $2/3$ or more, no treatment should be attempted.

Section 5.3

1. The observations are the test results: $S = \{$good, repairable, defective$\}$. The possible states of nature are the actual conditions of the blocks: $W = \{$good (w_1), repairable (w_2), defective $(w_3)\}$. The possible decisions are: $D = \{$use (d_1), repair (d_2), melt down $(d_3)\}$. The payoff function M is given by the matrix

	w_1	w_2	w_3
d_1	0	-50	-300
d_2	-5	-20	-5
d_3	-100	-100	0

3. The observations are the four possible classes which divides the men, $S = \{1, 2, 3, 4\}$. The states of nature are $W = \{$heart attack before 60, no heart

attack before 60}. The surgeon has two decisions, $D = \{(d_1)$ choose the man, (d_2) do not choose the man}. The payoff function is given in the matrix

	w_1	w_2
d_1	-100	5
d_2	5	0

5. The observations are the proportions of grade 1, grade 2, and scrap bearings. The states of nature are not available to us. The expected profit using the current process is \$1.80 per bearing. The expected profit using the new process is \$2.19 per bearing. The new process is better.

7. This is a decision problem. Using, the known probabilities of the six states of Nature, the expected gain if he holds the stock, is \$73,150. So the investor should hold the stock.

Section 5.4

1. a. $P(w_1) = 0.95$, $P(2_2) = 0.05$, $P(B \text{ only}|w_1) = 0.2$, and $P(B \text{ only}|w_1) = 0.1$. (See Example 5.4.) $P(w_1|B \text{ only}) = 0.97$. Now the expected payoff for the two decisions are—for d_1, -0.67; for d_2, 0. So the decision not to drill is the correct one.   b. When A and B are both observed (event N), we have $P(w_1|N) = 0.83$. Thus for decision d_1 the expected gain is 0.87, and just as in Example 5.3 we decide to drill.   c. For the event neither A nor B (event 0), we have $P(w_1|0) = 0.99$, and the expected gain with decision d_1 is -0.89, and he should not drill.

3. Now the physician must rely on his prior probabilities only. The expected payoffs are now for d_1, -2.16; for d_2, -2.24; for d_3, -2.40. The decision should be d_1 when no test results are available.

5. The event of interest here is the event E of observing 14 "cures," $P(p = 0.6|E) = 0.142$; $P(p = 0.7|E) = 0.441$; $P(p = 0.8|E) = 0.376$; $P(p = 0.9|E) = 0.041$. For decision d_1 the expected payoff is -2. For decision d_2 the expected payoff is $+9.19$, and d_2 is the correct decision.

7. Again we must find the posterior probabilities for each of the possible observed classes. Let A denote the event "heart attack" and 1, 2, 3, 4 stand for the events class 1, class 2, class 3, class 4, respectively. $P(A|1) = 0.027$; $P(A|2) = 0.053$; $P(A|3) = 0.182$; $P(A|4) = 0.182$. The expected payoff for each of these four classes is class 1 and $d_1 = 2.165$; class 1 and $d_2 = 0.135$; d_1 is preferred when class 1 is observed. Class 2 and $d_1 = -0.565$; class 2 and $d_2 = 0.265$; d_2 is preferred when class 2 is observed. Class 3 or 4 and $d_1 = -14.11$; class 3 or 4 and $d_2 = 0.91$; d_2 is preferred again. The surgeon should choose only men from class 1.

9. The prior probabilities for grade 1, grade 2, and scrap are 0.5, 0.4, and 0.1, respectively. We will write 1, 2, 3 to denote the inspection decisions grade 1, grade 2, scrap, respectively. $P(\text{grade } 1|1) = 0.71$; $P(\text{grade } 2|1) = 0.28$; $P(\text{scrap}|1) = 0.01$; $P(\text{grade } 1|2) = 0.25$; $P(\text{grade } 2|2) = 0.68$; $P(\text{scrap}|2) = 0.07$; $P(\text{grade } 1|3) = 0.18$; $P(\text{grade } 2|3) = 0.28$; $P(\text{scrap}|3) = 0.54$. Next we compute expected

payoffs for each decision and each observation. The results are

Observe	Decide	Expected payoff
1	grade 1	−0.89
1	grade 2	−1.45
1	scrap	−2.41
2	grade 1	−2.39
2	grade 2	−0.71
2	scrap	−1.43
3	grade 1	−3.54
3	grade 2	−1.98
3	scrap	−0.82

For observation 1 we decide grade 1; for observation 2 we decide grade 2; for observation 3 we decide scrap.

Section 5.5 1. To avoid the use of confusing numbers suppose that the possible returns are 1, 0, and −1. Then the table below shows probabilities which give results as desired here.

	Return		
	1	0	−1
Investment 1	0.6	0	0.4
Investment 2	0.4	0.6	0

INDEX